Solid Waste Management
Navigation from Fundamentals to Innovation

José María Ponce-Ortega
Chemical Engineering Department
Universidad Michoacana de San Nicolás de Hidalgo
Michoacan, Mexico

Rogelio Ochoa-Barragán
Chemical Engineering Department
Universidad Michoacana de San Nicolás de Hidalgo
Michoacan, Mexico

César Ramírez-Márquez
Chemical Engineering Department
Universidad Michoacana de San Nicolás de Hidalgo
Michoacan, Mexico

José Ezequiel Santibañez-Aguilar
Professor Chemical Engineering
Tecnologico de Monterrey
Monterrey, Mexico

Aurora del Carmen Munguía-López
University of Wisconsin
Madison, USA

CRC Press
Taylor & Francis Group
Boca Raton London New York

CRC Press is an imprint of the
Taylor & Francis Group, an **informa** business
A SCIENCE PUBLISHERS BOOK

First edition published 2026
by CRC Press
2385 NW Executive Center Drive, Suite 320, Boca Raton FL 33431

and by CRC Press
4 Park Square, Milton Park, Abingdon, Oxon, OX14 4RN

Library of Congress Cataloging-in-Publication Data (applied for)

ISBN: 978-1-032-82635-6 (hbk)
ISBN: 978-1-032-85061-0 (pbk)
ISBN: 978-1-003-51632-3 (ebk)

DOI: 10.1201/9781003516323

Typeset in Times New Roman
by Prime Publishing Services

Preface

Solid waste management is one of our time's most pressing and difficult issues, particularly in a global setting that increasingly demands sustainable and responsible practices. Our books, Solid Waste Management: Navigating from Fundamentals to Innovation offers a comprehensive view of this field, presenting both fundamental concepts and advanced, innovative approaches to waste management. This work is designed to serve as a key resource for students, researchers, and professionals who seek to improve and optimize waste management systems in pursuit of sustainable development.

The book is organized into three sections. Section I: Fundamentals of Solid Waste Management covers essential concepts and technical foundations. Chapter 1: Introduction to Solid Waste Management presents the historical background of solid waste management and introduces key concepts that establish a framework for the discipline. Chapter 2: Processing Technologies for Municipal Solid Waste describes methods and technologies used in waste treatment and processing, focusing on approaches that transform and manage various waste types to minimize environmental impact. Chapter 3: Waste Collection and Transportation Systems explores the logistics, challenges, and strategies involved in efficiently collecting and transporting solid waste.

Section II: Sustainable Waste Utilization and Planning emphasizes the importance of strategic planning for achieving sustainable waste management. Chapter 4: Optimal Planning for Sustainable Solid Waste Utilization discusses methods that maximize the reuse and sustainable utilization of waste resources. Chapter 5: Planning for Reuse of Municipal Solid Waste explores strategies for reusing municipal waste to support a circular economy model. Chapter 6: Strategic Planning for Municipal Solid Waste Management examines approaches for creating long-term plans that incorporate sustainable policies, yielding positive outcomes in waste reduction. Chapter 7: Financial Risk Analysis in Waste Management focuses on evaluating financial risks related to waste management, providing a detailed view of the economic dimensions of the field. Finally, Chapter 8: Optimization Framework for Solid Waste Management introduces an optimization framework that aids in making informed, effective decisions through mathematical and programming models.

Section III: Challenges and Innovations in Waste Management addresses contemporary challenges and technological innovations transforming the field. Chapter 9: Optimal Waste Management during the COVID-19 Pandemic investigates the unique challenges faced during the pandemic, highlighting practices that enabled efficient and safe waste management. Chapter 10: Hybrid Machine Learning-Mathematical Programming Approach combines machine learning and mathematical programming to introduce a hybrid approach, uniting artificial intelligence with traditional optimization techniques to develop advanced solutions for waste management.

This book is the result of years of research and collaboration at the Universidad Michoacana de San Nicolás de Hidalgo, supported by the contributions of numerous institutions and leading experts. Special acknowledgments are extended to Professor Fabricio Nápoles-Rivera and Professor Luis Fernando Lira Barragán, both of the Universidad Michoacana de San Nicolás de Hidalgo, whose leadership and expertise have been instrumental in the success of this work. Recognition is also due to Professor Mahmoud M. El-Halwagi from Texas A&M University for his ongoing guidance and encouragement, fostering an interdisciplinary perspective on the global complexities of waste management.

Further appreciation is extended to Professor Víctor M. Zavala, M.S. Alicia Danae Díaz-Barriga-Fernández, Professor Neyara Radwan, Professor J. Betzabe González-Campos, and Professor Juan Martínez-Gómez, whose contributions in key areas such as mathematical modeling, financial risk analysis, sustainability, and processing technologies have enriched this work, lending it a broad and innovative outlook on solid waste management's challenges and opportunities.

This book has been prepared with the objective of transforming waste management into a discipline that not only mitigates environmental impact but also promotes a circular and sustainable approach at every level. It is hoped that this work will inspire further research and contribute to the development of a cleaner, more sustainable future.

Contents

Section I

Fundamentals of Solid Waste Management

1

Introduction to Municipal Solid Waste Management

This chapter delivers a comprehensive analysis of solid waste management (SWM), addressing its historical advancement and development along with the progressive incorporation of optimization towards efficiency gains. Developments in SWM are becoming increasingly important as populations in urban contexts increase and continue to generate more solid waste. For example, global waste should rise from 2.01 billion tons/year in 2020 to 3.40 billion tons in 2050. The historical context has shifted from waste disposal methods to a more systematic and sustainable means of management. The complementary optimization methods are key tools for working towards a more efficient SWM system. The objectives of optimization methods are to reduce costs for managing waste while increasing resources for recovery, thereby optimistically moving in the direction of environmental sustainability. The chapter also investigates the challenges of employing optimization, such as data visibility and accessibility, things involved with sorting informal waste systems and movement toward technology. However, the potential services optimization might provide as waste systems globally are becoming increasingly difficult to manage is incredibly beneficial and leads to innovation towards decreasing the complexities that are noticeable factors of a future of sustainability.

1.1 Introduction to the Significance of Solid Waste Management

Solid Waste Management (SWM) is the planning process of handling waste from creation to disposal with the intent of reducing the impact of waste on public health, quality of life, and the environment, which includes collection, transport, treatment and disposal of waste, and resource recycling and recovery to manage waste sustainably (Oyebode 2024) (see Fig. 1). SWM is an essential urban infrastructure component and integral to public health services, pollution reduction, and resource sustainability in both developed and developing countries (Cheela et al. 2021).

Accruing urban populations exacerbates the challenges facing SWM globally, especially with greater waste impacts wherein consumption is increasing (in metropolitan and urban areas) and populations are growing, resulting in extreme increases in MSW (Municipal Solid Waste) from non-degradable (glass, plastics, metals, certain chemicals) (Hayat 2023). Global trends indicate that MWR (Municipal Wastewater Regulation) inappropriate or ineffective management will have significant consequences on the environment, health, and the economy (Yıldız et al. 2024).

1.1.1 Health and Environmental Challenges

Public health is arguably one of the most significant effects of poor SWM. When waste management practices fail to collect or require adequate disposal of waste, there is an increased potential for mosquitoes, rodents, and flies to breed and become vectors for transmission of a range of diseases,

Fig. 1. Solid Waste Management Process.

including malaria, cholera, and dysentery (Ravichandran and Venkatesan 2021). The effects of improper waste management are often heightened in high-density urban areas, as waste is often found in close or residential proximity to neighborhoods. Anecdotal evidence from the National Institute of Urban Affairs in many Indian cities suggests that the formally managed waste management system collects only 60–70% of what is generated, leaving significant amounts unprocessed which poses a risk to public health (Hazra and Goel 2009). Similarly, Kumar et al. (2024) highlight the urgent need for waste collection and recycling systems to minimize the environmental and health risks associated with urban waste.

The environmental effects of poor SWM are equally troubling. The practice of open dumping, which has remained prevalent in many cities, creates a risk of contamination of soil and groundwater as hazardous chemicals and leachates from the waste infiltrate the water that we consume (Siddiqua et al. 2022). For example, the Bakoteh dump site in Gambia has resulted in groundwater contamination as well as severe air pollution as a result of burning waste products (Sanneh et al. 2011). Untreated waste also contributes to air pollution, especially if it is burned which releases particulate matter, carbon dioxide, and methane into the air. This not only deteriorates urban air quality but also contributes to global climate change. According to Masud et al. (2023), the mismanagement of MSW in China has caused different kinds of burdens on the environment, a growing concern in many developing countries where open dumping is also prevalent.

1.1.2 Economic and Social Dimensions

The practice of effective SWM is not just important to safeguard environmental well-being but is recognized to be a fundamental pillar of economic and social systems, particularly in the informal sector. This informal recycling sector comprises waste pickers, itinerant buyers, and scrap dealers that significantly contribute towards waste diversion efforts by recycling anywhere from 10% to 15% of all waste generated with cities like Kolkata, Bangalore, and Chennai being exemplars (Colenbrander et al. 2017). The informal sector also provides livelihood and income for thousands of individuals who collect, sort, and sell recyclable goods to prevent what could have been large amounts of waste diverted into landfills (Omar 2018). Lissah et al. (2021) illustrate how important the role of informal sectors is to urban solid waste management systems given that the municipal authorities have failed to recognize them. Furthermore, Wilson et al. (2009) argue that informal workers should continue to be included in recycling activities to increase recycling levels in solid waste management systems in developing countries.

While that may be the case, the contributions made by the informal sector are often looked over when discussing SWM where workers are exposed to poorly regulated conditions. Waste pickers often work in precarious and hazardous settings in dealing with waste which are potentially toxic, where they will expose themselves (i.e., handling unsorted garbage, hazardous waste) without any proper protective gear or legal protection. Yet, again the informal sector is tied to the ways that SWM work in cities. One of the improved ways to accomplish better solid waste management systems with increased efficiency and fairness would be to support the informal sector (Aparcana 2017). That is, managing or writing policies that would work to informally or formally recognize the informal sector's practices and roles would create better, more effective solid waste management systems.

1.1.3 SWM as a Catalyst for Sustainability

The sustainable handling of solid waste is central to broader goals for environmental and economic sustainability. Recycling and recovery of resources from waste is key to not only decreasing the environmental impact of disposing of waste but also creating economic opportunities. Gaur et al. (2022) argue that recycling contributes to both reducing the total amount of waste sent to landfills and recovering valuable resources for the production process. Instead of a waste-centric disposal model, recovery rather than disposal is crucial for sustainable development in urban development. For example, several researchers point to the need to embrace the 3Rs-Reduce, Reuse, Recycle-as the basis for waste management again, moving past a disposal-oriented framework (Mohan et al. 2024).

In India, Dasgupta et al. (2017) proposed a system dynamics model to better understand waste generation, informal recycling and formal waste collection systems and how these interact as they seek to understand how policy and intervention would lead to more sustainable SWM, particularly in terms of improving solid waste segregation and resource recovery. The initial step to improve recycling is to have households separate organics from recyclables from trash - creating a source separation program. Abdel-Shafy and Mansour (2018), argue that not only should households separate organics from trash and recyclables fairly easily contribute to improved recycling rates but would also lessen the environmental burden of waste (with more well-planned collection systems potentially capturing more and better recyclables from households).

1.1.4 Technological and Policy Interventions

To ensure sustainable SWM, cities should invest in both technological and policy responses to promote sustainable and efficient waste collection, processing, and recycling. Technological innovations, including waste-to-energy plants and composting facilities, provide adequate means for managing waste rather than following conventional waste disposal practices. For instance, the city of Chennai, has experimented with Refuse-Derived Fuel (RDF) pellets to turn waste into energy, thereby replacing landfill disposal with renewable energy production (Kalyani and Pandey 2014). Composting is also an effective way of managing organic waste, which is often an important part of the MSW stream in many cities (Manea et al. 2024). The centralized construction of composting facilities can remove a stream of waste from the waste stream and produce a much-needed product, compost for agriculture.

As well as technological interventions, policy responses must be undertaken to improve SWM systems. To achieve sustainable SWM funding is critical, but also local community involvement, the participation of private sector players and informal waste workers, is also key. Strong institutional structures, appropriate regulations, and financial incentives that promote reduction and reuse are required to meet SWM sustainably (Batista et al. 2021). Furthermore, as noted by Maryanti et al. (2017), suggestions for stakeholder engagement, including community-based organizations and NGOs, are necessary to achieve significant improvements within SWM systems.

1.1.5 Global Perspectives on SWM

Around the world, the ongoing need for enhanced SWM has come to be recognized as a pillar for sustainable development. The United Nations has operationalized sustainable waste management under its umbrella for achieving the Sustainable Development Goals (SDGs). Namely, SDG 11, which is committed to making cities inclusive, safe, resilient, and sustainable, identifies SWM as a significant contributing factor to decreasing the negative environmental implications of cities when implemented properly. The Millennium Development Goals (MDGs) similarly acknowledged that improved waste management practices could enhance urban livability while eradicating poverty (Abubakar and Aina 2019).

The importance of SWM, however, extends beyond simply dealing with waste within urban areas. SWM is fundamentally important to public health, environmental sustainability, and economic resilience. Therefore, an effective SWM system should take a 'whole of system' approach, consider both formal and informal arrangements, help build a circular economy based on recycling and resource recovery, and engage stakeholders at every level. In adopting sustainable waste management practices to preside within urban areas, cities can lower their environmental impacts, positively impact public health outcomes, and create an economic bottom line for marginalized populations. The process of implementing the necessary technological, policy, and behavioral changes for effective, sustainable SWM systems is the challenge.

1.2 Historical Evolution of Solid Waste Management

The history of solid waste management reflects the evolution of societies and their interaction with the environment. In its simplest form waste is any material that an owner intends to abandon. The sense that its understanding has had historically, however, has differed greatly (see Fig. 2). In ancient and medieval cities, the streets were often used as receptacles for all kinds of waste including the excreta of both people and animals', organic materials from households and artisanal activities, as well as rubble from demolitions (Barles 2014). The inadequacy of waste collection methods resulted in the occurrence of serious public health problems, contributing to soaring urban mortality levels.

In the 18th century, the emergence of Neo-Hippocratic medicine brought attention to the connection between environmental conditions and public health, which helped to create systems for waste management (Mann 2012). In Europe, cities started many policies to enhance urban sanitation, especially cleaning the streets of waste and implementing more systematic collections (Leal et al. 2016). At the same time waste also was of economic value, especially organic waste as fertilizer. H uman and animal excreta collected were returned to agricultural lands, closing the loop between urban environments and the countryside. There was a rise in recycling during this period, particularly due to the demand for raw materials created by new industries (Barles 2014).

From the end of the 19th century, urbanization and industrialization further added complexity to waste management (Zohoori and Ghani 2017). The availability of synthetic fertilizers and industrial raw materials, such as coal and other petroleum by-products, diminished the market value of urban waste (Odejobi et al. 2024). For example, animal bones, which were once important for constructing

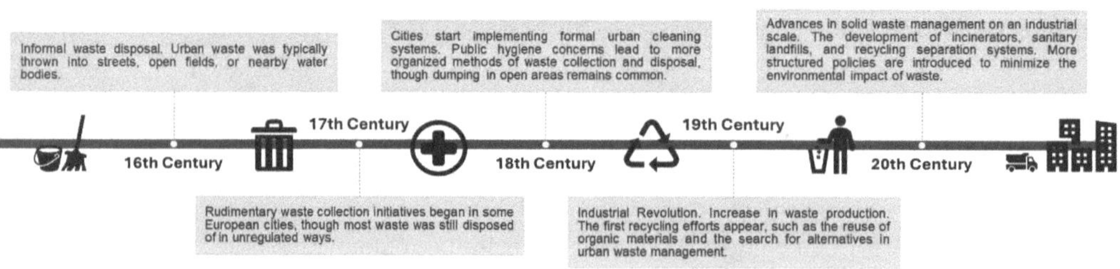

Fig. 2. Historical Evolution of Solid Waste Management from the 16th to 20th Century.

items or for use as fertilizer, were gradually supplanted by items made from synthetic materials such as plastics or artificial chemicals. With consumption patterns normalized around standardized, mass-produced products, and costs reduced by outsourcing/to Asian offshore plants, larger amounts of urban waste material were generated, and waste management became more difficult. In parallel, incineration was adopted in some cities, notably in Great Britain, to reduce the volume of waste, but this posed a new challenge as incineration schemes would often eliminate valuable recyclable materials (Cole et al. 2019).

The 20th century brought a significant change in how waste was perceived and addressed, particularly with the emergence of environmental consciousness in the 1960s and 1970s (Marshall and Farahbakhsh 2013). Waste was emblematic of the unsustainable nature of a consumer-driven society and policies to resolve the increasingly problematic waste crisis began to emerge. Legislation was enacted in the United States and Europe valuing source reduction rather than waste management in itself, followed by the introduction of recycling and waste-to-energy policies (Malinauskaite et al. 2017). The "polluter pays" concept was crafted to contribute to addressing waste, enabling people to internalize the costs of the environmental impact of waste, in terms of cost and merit, enabling waste generators to assist in covering waste disposal management costs (Gachenga 2022).

Notwithstanding these enhancements, the waste management apparatus remained under pressure. The level of municipal and industrial waste continued to increase, with waste from packaging and new consumer products significantly exacerbating the situation. Local authorities struggled to maintain inter-generational political equity concerning services being offered to citizens, especially with the costs of waste collection, waste treatment and waste recycling. For example, although recycling became more ubiquitous, the demand for and interactions within the sometimes-limited markets for secondary raw materials, like recycled plastics, often developed only tentatively and in response to cheaper material prices derived from fossil fuels.

In developing countries, this was markedly worse. With many levers and regulations for generating value through wealth wasted or considered waste subject to fees only before disposal, such as fines for creating noise, an underfunded waste collection and treatment system and massive urbanization increasing risks to human and environmental health were now often outpacing collection or implementation activities. While informal waste pickers can save valuable recyclables, they face many obstacles presented by not being recognized, and not having access or support from local officials. Informal waste pickers are often considered vital to recycling, in many parts of the world, while also working under unsafe environments that need to be improved for their safety.

Currently, solid waste management continues to have the dual focus of protecting environmental concerns while optimizing existing systems. This means generating any possible efficiencies from waste management systems to not only treat as well as dispose of waste-to-energy but to address source reduction and re cycling. Technology solutions, like automated sorting, and all waste to energy conversions can play an immediate role in reducing the amount of material exposed as landfilled. Lastly, increasing public knowledge on the need for community-level participation in their re cycling and/or source reduction efforts, all with the potential rewards or prices in the circular economy is equally as valuable.

In addition to the above, the goal of optimizing waste management includes closing material loops through product and package design innovations that will lead to easier recycling or reuse. Industries are looking at ideas, like industrial symbiosis, where the waste or by-products of one industry may be considered raw material for a different industry minimizing waste. Innovative technologies such as advanced biological and chemical treatments of organic waste and plastics will enable us to recover valued materials that before innovation appeared unrecoverable.

Furthermore, public policy is also critical in advancing the capacity to develop a viable and sustainable waste management infrastructure, which may include improving the waste collection capacity in urban and rural are as, disposing of hazardous waste properly, and incorporating urban planning needs with long-term waste management needs. A systems approach showcasing a combination of technology, regulation, and citizen action will offer solid waste initiatives the best opportunity to optimize waste management and much more holistically a sustainable model.

1.3 Emergence of Optimization in Solid Waste Management

Optimization in solid waste management has become an important strategy to deal with environmental and economic issues related to managing large amounts of waste. Using mathematical models, and multi-objective approaches, a system can be designed with the requisite efficiencies to reduce costs of operation and adverse environmental impacts. Optimization methodologies also enable better adoption of more advanced technologies like Waste-to-Energy (WtE) technologies, landfill biogas recovery, as well as Material Recovery Facilities (MRFs), and thus lead to a more efficient solid waste management process as a whole (Vlachokostas et al. 2021, Abdallah et al. 2021, Ooi et al. 2021).

An important space for optimization in solid waste management is the use of WtE conversion technologies, which change non-recyclable waste into electricity, or heat by thermochemical processes such as incineration or gasification. However, the energy efficiency of WtE conversion depends on the type of technology used, as well as the type of solid waste used as the input stream. It is here that optimization can play a valuable role in the enhancement of energy efficiency by the selection of technologies, as well as optimal operating parameters. Optimization techniques have been shown to improve the efficiency of WtE processes on the order of 12% to 27% making the facilities far more viable by incorporating less fossil fuel and producing less greenhouse gas emissions (Salem et al. 2023.)

Another application of optimization in solid waste management is in biogas recovery systems at landfills. Biogas systems have only been able to recover a small percentage of the methane generated. Optimization can greatly increase methane recovery and recovery utilization. Mathematically based models and scenario exploration can be employed to evaluate the best locations and operations of biogas recovery, maximizing methane recovery while minimizing losses. A follow-up localization study in Spain demonstrated that using profitability optimization studies, the biogas recovery systems could have a positive pay back in 3 years of operation, whereas globally landfill methane captures are less than 10% of emitted methane (Vidal-Barrero et al. 2022).

MRFs are an area where optimization is invaluable as well. MRFs allow for the various mixtures of waste to be sorted out into a maximized recovery of recyclable materials which is important for waste to be diverted from landfills. Therefore, optimization applies to evaluating the optimal location of MRF facilities and the operational design of automated sorting systems for operational efficiency. New design sensors have improved sorting accuracy such as Laser-Induced Breakdown Spectroscopy (LIBS) and machine vision type sensors, which can reduce operational costs while substantially improving the recovery of valuable recyclables (Araujo-Andrade et al. 2021).

Among the most effective approaches to optimizing these systems is using multi-objective models that can strike a balance between various aspects, ranging from costs to emissions to operational efficiency. This is especially effective in WtE systems and MRFs in which optimizing is not just about cutting costs, but also minimizing emissions while improving the efficiency of energy recovery and material recycling. Multi-objective models also allow waste managers to evaluate different alternatives and select the option that best fits specific sustainability and profitability goals (Ooi et al. 2021).

Table 1 summarizes various optimization methodologies considered in solid waste management technologies.

Optimization allows solid waste management systems to improve to become more efficient and sustainable. New models incorporating economic, environmental, and operational attributes can enhance the performance of key technologies, including WtE conversion, biogas recovery, and MRFs, concerning the environmental impact of disposal and sustainability of the overall system. Optimization of technology can substantially reduce the need to send solid waste materials to landfills.

Table 1. Optimization Approaches in Solid Waste Management Technologies.

Technology	Optimization Objective	Impact
Waste-to-Energy (WtE) Conversion	Maximize energy efficiency and reduce emissions	Increased energy production and reduced landfill waste
Landfill Biogas Recovery	Maximize methane capture and minimize losses	Reduced methane emissions and improved profitability
Material Recovery Facilities (MRFs)	Optimize recyclable material separation	Higher resource recovery rates and reduced disposal costs

1.4 Contributions of Optimization to Solid Waste Management

The application of optimization techniques in SWM plays a critical role in streamlining operations, improving resource allocation, and reducing environmental footprints. Beyond the more commonly discussed areas like WtE and landfill biogas recovery, several under-explored areas reveal how optimization can enhance SWM in novel ways.

One significant area is the optimization of collection and transportation logistics, where route planning for waste collection vehicles can dramatically cut costs and emissions. By employing algorithms such as the Vehicle Routing Problem (VRP) and its variants, municipalities can optimize the number of vehicles used, the sequence of pick-ups, and the distances traveled, reducing fuel consumption and operational expenses. For example, adaptive and dynamic routing systems consider real-time factors like traffic conditions and waste bin fill levels to adjust routes on the fly, significantly improving collection efficiency (Yang et al. 2022). This not only enhances the cost-effectiveness of SWM operations but also reduces the environmental impact by minimizing the carbon footprint associated with waste transport (Das et al. 2019).

Optimization also contributes to the design of recycling systems. Sophisticated optimization models help determine the best allocation of recycling resources, such as the number and location of recycling centers, the type of materials processed, and the sorting methods used. For instance, Multi-Criteria Decision-Making (MCDM) models can weigh environmental, economic, and social factors to optimize recycling processes and ensure that the most sustainable options are prioritized. These techniques enhance the effectiveness of recycling by improving recovery rates and reducing contamination, thereby promoting a circular economy (Garcia-Garcia 2022).

Another critical yet less frequently discussed contribution is in the optimization of landfill design and lifespan extension. By employing models that optimize the layering and compaction of waste, the volume of landfill space used can be minimized, extending the operational life of landfills. Additionally, landfill site selection optimization helps municipalities identify suitable locations for new landfills, balancing environmental risks (e.g., proximity to water sources) with logistical considerations such as accessibility and distance from urban centers (Ayyildiz and Erdogan 2024). Advanced geospatial analysis tools integrated with optimization models are increasingly used for this purpose, helping planners reduce the ecological impact of landfill placement (Spigolon et al. 2018).

Resource recovery through advanced waste sorting technologies is another area where optimization has contributed significantly. For instance, the use of linear programming and other optimization techniques in designing Material Recovery Facilities (MRFs) helps maximize the efficiency of waste segregation and material recovery processes. These models determine the optimal mix of manual and automated processes, leading to more cost-effective operations with higher throughput (Dubanowitz 2000). Optimization has also been applied to minimize energy consumption in these facilities, making recycling processes more sustainable (Rizwan et al. 2020).

Optimization plays a critical role in policy design and implementation. Municipalities and governments use optimization models to evaluate the impacts of various waste management policies, such as pay-as-you-throw schemes or recycling incentives. These models simulate different policy scenarios, helping decision-makers understand how changes in waste tariffs, regulations, or public awareness campaigns might affect waste generation and diversion rates (Zhang et al. 2021). Optimization in policy design ensures that resources are directed toward the most effective and equitable solutions, promoting sustainable waste management practices that are also economically viable (Roy et al. 2023).

The contributions of optimization to SWM extend beyond traditional methods, integrating innovative approaches such as Artificial Intelligence (AI), Machine Learning (ML), and the Internet of Things (IoT) to enhance the efficiency of collection, sorting, and waste treatment. These advancements offer significant potential to improve resource recovery, reduce costs, and mitigate environmental impacts.

AI and ML have become instrumental in the automation of the waste sorting processes. Modern SWM systems increasingly utilize visual recognition technology powered by deep learning algorithms such as convolutional neural networks (CNNs) to identify and classify various waste materials (Shahab et al. waste-sorting 2022). This method allows for the precise separation of plastics, metals, glass, and organic waste, dramatically increasing sorting accuracy. By leveraging these technologies, facilities can significantly improve the recovery of recyclable materials while reducing manual labor and associated costs (Kurniawan et al. 2023).

In addition to visual recognition, the use of robotic systems in SWM has revolutionized waste sorting. These systems, equipped with sensors and AI-driven algorithms, are capable of handling complex waste streams and making real-time decisions about material classification. Robotic arms, often combined with AI-based models such as "You Only Look Once" (YOLO), enable faster and more efficient waste processing, reducing the volume of waste sent to landfills (Farghali and Osman 2024). The integration of these systems not only optimizes sorting but also contributes to more sustainable SWM practices by enhancing resource recovery.

Another crucial area in which AI contributes is predictive analytics. ML models can analyze historical waste data, seasonal trends, and demographic shifts to predict waste generation patterns. This enables municipalities to optimize waste collection routes and schedules, ensuring that resources are deployed efficiently while minimizing fuel consumption and greenhouse gas emissions. For example, by analyzing real-time data from smart bins, AI-driven algorithms can notify waste collection teams when bins are full, leading to more effective collection and reduced operational costs (Andeobu et al. 2022).

The role of IoT in SWM optimization cannot be overlooked. Smart bins embedded with IoT sensors and AI algorithms provide continuous monitoring of waste levels, composition, and bin conditions. This real-time data is transmitted to a central system, allowing waste management operators to track bin usage and optimize collection services based on demand. Additionally, these systems can detect hazardous materials, enabling immediate response and preventing potential environmental harm (Sosunova and Porras 2022).

Optimization techniques also enhance the efficiency of WtE processes. By integrating AI-based models, WtE facilities can optimize the incineration process, improving energy recovery rates while minimizing emissions. AI helps in the dynamic adjustment of operating conditions to ensure maximum energy extraction from waste, resulting in a cleaner and more viable solution for waste disposal (Huang and Koroteev 2021).

1.5 Challenges and Advantages of Applying Optimization in Solid Waste Management

The application of optimization techniques in SWM brings both significant advantages and unique challenges, particularly as cities strive to address increasing waste volumes, environmental concerns, and resource limitations. By analyzing the complexities of implementing optimization solutions, this section highlights both the potential and obstacles that cities and municipalities face when integrating advanced technologies into their waste management systems.

1.5.1 Challenges in Implementing Optimization in SWM

1. **Technological Limitations and Infrastructure Gaps:** One of the primary challenges in applying optimization techniques to SWM is the technological infrastructure required to support sophisticated systems such as Artificial Intelligence (AI) and the Internet of Things (IoT). Many cities, especially in developing countries, lack the necessary technological foundation for real-time data collection and processing, which are critical to advanced waste management systems. Without smart bins, sensors, and high-speed data networks, cities are unable to fully leverage the potential of optimization techniques (Gatti et al. 2024).
2. **Financial and Resource Constraints:** Implementing optimization solutions, particularly those relying on AI and robotics, requires significant initial investments. This includes the costs of installing advanced sorting systems, developing predictive models, and integrating IoT technology for real-time waste monitoring. Many municipalities face budget constraints, which may limit their ability to adopt these technologies on a wide scale. Furthermore, the ongoing costs of maintaining these systems, including regular updates and technical support, add to the financial burden (Das et al. 2019). These factors can deter cities from committing to large-scale optimization programs.
3. **Data Availability and Quality:** The effectiveness of optimization in SWM depends heavily on the availability and quality of data. AI and Machine Learning (ML) algorithms require substantial datasets to accurately predict waste patterns, optimize collection routes, and improve sorting efficiency. In many cases, data is either insufficient or not properly organized, which undermines the potential benefits of optimization efforts (Nishant et al. 2020). Additionally, waste generation can be highly variable, influenced by cultural, seasonal, and regional factors, making it challenging to create universally applicable optimization models.
4. **Integration with Existing Systems:** Another challenge is the integration of optimization technologies with existing waste management infrastructure. Cities that already have established systems may face difficulties incorporating new technologies without disrupting their current operations. This includes the challenge of retraining personnel, modifying logistics, and adapting to new protocols that automation and AI-driven processes introduce. Additionally, legacy waste management systems may be incompatible with new technologies, necessitating significant structural changes.
5. **Public and Stakeholder Engagement:** The success of optimization initiatives in SWM also depends on the active participation of the public and other stakeholders. Policies that encourage waste segregation at the source, for example, are critical to the success of AI-driven sorting technologies. However, public resistance or lack of awareness can hinder the effectiveness of such programs. Further, involving the informal waste sector, which plays a significant role in many cities' recycling efforts, requires careful consideration to ensure that technological advancements do not marginalize these workers (Aparcana 2017).

1.5.2 Advantages of Applying Optimization in SWM

1. **Enhanced Efficiency and Cost Savings:** These are o ne of the greatest advantages of applying optimization techniques in SWM is the potential for enhanced efficiency. AI-powered systems improve the sorting of waste materials, maximizing resource recovery and reducing the amount of waste sent to landfills. Predictive analytics can optimize collection routes, reducing fuel consumption and labor costs. According to Santibañez-Aguilar et al. (2013), cities that have integrated optimization techniques have seen a significant reduction in operational costs, particularly in waste collection and sorting.
2. **Environmental Benefits Optimization:** These not only improve operational efficiency but also contribute to environmental sustainability. By optimizing WtE processes and improving

recycling rates, cities can reduce their reliance on landfills, which are a major source of methane emissions. Additionally, better resource recovery helps conserve natural resources and reduce the environmental footprint of waste disposal. The implementation of smart waste management systems, supported by AI and IoT, also allows for more accurate monitoring and regulation of waste-related environmental impacts (Czekała et al. 2023).

3. **Improved Public Health Effective:** SWM, supported by optimization technologies, directly impacts public health by ensuring timely and efficient waste collection and disposal. In urban areas where waste often accumulates and leads to health hazards, optimization systems can minimize the risk of disease transmission by reducing the amount of unmanaged waste. AI-driven technologies enable better monitoring of hazardous waste, ensuring that dangerous materials are identified and disposed of correctly (Reza and Hassan 2023).
4. **Scalability and Adaptability:** Another advantage of optimization in SWM is its scalability. Optimization models can be adapted to different urban environments, from small cities to large metropolitan areas. Technologies like IoT and AI can adjust to the specific waste generation patterns of a city, whether it involves seasonal variations, demographic shifts, or industrial waste (Maiurova et al. 2022). This adaptability makes optimization a valuable tool for cities looking to continuously improve their waste management practices.
5. **Supporting Circular Economies:** Finally, the use of optimization technologies in SWM supports the development of circular economies, where waste is treated as a resource rather than a byproduct. AI-powered sorting systems improve the recovery of valuable materials like metals, plastics, and glass, which can be reintroduced into production cycles. By promoting recycling and resource recovery, optimization technologies contribute to a more sustainable economic model where waste management plays a key role in resource conservation (Nižetić et al. 2019).

While the implementation of optimization techniques in SWM presents numerous challenges, including technological limitations, financial constraints, and data quality issues, the advantages far outweigh the obstacles. The integration of AI, IoT, and predictive analytics offers promising opportunities to improve the efficiency, sustainability, and environmental impact of solid waste management systems. As cities continue to grow and waste generation increases, optimization technologies will play an increasingly critical role in shaping the future of urban waste management.

By leveraging these technologies, cities can move towards more sustainable and efficient SWM practices, ultimately improving public health, environmental outcomes, and economic resilience.

1.6 References

Abdallah, M., Rahmat-Ullah, Z., and Metawa, A. 2021. Dynamic multi-objective optimization of integrated waste management using genetic algorithms. In *Soft Computing Techniques in Solid Waste and Wastewater Management* (pp. 257–274). Elsevier.

Abdel-Shafy, H. I., and Mansour, M. S. 2018. Solid waste issue: Sources, composition, disposal, recycling, and valorization. *Egyptian Journal of Petroleum*, *27*(4), 1275–1290.

Abubakar, I. R., and Aina, Y. A. 2019. The prospects and challenges of developing more inclusive, safe, resilient and sustainable cities in Nigeria. *Land Use Policy*, *87*, 104105.

Andeobu, L., Wibowo, S., and Grandhi, S. 2022. Artificial intelligence applications for sustainable solid waste management practices in Australia: A systematic review. *Science of The Total Environment*, *834*, 155389.

Aparcana, S. 2017. Approaches to formalization of the informal waste sector into municipal solid waste management systems in low-and middle-income countries: Review of barriers and success factors. *Waste Management*, *1*, 593–607.

Araujo-Andrade, C., Bugnicourt, E., Philippet, L., Rodriguez-Turienzo, L., Nettleton, D., Hoffmann, L., and Schlummer, M. 2021. Review on the photonic techniques suitable for automatic monitoring of the composition of multi-materials wastes in view of their posterior recycling. *Waste Management & Research*, *39*(5), 631–651.

Ayyildiz, E., and Erdogan, M. 2024. Literature analysis of the location selection studies related to the waste facilities within MCDM approaches. *Environmental Science and Pollution Research*, 1–22.

Barles, S. 2014. History of waste management and the social and cultural representations of waste. In *The Basic Environmental History* (pp. 199–226). Cham: Springer International Publishing.

Batista, M., Caiado, R. G. G., Quelhas, O. L. G., Lima, G. B. A., Leal Filho, W., and Yparraguirre, I. T. R. 2021. A framework for sustainable and integrated municipal solid waste management: Barriers and critical factors to developing countries. *Journal of Cleaner Production*, *312*, 127516.

Cheela, V. S., Ranjan, V. P., Goel, S., John, M., and Dubey, B. 2021. Pathways to sustainable waste management in Indian Smart Cities. *Journal of Urban Management*, *10*(4), 419–429.

Cole, C., Gnanapragasam, A., Cooper, T., and Singh, J. 2019. An assessment of achievements of the WEEE Directive in promoting movement up the waste hierarchy: experiences in the UK. *Waste Management*, *87*, 417–427.

Colenbrander, S., Gouldson, A., Roy, J., Kerr, N., Sarkar, S., Hall, S., Sudmant, A, Ghatak, A, Chakravarty, D, Gangly, D., and Mcanulla, F. 2017. Can low-carbon urban development be pro-poor? The case of Kolkata, India. *Environment and Urbanization*, *29*(1), 139–158.

Czekała, W., Drozdowski, J., and Łabiak, P. 2023. Modern technologies for waste management: A review. *Applied Sciences*, *13*(15), 8847.

Das, S., Lee, S. H., Kumar, P., Kim, K. H., Lee, S. S., and Bhattacharya, S. S. 2019. Solid waste management: Scope and the challenge of sustainability. *Journal of Cleaner Production*, *228*, 658–678.

Dasgupta, D., Debsarkar, A., Hazra, T., Bala, B. K., Gangopadhyay, A., and Chatterjee, D. 2017. Scenario of future e-waste generation and recycle-reuse-landfill-based disposal pattern in India: a system dynamics approach. *Environment, Development and Sustainability*, *19*, 1473–1487.

Dubanowitz, A. J. 2000. Design of a materials recovery facility (MRF) for processing the recyclable materials of New York City's municipal solid waste. *Columbia University*.

Farghali, M., and Osman, A. I. 2024. Revolutionizing waste management: unleashing the power of artificial intelligence and machine learning. In *Advances in Energy from Waste* (pp. 225–279). Woodhead Publishing.

Gachenga, E. 2022. Linear to circular waste policies: Breathing life into the polluter pays principle?. *International Journal for Crime, Justice and Social Democracy*, *11*(2), 61–73.

Garcia-Garcia, G. 2022. Using Multi-Criteria Decision-Making to optimise solid waste management. *Current Opinion in Green and Sustainable Chemistry*, *37*, 100650.

Gatti, A., Barbierato, E., and Pozzi, A. 2024. Toward Greener Smart Cities: A Critical Review of Classic and Machine-Learning-Based Algorithms for Smart Bin Collection. *Electronics*, *13*(5), 836.

Gaur, A., Gurjar, S. K., and Chaudhary, S. 2022. Circular system of resource recovery and reverse logistics approach: key to zero waste and zero landfill. In *Advanced Organic Waste Management* (pp. 365–381). Elsevier.

Hayat, P. 2023. Integration of advanced technologies in urban waste management. In *Advancements in Urban environmental studies: Application of geospatial technology and artificial intelligence in Urban studies* (pp. 397–418). Cham: Springer International Publishing.

Hazra, T., and Goel, S. 2009. Solid waste management in Kolkata, India: Practices and challenges. *Waste management*, *29*(1), 470–478.

Huang, J., and Koroteev, D. D. 2021. Artificial intelligence for planning of energy and waste management. *Sustainable Energy Technologies and Assessments*, *47*, 101426.

Kalyani, K. A., and Pandey, K. K. 2014. Waste to energy status in India: A short review. *Renewable and Sustainable Energy Reviews*, *31*, 113–120.

Kumar, G., Vyas, S., Sharma, S. N., and Dehalwar, K. 2024. Challenges of Environmental Health in Waste Management for Peri-urban Areas. In *Solid Waste Management: Advances and Trends to Tackle the SDGs* (pp. 149–168). Cham: Springer Nature Switzerland.

Kurniawan, T. A., Meidiana, C., Othman, M. H. D., Goh, H. H., and Chew, K. W. 2023. Strengthening waste recycling industry in Malang (Indonesia): Lessons from waste management in the era of Industry 4.0. *Journal of Cleaner Production*, *382*, 135296.

Leal Filho, W., Brandli, L., Moora, H., Kruopienė, J., and Stenmarck, Å. 2016. Benchmarking approaches and methods in the field of urban waste management. *Journal of Cleaner Production*, *112*, 4377–4386.

Lissah, S. Y., Ayanore, M. A., Krugu, J. K., Aberese-Ako, M., and Ruiter, R. A. 2021. Managing urban solid waste in Ghana: Perspectives and experiences of municipal waste company managers and supervisors in an urban municipality. *PloS one*,*16*(3), e0248392.

Maiurova, A., Kurniawan, T. A., Kustikova, M., Bykovskaia, E., Othman, M. H. D., Singh, D., and Goh, H. H. 2022. Promoting digital transformation in waste collection service and waste recycling in Moscow (Russia): Applying a circular economy paradigm to mitigate climate change impacts on the environment. *Journal of Cleaner Production*, *354*, 131604.

Malinauskaite, J., Jouhara, H., Czajczyńska, D., Stanchev, P., Katsou, E., Rostkowski, P., Thorne, R. J.,Colón, J., Ponsá, S., Al-Mansour, F., Anguilano, L., Krzyżyńska, R., López, I. C., Vlasopoulos, A., and Spencer, N. 2017. Municipal solid waste management and waste-to-energy in the context of a circular economy and energy recycling in Europe. *Energy*, *141*, 2013–2044.

Manea, E. E., Bumbac, C., Dinu, L. R., Bumbac, M., and Nicolescu, C. M. 2024. Composting as a Sustainable Solution for Organic Solid Waste Management: Current Practices and Potential Improvements. *Sustainability*, *16*(15), 6329.

Mann, A. 2012. Waste management: Tobias Smollett and remediation. *Eighteenth-Century Fiction*, *25*(2), 359–382.

Marshall, R. E., and Farahbakhsh, K. 2013. Systems approaches to integrated solid waste management in developing countries. *Waste Management*, *33*(4), 988–1003.

Maryanti, D. F. 2017. *Performance of community-based solid waste management for integrated and sustainable solid waste management: The case of Bogor City, Indonesia* (Doctoral dissertation, UNESCO-Ihe).

Masud, M. H., Mourshed, M., Hossain, M. S., Ahmed, N. U., and Dabnichki, P. 2023. Generation of waste: problem to possible solution in developing and underdeveloped nations. In *Waste Management and Resource Recycling in the Developing World* (pp. 21–59). Elsevier.

Mohan, J., Mohanty, C. R. C., Cross, J. S., and Khajuria, A. 2024. Mapping 3R and Circular Economy Policy Implementation in Asia and the Pacific. *Circular Economy and Sustainability*, *4*(1), 671–692.

Nishant, R., Kennedy, M., and Corbett, J. 2020. Artificial intelligence for sustainability: Challenges, opportunities, and a research agenda. *International Journal of Information Management*, *53*, 102104.

Nižetić, S., Djilali, N., Papadopoulos, A., and Rodrigues, J. J. 2019. Smart technologies for promotion of energy efficiency, utilization of sustainable resources and waste management. *Journal of Cleaner Production*, *231*, 565–591.

Odejobi, O. J., Ajala, O. O., and Osuolale, F. N. 2024. Review on potential of using agricultural, municipal solid and industrial wastes as substrates for biogas production in Nigeria. *Biomass Conversion and Biorefinery*, *14*(2), 1567–1579.

Omar, H. M. 2018. Sustainability of informal solid waste collection livelihood in urban areas: A case of Kinondoni municipality, Dar Es Salaam, City. Tanzania. *Unpublished PhD Thesis of the Open University of Tanzania*.

Ooi, J. K., Woon, K. S., and Hashim, H. 2021. A multi-objective model to optimize country-scale municipal solid waste management with economic and environmental objectives: A case study in Malaysia. *Journal of Cleaner Production*, *316*, 128366.

Oyebode, O. J. 2024. Promoting integrated and sustainable solid waste management system in A developing city for public health and cleaner environment. *Journal of Harbin Engineering University*, *45*(01).

Ravichandran, C., and Venkatesan, G. 2021. Toward sustainable solid waste management–challenges and opportunities. *Handbook of Advanced Approaches Towards Pollution Prevention and Control*, 67–103.

Reza, M., and Hassan, M. 2023. AI-Driven solutions for enhanced waste management and recycling in urban areas. *International Journal of Sustainable Infrastructure for Cities and Societies*, *8*(2), 1–13.

Rizwan, M., Saif, Y., Almansoori, A., and Elkamel, A. 2020. A multiobjective optimization framework for sustainable design of municipal solid waste processing pathways to energy and materials. *International Journal of Energy Research*, *44*(2), 771–783.

Roy, S., Rautela, R., and Kumar, S. 2023. Towards a sustainable future: Nexus between the sustainable development goals and waste management in the built environment. *Journal of Cleaner Production*, *415*, 137865.

Salem, K. S., Clayson, K., Salas, M., Haque, N., Rao, R., Agate, S., Singh, A., Levis, J., Mittal, A., Yarbrough, J., Venditti, R., Jameel, H., Lucia, L., and Pal, L. 2023. A critical review of existing and emerging technologies and systems to optimize solid waste management for feedstocks and energy conversion. *Matter, 6*, 3348–3377.

Sanneh, E. S., Hu, A. H., Chang, Y. M., and Sanyang, E. 2011. Introduction of a recycling system for sustainable municipal solid waste management: a case study on the greater Banjul area of the Gambia. *Environment, Development and Sustainability*, *13*, 1065–1080.

Santibañez-Aguilar, J. E., Ponce-Ortega, J. M., González-Campos, J. B., Serna-González, M., and El-Halwagi, M. M. 2013. Optimal planning for the sustainable utilization of municipal solid waste. *Waste Management*, *33*(12), 2607–2622.

Shahab, S., Anjum, M., and Umar, M. S. 2022. Deep learning applications in solid waste management: A deep literature review. *International Journal of Advanced Computer Science and Applications*, *13*(3).

Siddiqua, A., Hahladakis, J. N., and Al-Attiya, W. A. K. 2022. An overview of the environmental pollution and health effects associated with waste landfilling and open dumping. *Environmental Science and Pollution Research*, *29*(39), 58514–58536.

Sosunova, I., and Porras, J. 2022. IoT-enabled smart waste management systems for smart cities: A systematic review. *IEEE Access*, *10*, 73326–73363.

Spigolon, L. M., Giannotti, M., Larocca, A. P., Russo, M. A., and Souza, N. D. C. 2018. Landfill siting based on optimisation, multiple decision analysis, and geographic information system analyses. *Waste Management & Research*, *36*(7), 606–615.

Vidal-Barrero, F., Baena-Moreno, F. M., Preciado-Cárdenas, C., Villanueva-Perales, A., and Reina, T. R. 2022. Hydrogen production from landfill biogas: Profitability analysis of a real case study. *Fuel*, *324*, 124438.

Vlachokostas, C., Michailidou, A. V., and Achilles, C. 2021. Multi-criteria decision analysis towards promoting waste-to-energy management strategies: a critical review. *Renewable and Sustainable Energy Reviews*, 138, 110563.

Wilson, D. C., Araba, A. O., Chinwah, K., and Cheeseman, C. R. 2009. Building recycling rates through the informal sector. *Waste Management*, *29*(2), 629–635.

Yang, J., Tao, F., and Zhong, Y. 2022. Dynamic routing for waste collection and transportation with multi-compartment electric vehicle using smart waste bins. *Waste Management & Research*, *40*(8), 1199–1211.

Yıldız, T. D., Güner, M. O., and Kural, O. 2024. Effects of EU-compliant mining waste regulation on Turkish mining sector: A review of characterization, classification, storage, management, recovery of mineral wastes. *Resources Policy*, 90, 104836.

Zhang, J., Qin, Q., Li, G., and Tseng, C. H. 2021. Sustainable municipal waste management strategies through life cycle assessment method: A review. *Journal of Environmental Management*, *287*, 112238.

Zohoori, M., and Ghani, A. 2017. Municipal solid waste management challenges and problems for cities in low-income and developing countries. *International Journal of Science and Engineering Applications*, *6*(2), 39–48.

2

Processing Technologies for Municipal Solid Waste

Solid waste generation is a consequence of human activity that represents a problem that produces adverse effects on health and the environment. As urban populations increase in size, waste management becomes increasingly complicated. In this context, to limit the negative effects of trash accumulation, this chapter investigates the application of various technologies to treat the many forms of solid waste produced. This involves analyzing the use of technologies aimed at the generation of new materials and technologies focused on the conversion of waste to energy according to the physicochemical, biological, and even economic characteristics associated with each type of waste. This chapter provides insights into the diverse approaches available for effectively managing solid waste and supporting a transition toward a circular economy.

2.1 Introduction

Current technological advances have introduced new and efficient technologies considering different economic and environmental aspects, the tendency of which is to displace traditional technologies such as landfilling and incineration, which produce strong environmental impacts, within the next decades. Although there is a wide range of processing methods for solid waste, the technology to be adopted will rely on various criteria, such as the nature of the waste, its economic potential, or even the environmental impact generated by its processing. For example, organic waste may be appropriate for technologies such as composting or anaerobic digestion, since they allow the production of bioproducts such as compost or biogas, while when it comes to mixtures of waste or hazardous waste, a treatment such as incineration or pyrolysis can help mitigate the risks of this waste while its volume is significantly reduced and electricity and heat are generated in the process. For this reason, the urban and industrial sectors need to understand which technologies are suitable for each type of waste generated, in such a way that it allows them to minimize the environmental and economic impact in favor of a sustainable society.

In the following subsections of this chapter, readers will be introduced to some of the most common technologies for solid waste treatment. These technologies vary widely in approach and application, reflecting the diversity of waste materials and the specific needs of different environments. By exploring the available options, this chapter aims to highlight the importance of selecting appropriate treatment methods to enhance resource recovery, minimize environmental impact, and support sustainable waste management practices.

2.2 Overview of Solid Waste Processing

Effective methods such as physical, thermal, and chemical recycling technologies are needed to recycle plastics (Kalantar et al. 2012, Hur et al. 2003) (Table 1). Gasification is a proper alternative to the thermal recycling of plastic waste to produce a synthesis gas (syngas). Also, plastic waste can be transformed into carbon nanotubes (Alireza and Gordon 2012, Wu et al. 2012).

There are several routes for recycling paper wastes, including thermal processing and landfilling, which do not typically require sorting, while others involve the separation of paper; for example, paper digestion can be used for papermaking (Hanan et al. 2012) and fermentation to yield bioethanol (Wang et al. 2012), mixed alcohols, acids, ketones, or fuel (Pham et al. 2010).

Biowastes (food and garden wastes) can be used to obtain bioproducts through composting, anaerobic digestion, or another processing technology; or even to obtain biofuels like biogas or bioethanol; however, their composition can be very different and variable and the determination of production factors can be complicated, this type of wastes are taken into account with the mix of the remaining waste after the separation of plastics, glasses, paper, and metal.

Thermal pathways such as incineration, gasification, and pyrolysis have been proposed to treat the wastes as a mixture (Grieco and Baldi 2012). Incineration is a combustion process that uses an excess of oxygen to burn municipal solid waste. Pyrolysis is a thermal decomposition of carbon-based materials in an atmosphere without oxygen using heat to produce syngas, the process is endothermic. For the pyrolysis process, there are several variations, for example, there is pyrolysis-gasification, where another reactor is added to gasify the produced liquids from the first step of pyrolysis. Conventional gasification is a thermal process, which converts carbonaceous materials into syngas using a limited quantity of oxygen, where steam may be injected to promote CO and hydrogen production. Another variation of the gasification process is plasma arc gasification, which is a high-temperature pyrolysis process, where organic compounds are converted into syngas and inorganic materials and minerals (Young 2010).

2.3 Mechanical Sorting and Separation Technologies

Sorting and separating waste may seem like a simple task; however, the reality is much more complex. Although there are widely studied technologies for these activities, it is common to find communities where inhabitants do not perceive the potential for revalorization of the waste they generate. This causes waste to not be separated properly. If the separation is not carried out correctly from the collection stage, it can significantly affect the efficiency and costs of subsequent processes (Laurieri et al. 2020). Waste separation and classification are part of mechanical treatment processes. However, understanding the differences between the two processes can be confusing. Therefore, they are defined as follows, as presented in the work of Lahtela and Kärki (2018):

- **Classification:** Mechanical separation based on the geometric properties of the waste generated, which precedes sorting.
- **Sorting:** Mechanical separation based on the physicochemical properties of the waste generated, which is carried out after classification.

Regardless of whether it is classification or sorting, most of the available technologies require pretreatment of the waste. This pretreatment is usually based on reducing the size of the waste to facilitate its handling. Depending on the source of the waste, another common pretreatment is cleaning to remove impurities. These pretreatments are vital since excessive size reduction can generate toxic dust, which puts the health of workers at risk. In addition, there is a greater risk of loss of valuable elements in the later stages of the process (Bachér and Kaartinen 2017). Techniques used for size reduction include cutting, tearing, and pulverizing.

There is a wide catalog of sorting techniques, which have been widely studied in the literature. These techniques seek to take advantage of various physicochemical properties of materials, such as magnetism, electrical conductivity, hydrophobic properties, differences in melting points, and density, among others.

Table 1. Efficiency factors for the processing technologies to treat municipal solid waste.

Waste	Sub-waste	Technology	Product	Factor kg Product/kg Waste
Plastic	PP	Technology 1 for nanotubes	Nanotubes of 140um diameter	0.11
Plastic	PE	Technology 1 for nanotubes	Nanotubes of 140um diameter	0.16
Plastic	PET	Technology 1 for nanotubes	Nanotubes of 140um diameter	0.19
Plastic	PE	Technology 2 for nanotubes	Nanotubes of 252um diameter	0.32
Plastic	PP	Technology 2 for nanotubes	Nanotubes of 252um diameter	0.51
Plastic	PE	Material Recycling 1	Pellet	0.672
Plastic	PS	Material Recycling 1	Pellet	0.672
Plastic	PP	Material Recycling 1	Pellet	0.672
Plastic	PE	Thermal Recycling	Gasoline	0.299
Plastic	PE	Thermal Recycling	Diesel	0.089
Plastic	PE	Thermal Recycling	Heavy Oil	0.068
Plastic	PE	Chemical Recycling	Heat Energy	19.44*
Plastic	HDPE	Pyrolysis	n-Paraffins	0.432
Plastic	HDPE	Pyrolysis	n-Olefines	0.446
Plastic	PP	Pyrolysis	Branched Paraffins	0.304
Plastic	PP	Pyrolysis	Branched Paraffins	0.333
Paper	Paper	Material Recycling 2	New Paper	0.84
Metal	Aluminum	Material Recycling 3	Useful Aluminum	1
Glass	Clear Glass	Material Recycling 4	New Glass	1
Glass	Green Glass	Material Recycling 4	New Glass	1
Glass	Brown Glass	Material Recycling 4	New Glass	1
MSW	MSW	Mass Burn (Incineration)	Energy	1.9584*
MSW	MSW	Pyrolysis	Energy	2.0556*
MSW	MSW	Pyrolysis/Gasification	Energy	2.466*
MSW	MSW	Plasma Arc Gasification	Energy	2.9376*
MSW	MSW	Conventional gasification	Energy	2.466*

* MJ Energy/kg Waste

2.3.1 Well-established Technologies for Mechanical Sorting

One of the most common methods employed is manual sorting, which consists of manually separating waste, usually carried out in homes. However, despite its simplicity, this technique has been reported to have low efficiency (Yuan et al. 2015). On the other hand, there is the screen sorting technique, which takes advantage of the difference in the size of the waste, making it pass through a perforated screen, known as a trommel screen. The holes in this screen gradually vary in size, allowing the smallest waste to pass through first, while the largest remains inside. Due to its simplicity and low maintenance costs in comparison to other technologies, this technology offers substantial economic advantages. Given the large problem that plastic waste represents, many techniques have been developed specifically for its treatment. An example is magnetic separation, which allows the removal of metallic contaminants. Magnetic materials adhere to a magnetic surface, while non-magnetic materials fall by gravity. Despite its simple operating principle, its feed rate has been determined to be a key factor in ensuring the efficiency of the separation process (Ruan and Xu 2011).

2.3.2 Emerging Technologies for Mechanical Sorting

More sophisticated methods, like triboelectric, electrostatic corona, and eddy current sorting, are based on electrical conductivity and are intended to separate plastic trash from metals. However, it has been noted that when processing volume is great, these methods can result in high costs (Bakker et al. 2009).

On the other hand, separation techniques based on optical principles and images have gained popularity in the industry thanks to their versatility (Tatzer et al. 2005), since they are not limited only to the separation of plastics and metals. However, these technologies have limitations related to the size and shape of the samples, in addition to their high costs. An example of these techniques is X-ray fluorescence (XRF), whose operating principle is based on the wavelength of the X-rays emitted by the elements. Each element has its own "X-ray signature", which allows for identifying the elemental compositions of the waste. These sorting methods are highly versatile and capable of identifying and separating wood treated with hazardous preservatives, such as chromated copper arsenate (CCA), used as a pesticide. They can also separate Waste Electrical and Electronic Equipment (WEEE) by identifying flame retardants present in electronic components, which pose a significant risk to human health (GSPI 2024). XRF technology has the advantage of being highly efficient and non-destructive in separating waste. However, its application in regions with high temperatures is limited, as the detection system can overheat, affecting its performance. These kinds of technology can be combined with cutting-edge tools like machine learning algorithms, which are revolutionizing mechanical sorting by enhancing the process's accuracy and efficiency When it comes to automated sorting, convolutional neural networks (CNNs) show enormous potential for real-time material recognition, allowing plastics, metals, or paper to be identified on conveyor belts using computer vision. Furthermore, by integrating cameras and sensors, it is possible to detect features that traditional systems would not be able to easily capture (Ren 2021, Choudhary 2023). The scope of these tools is not limited solely to identification tasks, since, in terms of material flow optimization, machine learning algorithms dynamically adjust parameters such as belt speed and separator configuration, which helps to avoid bottlenecks and maximize operational efficiency (Choudhary 2023).

2.4 Biological Treatment Methods

The processing of solid waste through biological processes is one of the many alternatives for converting waste to energy (WtE), often through the production of methane or biogas. These treatments allow the revalorization of solid waste with low economic value, such as organic waste, reducing the amount destined for landfills and generating energy in the process (Istrate et al. 2021).

2.4.1 Well-established Technologies for Biological Treatment Methods

The most common methods of biological waste treatment are activated sludge processes, which are based on the use of aerobic bacteria suspended in an aerated environment to eliminate organic pollutants. This process has evolved over time, allowing the integration of new technologies that facilitate, for example, the removal of phosphorus and simultaneous nitrification-denitrification, increasing its efficiency in removing nutrients.

There is also anaerobic digestion, applied mainly to wastewater, but also compatible with solid waste such as food waste, agricultural waste and residual sludge from treatment plants. This method is based on a process in which microorganisms break down organic matter in the absence of oxygen, generating biogas and digestate as by-products.

2.4.2 Emerging Technologies for Biological Treatment Methods

Currently, innovative treatment methods are being proposed that, although still in development or being studied for scale-up, offer promising alternatives for municipal solid waste (MSW) management. Bioconversion using the Black Soldier Fly (BSF) is a prominent example; in this process, BSF larvae are used to consume various organic wastes. These larvae can then be processed into animal feed or oils, contributing to a circular economy (Van Huis 2020). This technology has gained attention in food waste management due to its ability to significantly reduce the amount of waste.

Another important advance has been in the field of anaerobic digestion, where biogas purification systems have recently been developed to convert it into biomethane, allowing its use as a substitute for natural gas (Haldar et al. 2023). Within composting, Enzyme-based Accelerated Composting stands out, a process that uses enzymes or microbial inoculants to accelerate the decomposition of organic waste. This allows composting, which could normally take months, to be notably reduced depending on the type of waste treated, which represents a considerable advance in the rapid treatment of organic waste (Gattupalli et al. 2024).

2.5 Thermal Treatment Techniques

Thermal treatment techniques are based on altering different physicochemical and biological properties of matter by adding heat and covering a wide range of technologies mainly intended for WtE.

2.5.1 Well-established Technologies for Thermal Treatment Techniques

The most common thermal treatment process for Municipal Solid Waste (MSW) management is incineration, which involves burning the waste at high temperatures (usually over 850 degrees Celsius). The great advantage of this technology is that it can reduce the volume of waste by up to 90%, and it is also capable of treating waste with associated biological risks (medical waste). In addition, the heat produced can produce steam and electricity. However, it also has the great disadvantage of generating a huge amount of emissions, which is why advanced emission controls are required to limit some pollutants such as dioxins and furans (Rogoff and Screve 2019).

2.5.2 Emerging Technologies for Thermal Treatment Techniques

With more advanced techniques in the field of thermal treatments, we have those technologies focused on maximizing energy and minimizing environmental impact. One example of these technologies is the gasification process, which produces syngas, a mixture of hydrogen and carbon monoxide. The resulting product can be used directly to produce energy or as a raw material for other processes. On the other hand, there is an advanced gasification process that uses high-energy plasma to vaporize waste into syngas (plasma arc technique) (Hazra et al. 2019), which allows for minimizing process waste and maximizing energy recovery. However, this technology is still under development, and the main challenges to overcome are the high cost of using these technologies and how complex their operation is (Sanjaya and Abbas 2023).

2.6 Processing Technologies for the Type of Solid Waste

2.6.1 E-waste Processing Technologies

Two technologies are considered for processing electronic waste, which corresponds to waste from small and large electronic devices (Streicer-Porte et al. 2005). Commonly, this type of waste contains a great variety of metals with great economic potential. However, due to the large number of materials they contain, the recycling process for this type of material is limited. The most common process consists of the dismantling and separation of the main pieces, followed by a reduction in size, to end with a magnetic separation that seeks to separate metals of plastics, the separated materials are ready for sale as the raw material used to produce new electronic or other products (Kang and Schoenung 2005).

2.6.2 Metal Waste Processing Technologies

There are several alternatives to treat metal waste with different associated costs. For example, hydrometallurgy is cheap in terms of investment and has a low environmental impact since it does not generate hazardous gases or dust, which makes it ideal for small and medium-scale applications, in addition to having a high percentage of metal recovery, using a two-stage process based on metal leaching (Tuncuk et al. 2012). Furthermore, pyrometallurgy has a great disadvantage due to the high temperatures that are handled in this process and materials such as plastic can be lost. It has a high energy requirement and is not environmentally friendly (Congcong et al. 2019); however, it is possible to use this technology for large quantities of material, in addition to requiring short processing times (Marugappan and Karthikeyan 2021). Remelting is the last option proposed for the processing of metal waste, it is the simplest process, which involves melting previously separated and compressed metals in large furnaces. Although the energy requirement is high, it is still lower than obtaining the metals from their main production route (Capuzzi and Temeli 2018). Afterward, purification is commonly carried out by electrolysis or simple magnetic separation to eliminate the imperfections present in the new material (Jirang and Roven 2010).

2.6.3 Paper Waste Processing Technologies

The technology used to recycle paper depends mainly on the type of paper to be recycled. For example, waterless technology is a relatively recent technology, and its full potential has not yet been reached, therefore, it is not yet possible to recycle all types of paper with this technology (EPSON 2021), in addition to being highly demanding in terms of electrical energy. On the other hand, recycling paper using water can be used for all types of paper, and energetically speaking is much cheaper than using waterless technology. Although it involves using large quantities of water resources, it is more attractive economically due to the low associated cost. The objective of both technologies is the same (to generate recycled paper), but from the economic and environmental points of view, they are completely different. The recycled paper from technologies without water generates a favorable environmental impact since it reduces the enormous amounts of water required for recycling paper (WNR 2015). However, the price to pay is quite high, since from an economic point of view it is 4.5 times more expensive to invest in waterless technology than in water technology, and the cost to produce a ton of paper is about 30 times higher, which is associated with the high consumption of electrical energy related to waterless recycling machines (EPSON 2021). This might not be environmentally friendly, so the simultaneous use of both technologies is more favorable.

2.6.4 Glass Waste Processing Technologies

Glass recycling can be controversial due to how expensive the process is because of the ease with which it fragments, the difficulty of its correct separation, and its low economic potential (Rogoff and Gardner 2016). It is not attractive from an economic point of view (Jacoby 2019), but its recycling is convenient from an ecological point of view. The technology to be used is remelting because of its process' simplicity and efficiency.

2.6.5 Plastic Waste Processing Technologies

Plastics have enormous potential as a source of chemicals and energy. The revaluation of this type of waste has been one of the most studied (Neena and Kurian 2014). The most common method for plastic waste recycling is mechanical recycling (Al-Salem et al. 2009). This process typically includes collection, sorting, washing, and grinding of the material. Steps may occur in a different order, multiple times or not at all, depending on the origins and composition of the waste (Ragaert et al. 2017). Chemical recycling is especially useful for plastic mixtures that are difficult to separate or when separation is expensive. This technique consists of converting polymers into small molecules to produce new fuels using a thermochemical process thanks to their high hydrocarbon content (Ragaert et al. 2017). An additional alternative to chemical recycling is the production of nanotubes through catalytic pyrolysis. This type of process is novel and produces high-value derivatives, although it also requires large investments (Williams 2021).

2.6.6 Medical Waste Processing Technologies

Although burning this kind of waste has been the method of choice in many parts of the world, other technologies are being developed that may be helpful for proper processing because of the waste's potential danger to humans. Ilyas et al. (2020) proposed the use of technologies such as incineration, chemical treatment, and alternative heat treatment for waste disinfection as follows:

- Treatment by incineration consists of incinerating the trash at a high temperature between 800 and 1200 degrees Celsius, killing the pathogen and destroying 90 percent of the organic materials. This form of treatment emits a lot of pollutants into the atmosphere, in addition to the pretreatment required for the waste, it may require considerable amounts of energy due to the humidity present in this type of waste (Montiel-Bohórquez 2019).
- Chemical treatment is based on disinfection, which is commonly employed as a pretreatment in COVID-19 waste and is then followed by mechanical crushing. This approach has good yields, fast action, and a wide sterilization spectrum that can kill the pathogen and the microorganisms present and destroy bacterial spores (Wang et al. 2020).
- Alternative treatments are mainly two processes, those based on high-temperature pyrolysis techniques and those based on the use of microwaves in medium-temperature ranges.

Any of the treatments described above seek to reduce the health risks presented by medical waste.

2.6.7 Organic Waste Processing Technologies

Organic waste has become of special interest at a global level due to its high potential to produce energy and fuels in a sustainable way (Miccio et al. 2020). For this type of waste, three WTE alternatives are presented: incineration, pyrolysis, gasification, and an alternative option to convert organic waste to compost. The incineration of organic waste can take advantage of the energy contained in organic waste; however, it increases the generation of SO_2, Nox, and other toxic emissions for humans (Cole-Hunter et al. 2020). In comparison, pyrolysis consists of the thermal degradation of substances without the need for combustion reactions to occur, which translates into a lower environmental impact, although its energy efficiency is lower than incineration (Huang and Tang 2007). Gasification converts the waste into a combustible gas using a gasifying agent, which makes it possible to take advantage of the energy potential of this generated gas, this process generates very low emissions compared with conventional coal combustion (Liping et al. 2019). However, the yields obtained, and the characteristics of the gases produced in gasification depend, mainly, on the conditions present in the reaction (temperature, pressure, gasifying agent, reaction time) so it is a more complex process than those previously mentioned (Gómez-Barea et al. 2009). As an alternative to energy treatments, it is possible to generate compost from organic waste, through the biological degradation of organic matter. This process is friendly to the environment, and although it produces carbon dioxide emissions, these emissions depend mainly on the type of waste and its composition. Producing high-quality compost from MSW can require a lot of energy due to the use of heavy machinery that generates carbon dioxide emissions (Luo et al. 2009). However, the use of compost can reduce the need for fertilizers and pesticide chemicals, which implies the reduction of greenhouse gas emissions associated with agricultural production (Sánchez et al. 2015).

2.7 References

Alireza, B., and Gordon, M. 2012. A review-synthesis of carbon nanotubes from plastic wastes. Chemical Engineering Journal, *195–196*, 377–391.

Al-Salem, S. M., Lettieri, P., and Baeyens, J. 2009. Recycling and recovery routes of plastic solid waste (PSW): A review. Waste Management, *29*(10), 2625–2643. DOI: 10.1016/j.wasman.2009.06.004

Bachér, J., and Kaartinen, T. 2017. Liberation of printed circuit assembly (PCA) and dust generation in relation to mobile phone design in a size reduction process. Waste Management, *60*, 609–617. https://doi.org/10.1016/j.wasman.2016.09.037

Bakker, E. J., Rem, P. C., and Fraunholcz, N. 2009. Upgrading mixed polyolefin waste with magnetic density separation. Waste Management, *29*(5), 1712–1717. https://doi.org/10.1016/j.wasman.2008.11.006

Capuzzi, S., and Temeli, G. 2018. Preparation and melting of scrap in aluminum recycling: A review. *Metals*, *8*(4), 249. DOI:10.3390/met8040249

Choudhary, B. 2023, December 29. Enhancing maintenance efficiency with ai & machine learning. FieldCircle. https://www.fieldcircle.com/articles/smart-maintenance-with-ai-and-machine-learning/

Cole-Hunter, T., Johnston, F. H., Marks, G. B., Morawska, L., Morgan, G. G., Overs, M., Porta-Cubas, A., and Cowie, C. T. 2020. The health impacts of waste-to-energy emissions: a systematic review of the literature. Environmental Research, *15*, 123006. DOI: 10.1088/1748-9326/abae9f

Congcong, Y., Jian, P., De-qing, Z., Zhengqi, G., and Xiao-Ming, L. 2019. Pyrometallurgical recycling of stainless steel pickling sludge: a review. Journal of Iron and Steel Research International, *26*, 547–557. DOI: 10.1007/s42243-019-00278-y.

EPSON. 2016. The sustainable way to securely destroy and upcycle paper. EPSON. https://www.epson.eu/paperlab#view-specifications (Accessed 29 September 2021).

Gattupalli, M., Dashora, K., Javed, Z., and Tripathi, G. D. 2024. Exploring garbage enzymes as novel biocatalyst for enhancing bioprocess performance in composting. Process Safety and Environmental Protection, *188*, 73–80. https://doi.org/10.1016/j.psep.2024.05.080

Gómez-Barea, A., Vilches, L.F., Leiva, C., Campoy, M., and Fernández-Pereira, C. 2009. Plant optimisation and ash recycling in fluidised bed waste gasification. Chemical Engineering Journal. *146*, 227–236. DOI: 10.1016/j.cej.2008.05.039

Grieco, E. M., and Baldi, G. 2012. Pyrolysis of polyethylene mixed with paper and wood: Interaction effects on tar, char and gas yields. Waste Management, *32*(5), 833–839.

GSPI. (2024). *Electronics - Green Science Policy Institute*. Green Science Policy. https://greensciencepolicy.org/our-work/electronics/

Haldar, D., Bhattacharjee, N., Shabbirahmed, A. M., Anisha, G. S., Patel, A. K., Chang, J.-S., Dong, C.-D., and Singhania, R. R. 2023. Purification of biogas for methane enrichment using biomass-based adsorbents: A review. Biomass and Bioenergy, *173*, 106804. https://doi.org/10.1016/j.biombioe.2023.106804

Hanan, D., Burnley, S., and Cooke, D. 2012. A multi-criteria decision analysis assessment of waste paper management options. Waste Management, *33*(3), 566– 573.

Hazra, A., Das, S., Ganguly, A., Das, P., Chatterjee, P. K., Murmu, N. C., and Banerjee, P. 2019. Plasma arc technology: A potential solution toward waste to energy conversion and of ghgs mitigation. In *Waste Valorisation and Recycling* (pp. 203–217). Springer Singapore. http://dx.doi.org/10.1007/978-981-13-2784-1_19

Huang, H., and Tang, L. 2007. Treatment of organic waste using thermal plasma pyrolysis technology. Energy Conversion and Management, *48*(4), 1331–1337. DOI: 10.1016/j.enconman.2006.08.01

Hur, T., Lim, S., and Lee, H. 2003. A study on the eco-efficiencies for recycling methods of plastics wastes. InLCA. 22.

Ilyas, S., Srivastava, R.R., and Kim, H. 2020). Disinfection technology and strategies for COVID-19 hospital and bio-medical waste management. Science of The Total Environment, 749, 141652. DOI: 10.1016/j.scitotenv.2020.141652

Istrate, I.-R., Medina-Martos, E., Galvez-Martos, J.-L., and Dufour, J. 2021. Assessment of the energy recovery potential of municipal solid waste under future scenarios. Applied Energy, *293*, 116915. https://doi.org/10.1016/j.apenergy.2021.116915

Jacoby, M. 2019. Why glass recycling in the US is broken. c&en Chemical & Engineering News. https://cen.acs.org/materials/inorganic-chemistry/glass-recycling-US-broken/97/i6 (Accessed 24 February 2022).

Jirang, C. U. I., and Roven, H. J. 2010. Recycling of automotive aluminum. Transactions of Nonferrous Metals Society of China, *20*(11), 2057–2063. DOI: 10.1016/S1003-6326(09)60417-9

Kalantar, Z.N., Karim, M. R., and Mahrez, A. 2012. A review of using waste and virgin polymer in pavement. Construction and Building Materials, *33*, 55–62.

Kang, H. Y., and Schoenung, J. M. 2005. Electronic waste recycling: A review of U.S. infrastructure and technology options. Resources, Conservation and Recycling, *45*(4), 368–400. DOI: 10.1016/j.resconrec.2005.06.0

Lahtela, V., and Kärki, T. 2018. Mechanical sorting processing of waste material before composite manufacturing—A review. Journal of Engineering Science and Technology Review, *11*(6), 35–46. https://doi.org/10.25103/jestr.116.06

Laurieri, N., Lucchese, A., Marino, A., and Digiesi, S. 2020. A door-to-door waste collection system case study: A survey on its sustainability and effectiveness. Sustainability, *12*(14), 5520. https://doi.org/10.3390/su12145520

Liping, L. V., Zhang, Z., and Li, H. 2019. SNG-electricity cogeneration through MSW gasification integrated with a dual chemical looping process. Chemical Engineering and Processing - Process Intensification, *145*, 107665. DOI: 10.1016/j.cep.2019.107665

Lou, X. F., and Nair, J. 2009. The impact of landfilling and composting on greenhouse gas emissions—A review. *Bioresource Technology*, *100*(16), 3792–3798. DOI: 10.1016/j.biortech.2008.12.00

Miccio, M., Pierri, R., Cuccurullo, G., Metallo, A., and Brachi, P. 2020. Process intensification of tomato residues drying by microwave heating: experiments and simulation. Chemical Engineering and Processing - Process Intensification, *156*, 108082. DOI: 10.1016/j.cep.2020.108082

Montiel-Bohórquez, N. D. 2019. Generación de energía a partir de residuos sólidos urbanos. Estrategias termodinámicas para optimizar el desempeño de centrales térmicas. Estrategias termodinámicas para optimizar el desempeño de centrales térmicas. Información Tecnológica, *30*(1), 273–284. DOI:10.4067/s0718-07642019000100273

Murugappan, R. M., and Karthikeyan, M. 2021. 4 - Microbe-assisted management and recovery of heavy metals from electronic wastes. Environmental Management of Waste Electrical and Electronic Equipment, 65–88. DOI: 10.1016/B978-0-12-822474-8.00004-0

Neena, G., and Kurian, T. 2014. Recent developments in the chemical recycling of postconsumer poly(ethylene terephthalate) waste. Industrial and Engineering Chemistry Research, *53*, 14185–14198. DOI: 10.1021/IE501995M

Pham, V., Holtzapple, M.T., and El-Halwagi, M.M. 2010. Techno-economic analysis of biomass to fuel via the MixAlco process. Journal of Industrial Microbiology and Biotechnology, *37*(11), 1157–1168.

Ragaert, K., Delva, L., and Van Geem, K. 2017. Mechanical and chemical recycling of solid plastic waste. Waste Management, *69*, 24–58. DOI:10.1016/j.wasman.2017.07.044

Ren, Y. 2021. Optimizing predictive maintenance with machine learning for reliability improvement. ASCE-ASME Journal of Risk and Uncertainty in Engineering Systems, Part B: Mechanical Engineering, *7*(3). https://doi.org/10.1115/1.4049525

Rogoff, M. J., and Screve, F. 2019. Energy from waste technology. In Waste-To-energy (pp. 29–56). Elsevier. http://dx.doi.org/10.1016/b978-0-12-816079-4.00003-7

Rogoff, M. J., and Gardner, R. B. 2016. High Costs and Limited Markets Put a Crack in Glass Recycling. MSW Management. https://www.mswmanagement.com/recycling/article/13023949/high-costs-and-limited-markets-put-a-crack-in-glass-recycling, (Accessed 19 November 2021).

Ruan, J., and Xu, Z. 2011. Environmental friendly automated line for recovering the cabinet of waste refrigerator. Waste Management, *31*(11), 2319–2326. https://doi.org/10.1016/j.wasman.2011.06.004

Sánchez, A., Artola, A., Font, X., Gea, T., Barrena, R., Gabriel, D., and Mondini, C. 2015. Greenhouse gas from organic waste composting: Emissions and measurement. CO_2 Sequestration, Biofuels and Depollution, 33–70. DOI: 10.1007/978-3-319-11906-9_2

Sanjaya, E., and Abbas, A. 2023. Plasma gasification as an alternative energy-from-waste (EFW) technology for the circular economy: An environmental review. Resources, Conservation and Recycling, *189*, 106730. https://doi.org/10.1016/j.resconrec.2022.106730

Streicher-Porte, M., Widmer, R., Jain, A., Bader, H. P., Scheidegger, R., and Kytzia, S. 2005. Key drivers of the e-waste recycling system: Assessing and modelling e-waste processing in the informal sector in Delhi. Environmental Impact Assessment Review, *25*(5), 472–491. DOI: 10.1016/j.eiar.2005.04.004

Tatzer, P., Wolf, M., and Panner, T. 2005. Industrial application for inline material sorting using hyperspectral imaging in the NIR range. Real-Time Imaging, *11*(2), 99–107. https://doi.org/10.1016/j.rti.2005.04.003

Tuncuk, A., Stazi, V., Akcil, A., Yazici, E.Y., and Deveci, H. 2012. Aqueous metal recovery techniques from e-scrap: Hydrometallurgy in recycling. Minerals Engineering, *25*(1), 28–37. DOI: 10.1016/j.mineng.2011.09.019

van Huis, A., Oonincx, D. G. A. B., Rojo, S., and Tomberlin, J. K. 2020. Insects as feed: House fly or black soldier fly? Journal of Insects as Food and Feed, *6*(3), 221–229. https://doi.org/10.3920/jiff2020.x003

Wang, J. Shen, J., Ye, D., Yan, X., Zhang, Y., Yang, W., Li, X., Wang, J., Zhang, L. and Pan, L. 2020. Disinfection technology of hospital wastes and wastewater: suggestions for disinfection strategy during coronavirus disease 2019 (COVID-19) pandemic in China. Environmental Pollution, 262, 114665. DOI: 10.1016/j.envpol.2020.114665.

Wang, L., Templer, R., and Murphy, R.J. 2012. A life cycle assessment (LCA) comparison of three management options for waste papers: bioethanol production, recycling and incineration with energy recovery. Bioresource Technology, *120*, 89–98.

Williams, P. T. 2021. Hydrogen and carbon nanotubes from pyrolysis-catalysis of waste plastics: A review. Waste and Biomass Valorization, *12*, 1–28. DOI:10.1007/s12649-020-01054-w

WNR 2015. New Waterless Paper Recycling Technology. The Water Network. https://thewaternetwork.com/article-FfV/new-waterless-paper-recycling-technology-lIbnktHl35s50iRrPLLVUg (Accessed 28 October 2021).

Wu, C., Wang, W. L., Williams, P. T., and Huang, J. 2012. Sustainable processing of waste plastics to produce high yield hydrogen rich synthesis gas and high quality carbon nanotubes. RSC Advances, *2*, 4045–4047.

Yuan, H., Fu, S., Tan, W., He, J., and Wu, K. 2015. Study on the hydrocyclonic separation of waste plastics with different density. Waste Management, *45*, 108–111. https://doi.org/10.1016/j.wasman.2015.01.037

3

Planning for Waste Management Optimization

Modern society has encouraged consumption patterns that produce a wide range of trash, thus it is essential to put management and planning techniques into place to lessen the impact on the environment, particularly in densely populated places. This chapter introduces Waste Management, addressing various approaches depending on the type of waste and formulating them as an optimization problem using linear programming. It details step by step how to pose the problem, develop the mathematical model, and its implementation using the General Algebraic Modeling System (GAMS) and Python. Upon completion, the reader will be able to formulate a planning determinist problem in the management of municipal solid waste and generate graphics that facilitate the interpretation of useful results to optimize waste management.

3.1 Introduction

Planning for waste management is crucial for ensuring resource sustainability, promoting public health, and lessening the impact on the environment. Authorities in areas where trash generation rates have risen over time have been compelled to put different plans into place in order to lessen the impact of this issue. Depending on the local circumstances, these tactics differ; for instance, some emphasize the revaluation of metal and electronic trash, while others aim to provide financial incentives for reducing plastic waste. Regardless of the strategy adopted, it is crucial to define the key stages of Municipal Solid Waste (MSW) management. Figure 1 presents the common stages described in the literature, highlighting the three most common destinations considered in these strategies, along with the economic and environmental impacts (EI) associated with each process.

Although waste generation represents a challenge in virtually every community worldwide, this problem is exacerbated in regions where MSW planning and management is poor or even non-existent (Guerrero et al. 2013).In this context, optimization becomes an invaluable tool to improve the planning, recycling, control, and logistics of the different processes involved in waste management. Designing an optimization strategy for MSW management requires, first of all, defining which part of the process is being optimized. For example, at the collection stage, multiple strategies have been developed to determine the appropriate number of collection units and design optimal routes that minimize fuel consumption and time, while reducing the impact on vehicular traffic. Another alternative is to use optimization tools to evaluate government economic incentives or develop new proposals based on needs. These decision-making systems integrate the participation of different actors, such as businesspersons, heads of households, public managers and government authorities (Nevrlý et al. 2019). In general, MSW management has been investigated from different perspectives, not necessarily limited to the stages of the process described above, since works and strategies have also been presented aimed at predicting waste generation, monitoring waste collection systems, or

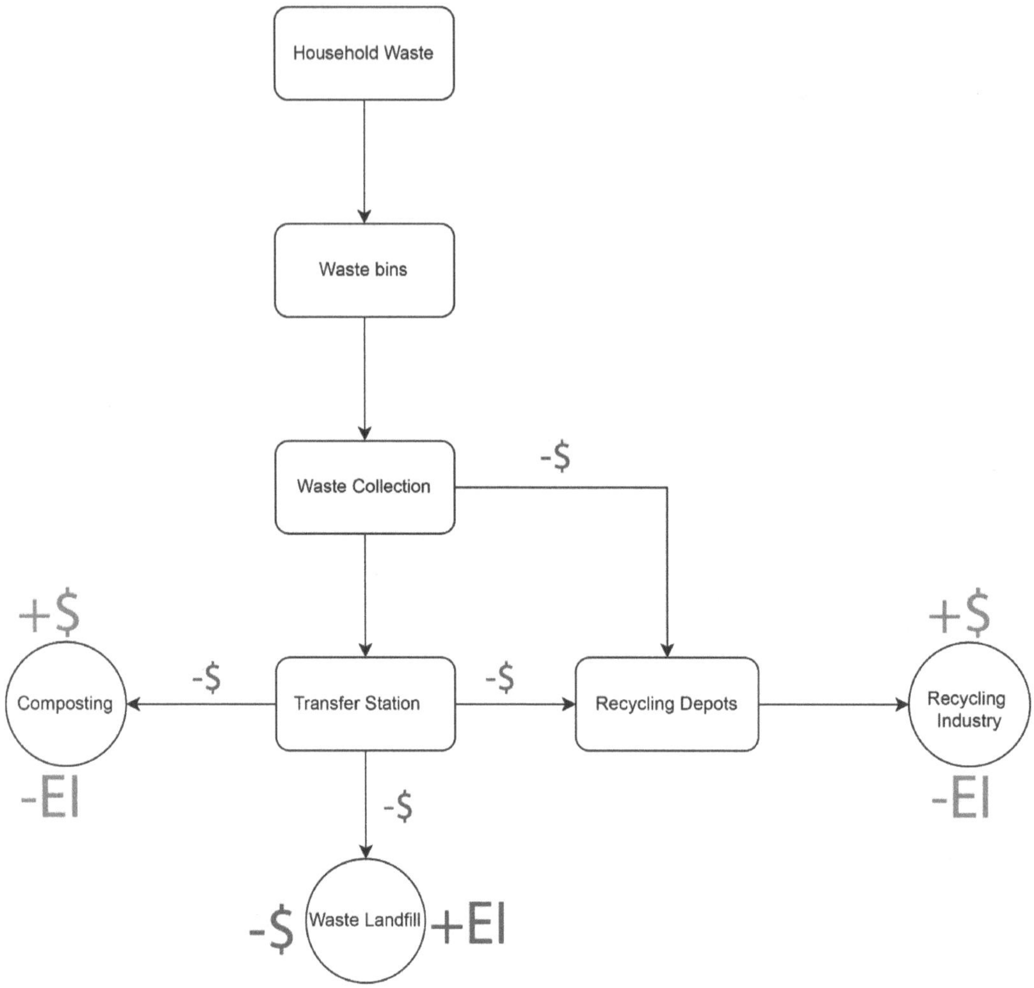

Fig. 1. MSW management steps.

even works aimed at new treatments for waste that requires special conditions such as those generated from the COVID-19 pandemic (Karimi et al. 2024, Ganguly and Chakraborty 2024).

This chapter introduces the reader to the different models proposed in the literature to address issues related to MSW management. The following sections focus on presenting different types of mathematical models and their application in MSW management, as well as exploring alternatives for recycling and revalorizing waste that promotes sustainable development. Finally, a step-by-step guide is offered for the implementation of deterministic models using Python and GAMS, providing a solid foundation for developing the topics covered in later chapters of this book.

3.2 Mathematical Models for Waste Management

As mentioned in the previous section, there are multiple methodologies proposed to optimize the various stages associated with MSW management. In this section, some methodologies found in the literature will be presented.

3.2.1 Linear Programming

Linear Programming (LP) can be understood as an optimization problem that seeks to maximize or minimize a linear objective function, subject to various linear equality or inequality constraints. An example of this type of formulation can be described by equations (3.1) to (3.6).

$$f(x) = -10x_1 - 3x_2 \tag{3.1}$$

Subject to:

$$x_1 + x_2 \leq 12 \tag{3.2}$$

$$x_1 \leq 6 \tag{3.3}$$

$$x_2 \leq 10 \tag{3.4}$$

$$x_1 \geq 0 \tag{3.5}$$

$$x_2 \geq 0 \tag{3.6}$$

The graphical representation of this problem is shown in Fig. 2, where the feasible region (shaded) indicates the possible values of the independent variables that meet the established constraints. It is important to note that, in this type of problem, the optimal solution is always located at one of the vertices of the feasible region. The reader is recommended to consult Chapter 3 of the book "Optimization of Chemical Processes: A Sustainable Perspective" (Ponce-Ortega et al. 2024) should the reader wish to deepen their theoretical understanding and practical application of linear programming.

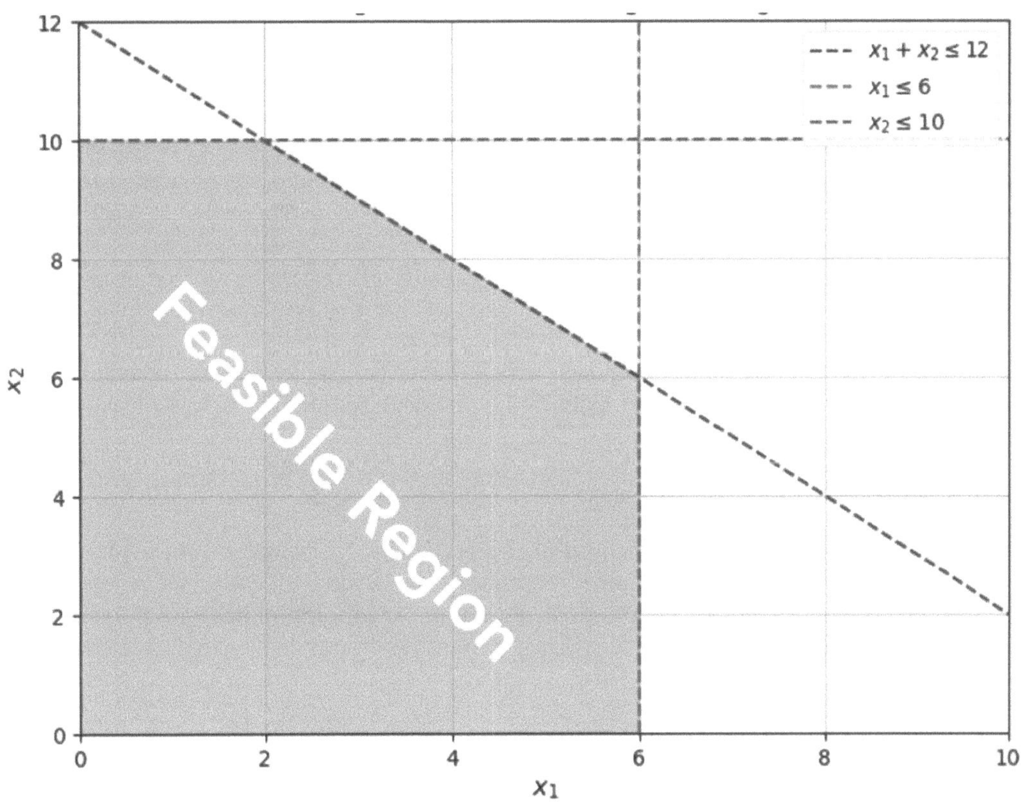

Fig. 2. Feasible Region for the linear programming problem.

Multiple MSW management strategies have been explored in the literature, which are commonly represented as mixed integer programming (MILP) models, either single-objective or multi-objective. In these models, it is common to find functions that minimize costs, environmental impact, or social impact, as well as those that maximize profits derived from waste revalorization, among others (Maity 2018). This type of model covers a wide range of applications, from the optimization of the route of MSW collection vehicles to the optimal selection of the technologies required for the treatment of a specific waste. Below are some works related to MSW management that have been solved using linear programming (LP):

In the work developed by Šomplák et al. (2019), a strategy was presented that seeks to reduce the amount of waste by maximizing the rate of recovered material and allocating a portion of the remaining waste to energy recovery. This model uses a multi-objective optimization methodology based on two criteria: cost minimization and greenhouse gas (GHG) minimization, while the composition of the waste generated is set randomly. The Czech Republic is the study region in this investigation. In addition, different restrictions associated with transporting the waste, the capacity of the facilities, and the waste sent to the landfill, among others, are proposed.

Munguía-López et al. (2022) addressed the problem generated by medical waste produced during the COVID-19 health emergency, using a Mixed Integer Programming (MILP) model for waste revalorization, following two routes: the conversion of waste into new materials and the conversion of waste into energy. This strategy presents a model capable of determining the optimal treatment technology for different types of waste (electronic, medical, plastic, organic, etc.), considering environmental and economic aspects. It was determined that, in order to promote the recycling of waste with low economic potential (either due to its high processing costs or the low obtainment of high-value materials), the incorporation of economic penalties in the strategy is required. The case study selected in this work was the New York region, recognized as one of the most affected by the generation of hazardous medical waste. In addition, restrictions associated with the processing capacities of the proposed technologies and the penalties assigned according to the rates of waste sent to landfills are presented. This strategy is explored in detail later in Chapter 9 of this book.

In general, Linear Programming (LP) strategies that address issues related to Municipal Solid Waste (MSW) management are characterized by having similar objective functions and constraints. Many of these models seek to reduce costs, minimize environmental impacts, or increase recycling rates. Depending on the stage of the MSW management process to be represented, it is common to find constraints related to transportation, processing, or even landfill space capacities.

The advantage of LP models lies in their ability to represent different stages of MSW management, whether on a large or small scale and in their ability to consider multiple objectives simultaneously. However, it is important to note that both objective functions and constraints must be linear. Therefore, incorporating complex processes whose models are non-linear may require the use of additional tools, such as reformulations (Big M, Convex Hull, etc.) or more advanced techniques, such as Machine Learning (ML).

3.2.2 Non-Linear Programming (NLP)

Similar to the strategies presented in the previous subsection, an NLP model can be used to reduce costs, environmental impact, or maximize recycling rates, however, these types of models are subject to objective functions and/or non-linear constraints. An example of this type of formulation is described in Eq (3.7) to (3.11).

$$min\, f(x_1, x_2) = (x_1 - 8)^2 - (\underline{x}_2 - 6)^2 \tag{3.7}$$

Subject to:

$$x_1^2 + x_2^2 \leq 25 \tag{3.8}$$

$$x_1 + x_2 \geq 1 \tag{3.9}$$

$$x_1 \geq 0 \tag{3.10}$$

$$x_2 \geq 0 \tag{3.11}$$

Figure 3 illustrates this problem, highlighting the objective function and the feasible region (shaded area), which represents the possible combinations of the independent variables that satisfy

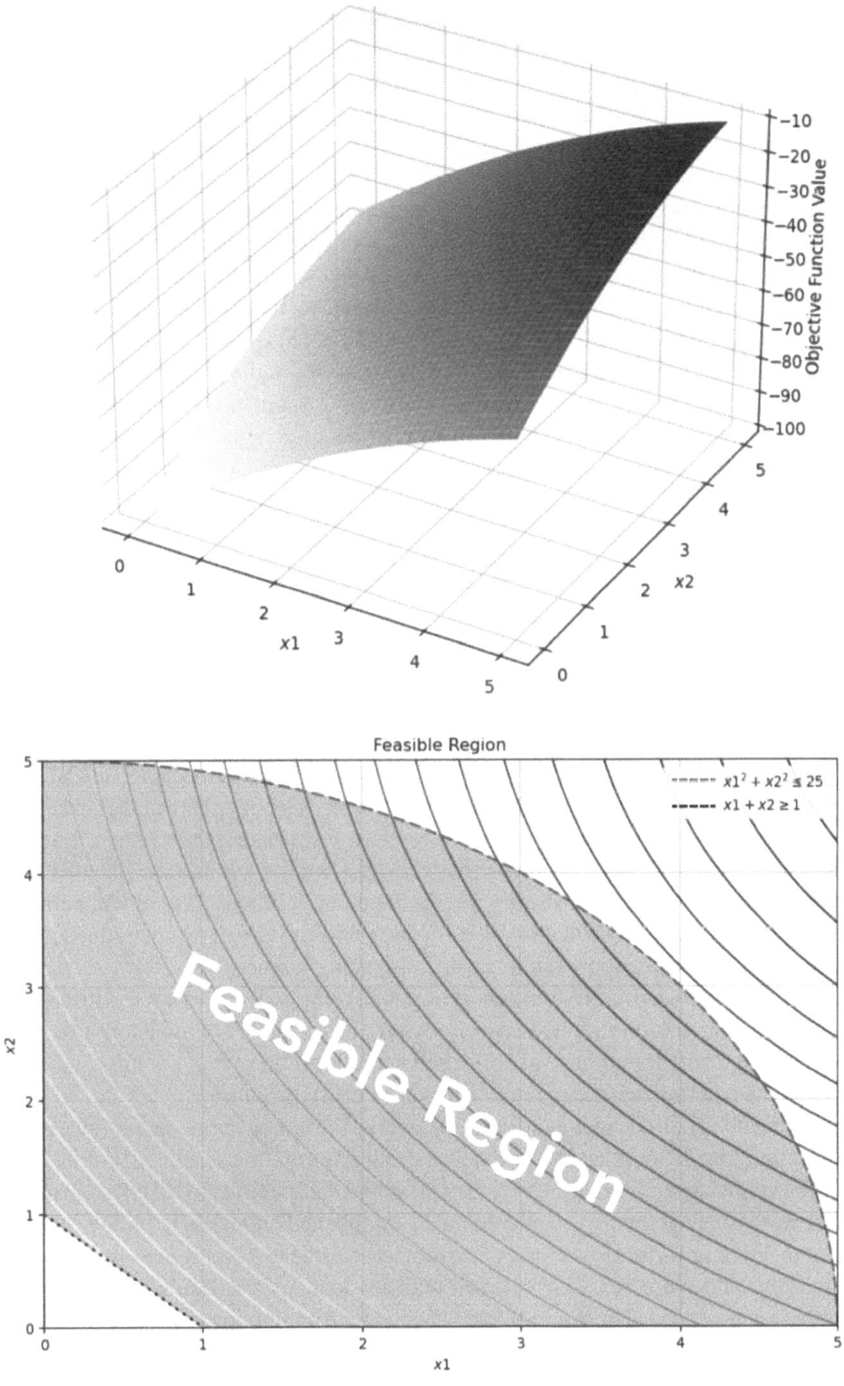

Fig. 3. Objective Function and Feasible Region for a nonlinear programming problem.

the defined constraints. Unlike an LP problem, where the optimal solution is located at one of the vertices of the feasible region, in a Nonlinear Programming (NLP) problem, the optimal solution can be located at any point within the feasible region, not necessarily at a vertex, due to the curved nature of the objective function or the constraints. This makes finding the optimal solution in NLP more complex since multiple local optima can exist.

In the context of municipal solid waste (MSW) management, NLP models have gained presence due to their ability to represent complex phenomena more accurately, which do not fit into a linear model. As mentioned in the work of Alshaikh and Abdelfatah (2024), some of the nonlinear relationships employed in NLP models include the representation of waste generation as a function of the economic growth of a region, as well as recycling rates in relation to the availability of recycling plants. For example, in the work of Zhao et al. (2023), a MINLP model is developed that uses variable treatment capacities, which involves setting constraints of a nonlinear nature due to the nonlinear relationships between waste rates and processing capacities of each technology. This work seeks to simultaneously optimize costs and CO_2 emissions. To compare decision-making results, this model was compared to its counterpart that employs fixed processing capacities, determining, through different scenarios, that adjustable processing capacities allow for more efficient decision-making.

Another use of NLP models proposed by Yadav et al. (2016) is to determine the optimal location of different waste collection systems in addition to the locations of the infrastructure necessary for the management and processing of waste. The non-linearity of this model is established in its restrictions since the data used to develop it present a high level of heterogeneity. The Indian region was taken into consideration when conducting this investigation.

3.2.3 Approximate models

All the strategies presented above belong to the category of exact models; however, various approximate model strategies have also been developed that employ stochastic methods or metaheuristic methodologies. These approaches allow considering some characteristics of the MSW management system as random variables, which facilitates the incorporation of variability and uncertainty in the model parameters and allows the model to be adjusted to new sources of information or changes in the environment.

An example of the use of approximate models is found in the work of Zaeimi et al. (2021), where a stochastic model is presented that considers its parameters as uncertain. In this case, values such as demand and volume of waste are considered random for a district of Qazvin, Iran.This model is solved as a MILP model that minimizes the total cost and pollution associated with the waste collection stage. The results suggest that by increasing the reliability of the model parameters (e.g., through better forecasting of waste demand), both costs and emissions can be reduced, highlighting the importance of a stochastic approach in waste collection planning.

On the other hand, the work of Moazzeni et al. (2022) employs Genetic Algorithms (GA) for the collection of electronic waste in the Edmonton region of Canada. This model allows the design of optimized routes for a smaller number of collection vehicles, minimizing the time and total distance traveled, as well as determining the optimal location of charging stations.

Although practically all the methodologies and works presented in this section share the common objective of optimizing MSW management considering environmental and economic aspects, each strategy may be viable depending on the region context and the specific purpose.Deterministic models are useful for waste management systems with well-established technologies, where the models can be reduced to their linear form. However, the current trend indicates that the use of hybrid techniques is becoming more and more relevant in areas with inadequate waste management or when new strategies are being proposed. These approaches enable the combination of deterministic models with other optimization tactics, including stochastic models or metaheuristic techniques, providing the opportunity to suggest novel approaches that have a high potential for enhancing MSW management decision-making.

3.3 Sustainable Utilization of Municipal solid Waste

Talking about sustainable use of MSW involves finding the optimal process routes at each of the stages associated with MSW management, from waste collection and transportation to processing and recycling. Typically, waste processing can be divided into two categories, Material Recovery and Waste to Energy (WtE):

- Material Recovery: This route focuses on revaluing waste through processes that transform it into new products or raw materials useful for other production cycles. First, collection and separation are carried out to identify and classify materials such as plastics, metals, or electronics. Then, the waste goes through compaction and crushing processes to facilitate its transport and handling. Finally, industrial reprocessing takes place, where waste is converted into new raw materials through techniques such as metal remelting, mechanical or chemical recycling of plastics, and paper transformation. Figure 4 illustrates these stages and their integration within the revaluation process.
- WtE: This route is designed to take advantage of waste with low potential for revalorization as new materials, transforming it into useful energy, mainly in the form of electricity or heat (Lam and Chase 2012, Kumar and Samadder 2019). The most common waste in this category includes organic waste, medical waste, and certain types of plastics.Key processes include incineration,

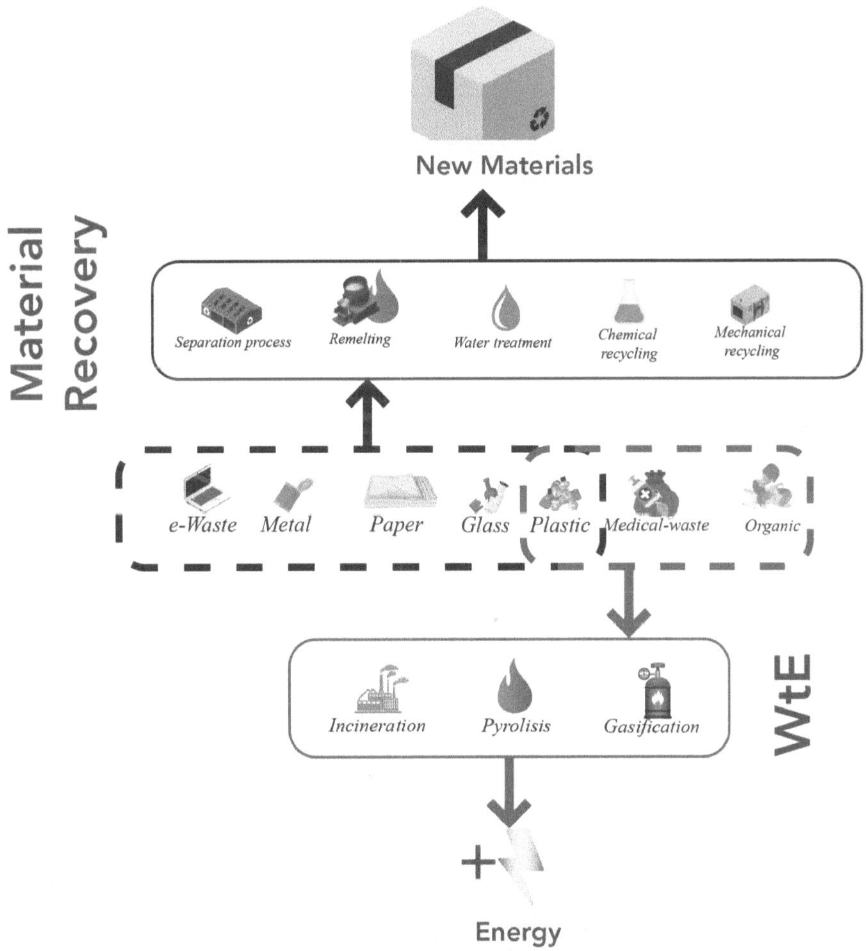

Fig. 4. Main MSW processing routes.

which generates heat to produce steam, electricity, or district heating. Pyrolysis and gasification are used to treat organic waste or plastics, allowing the production of combustible gases, such as syngas, or pyrolytic oils through thermal decomposition in the absence of oxygen. Figure 3.4 shows these stages and their integration within the WtE process.

With the processing routes established, a baseline example will now be established on solving MSW handling optimization problems.

3.3.1 Example 3.1

A waste management problem is posed that involves the collection and treatment of different types of solid waste generated by the community in specific quantities that must be processed to minimize the environmental impact and associated costs. With multiple technologies available for treatment, each presents different processing capacities, operating costs and greenhouse gas emissions. Develop an optimization model that is capable of efficiently assigning the collected waste to these technologies, ensuring that treatment capacities are not exceeded while minimizing both costs and emissions generated; in addition, the option of sending waste to landfills must be considered, which implies an additional cost and environmental footprint. Consider the Superstructure presented in Fig. 5 as a basis for the development of the mathematical model. The data associated with the case study are presented in Tables 1 to 4.

3.3.2 Example 3.1 Solution

The following model based on material balances is developed to manage the collection, treatment, and disposal of waste, optimizing costs, and Greenhouse Gas (GHG) emissions.

First Eq. (3.12), ensures that all collected waste is allocated to either treatment or landfill.

$$W_i = \Sigma_j^J \; f_{i,j}^{waste} + f_i^{landfill}, \forall i \in i \tag{3.12}$$

Table 1. Parameters associated with set i.

Waste type	W_i (tons) collected subwaste i (tons)	$EF_i^{landfill}$; emission factor from landfill disposal (tons of CO_2/ton disposed)
paper	12	0.2
metal	30	0.3
e-waste	40	0.25
glass	70	0.4
plastic	55	0.35
medical-waste	90	0.45
organic	20	0.5

Table 2. Parameters associated with set j.

Processing	$C_j^{procesing}$ processing cost (USD/ton)	$C_j^{transport}$ transportation cost (USD/ton)	$CA_j^{processing}$ maximum processing capacity (tons)	$EF_j^{procesing}$ emission factor for processing (tons of CO_2-eq/ton processed)
remelting	112	15	60	0.05
separation	2.82	20	50	0.1
chemical recycling	144	18	55	0.08
incineration	380	25	80	0.15
pyrolysis	354	22	75	0.12

Table 3. Efficiency Factors.

$h_{i,j}^{recov}$ **material recovery efficiency**				$h_{i,j}^{WtE}$ **WtE conversion efficiency (kWh/ton)**	
Waste type	**Remelting**	**Separation**	**chemical recycling**	**Incineration**	**Pyrolysis**
paper	0	0.6	0.85	750	0
metal	0.95	0.8	0	0	0
e-waste	0	0.5	0.9	0	0
glass	0.75	0.5	0	0	0
plastic	0	0.65	0.9	0	8000
medical-waste	0	0	0	900	0
organic	0	0.7	0	0	300

Table 4. Additional Parameters.

Additional Parameters	
$C^{landfill}$ landfill disposal cost (USD/ton)	30
CA^{truck} truck capacity (tons)	15

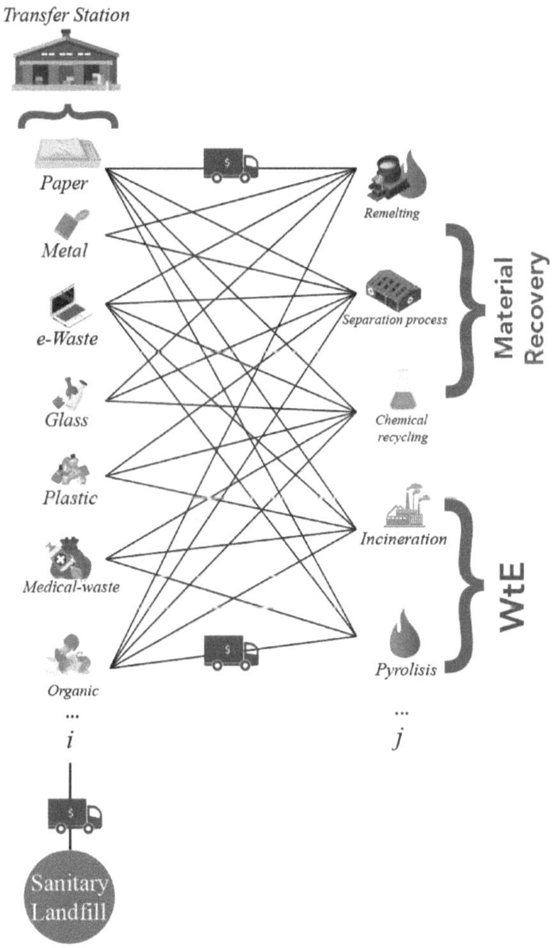

Fig. 5. Superstructure for MSW management of Example 3.1.

where denotes the total amount of waste collected of type *i* (in tons).The term indicates the quantity of waste of type *i* sent to treatment technology *j*, while represents the quantity of waste sent to the landfill.

Next, Eq. (3.13) estimates the total recovered material by each technology j that belongs to *JR*.

$$mat_j^{recov} = \Sigma_j^J \ (f_{i,j}^{waste}.\eta_{i,j}^{recov}), \forall j \in JR \tag{3.13}$$

where, mat_j^{recov} represents the total amount of material recovered through technology *JR*, and $h_{i,j}^{recov}$ denotes the recovery efficiency for waste type *i* using this technology

Following this, Eq.(3.14) calculates the total energy generated by each WtE technology, formulated as:

$$ene_j^{WtE} = \Sigma_j^J \ (f_{i,j}^{waste}.\eta_{i,j}^{WtE}), \forall j \in JE \tag{3.14}$$

where indicates the total energy produced using WtE technology *JE*, while $h_{i,j}^{WtE}$ represents the conversion efficiency from waste type *i* to energy (Note that when a waste type is not compatible with some technology, $\eta_{i,j}^{WtE}$ or $\eta_{i,j}^{recov}$ have a value of 0).

Eq. (3.15) defines the total cost of the system, represented as:

$$cost^{total} = cost^{procesing} + cost^{transport} \tag{3.15}$$

This equation calculates both processing and transportation costs.

Subsequently, Eq. (3.16) quantifies the total GHG emissions of the system as:

$$ghg^{total} = ghg^{procesing} + cost^{transport} \tag{3.16}$$

To further analyze costs, Eq. (3.17) estimates the total processing cost, given by the following relationship:

$$cost^{procesing} = \Sigma_j^J \ = \Sigma_j^J \ (f_{i,j}^{waste}.C_{i,j}^{procesing}) \tag{3.17}$$

where $C_{i,j}^{procesing}$ represents the cost per ton associated with processing waste using technology *j*.

Eq. (3.17) calculates the total transportation cost, including landfill disposal:

$$cost^{transport} = \Sigma_i^J \ \Sigma_j^J \ (f_{i,j}^{waste}.C_j^{transport}) + \Sigma_i^I \ (f_i^{landfill}.C^{landfill}) \tag{3.18}$$

This ensures that all expenses related to moving waste ($C_j^{transport}$ *and* $C^{landfill}$) to various destinations are accounted for.

Eq. (3.18) calculates the GHG emissions from processing, represented as follows:

$$ghg^{procesing} = = \Sigma_i^J \ \Sigma_j^J (f_{i,j}^{waste}.EF_j^{procesing}) \tag{3.19}$$

Here, $EF_j^{procesing}$ indicates the emissions factor for processing technology *j*.

Similarly, Eq. (3.19) focuses on GHG emissions from landfill disposal, defined as:

$$ghg^{landfill} = \Sigma_i^J \ (f_i^{landfill}.EF_j^{landfill}) \tag{3.20}$$

Where $EF_j^{procesing}$ represents the emissions factor for deposit waste *i* in the landfill.

To evaluate the logistical aspects of waste management, Eq.(3.21) calculates the number of trucks required (v_{number}), expressed as:

$$v^{number} = \frac{\Sigma_i^J \ \Sigma_i^J \ f_{i,j}^{wast}}{CA^{truck}} \tag{3.21}$$

Where CA^{truck} refers to the capacity of a truck (in tons per vehicle).

Lastly, Eq. (3.22) ensures that the processing capacity $CA_j^{procesing}$ is not exceeded, represented as:

$$\Sigma_j^J \ f_{i,j}^{waste} \leq CA_j^{procesing}, \forall j \in J \tag{3.22}$$

Multi-objective optimization can be a useful methodology for tackling problems that involve multiple, often conflicting objectives that need to be addressed simultaneously. In waste management scenarios, this approach is particularly useful for balancing economic costs with environmental impacts, such as GHG. In this example, the model proposes to minimize two primary objectives: the total GHG emissions, denoted as follows:

$$ghg^{total} \tag{3.23}$$

and the total cost associated with the waste management system is represented as:

$$cost^{total} \tag{3.24}$$

In this example, the epsilon-constraint technique is utilized to transform the multi-objective problem into a series of single-objective problems. By introducing a new parameter $GHG^{MINIMUM}$, epsilon constraints are added to limit one of the objectives, enabling the optimization process to focus on minimizing the other objective while adhering to predefined thresholds for the constrained objective. In other words, the following restriction is added to the model once the minimum value of ghg^{total} is known:

$$ghg^{total} \leq GHG^{MINIMUM} \tag{3.25}$$

To address this problem, increase the $GHG^{MINIMUM}$ value (for instance, by multiplying it by factors such as 1.01, 1.02, 1.03, …). This enables the identification of various Pareto optimal points. Pareto optimal points are solutions in a multi-objective optimization problem where you cannot improve one objective without making another worse. Once you identify these points, you can create a Pareto curve, also known as the Pareto front. This curve is a visual way to show all the Pareto optimal points and highlights the trade-offs between different objectives, making it easier to understand the choices available.

To solve this problem there are different alternatives, below in the following subsections the implementation in GAMS and Python will be shown. Note that the steps for implementing the mathematical model in GAMS and Python are included as comments within the code lines.

3.3.3 Using GAMS for waste management optimization

```
*****************************************************************
* -------------------------------------------------------
* Linear Programming Model for Waste Management
* -------------------------------------------------------
* Define sets.
Sets
        I subwaste /1*7/
*Types of waste collected.
        J technology /1*5/
*Available treatment technologies.
        JR(J) Material Recovery Tech /1, 2, 3/
*Material recovery technologies.
        JE(J) WtE Tech /4, 5/
*Waste-to-Energy technologies.
* -------------------------------------------------------
* Parameters: Input data for the model
* -------------------------------------------------------
* Quantity of collected waste per type (in tons).
Parameter W(i) collected subwaste i (tons)
/
```

```
1        12
2        30
3        40
4        70
5        55
6        90
7        20
/;
* Processing cost per technology (USD/ton).
Parameter C_processing(J) processing cost (USD\ton)
/
1        112
2        2.82
3        144
4        380
5        354
/;
* Transportation cost per technology (USD/ton).
Parameter C_transport(J) transportation cost (USD\ton)
/
1        15
2        20
3        18
4        25
5        22
/;
* Cost of landfill disposal (USD/ton).
Parameter C_landfill landfill disposal cost (USD\ton) /30/;
* Truck capacity (in tons per trip).
Parameter CA_truck truck capacity (tons) /15/;
* Emission factors for processing (tons of CO2/ton processed).
Parameter EF_processing(J) emission factors for processing (tons of CO2\ton processed)
/
1        0.05
2        0.10
3        0.08
4        0.15
5        0.12
/;
* Maximum processing capacity per technology (tons).
Parameter CA_processing(J) maximum processing capacity (tons)
/
1        60
2        50
3        55
4        80
5        75
/;
* Emissions from landfill disposal (tons of CO2/ton disposed).
Parameter EF_landfill(I) emissions from landfill disposal (tons of CO2\ton disposed)
```

```
/
1       0.20
2       0.30
3       0.25
4       0.40
5       0.35
6       0.45
7       0.50
/;
* Material recovery efficiency by type and recovery technology.
Table eta_recov(i,JR) material recovery efficiency
        1       2       3
1       0.00    0.60    0.85
2       0.95    0.80    0.00
3       0.00    0.50    0.90
4       0.75    0.50    0.00
5       0.00    0.65    0.90
6       0.00    0.00    0.00
7       0.00    0.70    0.00;
* WtE conversion efficiency by type and technology (kWh/ton).
Table eta_WtE(i,JE) WtE conversion efficiency (kWh\ton)
        4       5
1       750     0
2       0.00    0.00
3       0.00    0.00
4       0.00    0.00
5       0.00    8000
6       900     0
7       0       300;
* -------------------------------------------------------
* Variables: Defining the decision variables
* -------------------------------------------------------
variables
 ghg_total        * Total GHG emissions (tons CO2).
 cost_total       * Total cost of waste management (USD).
positive variables
 f_waste(i,j)     * Amount of waste allocated to each technology (tons).
 f_landfill(i)    * Amount of waste sent to landfill (tons).
 mat_recov(JR)    * Recovered material by recovery technology (tons).
 ene_WtE(JE)      * Energy generated by WtE technology (kWh).
 cost_processing  * Total processing cost (USD).
 cost_transport   * Total transportation cost (USD).
 ghg_processing   * GHG emissions from processing (tons CO2).
 ghg_landfill     * GHG emissions from landfill (tons CO2).
 v_number         * Number of truck trips required (trips).
* -------------------------------------------------------
* Equations: Model constraints and objective functions
* -------------------------------------------------------
* Constraint 1: Ensuring all collected waste is allocated to treatment or landfill.
Equation EQ1(I);
```

```
EQ1(I)..W(i) =E= sum(j, f_waste(i,j)) + f_landfill(i);
* Constraint 2: Calculating total recovered material by each recovery technology.
Equation EQ2(JR);
EQ2(JR)..mat_recov(JR) =E= sum(i, f_waste(i,JR) * eta_recov(i,JR));
* Constraint 3: Calculating total energy generated by each WtE technology.
Equation EQ3(JE);
EQ3(JE)..ene_WtE(JE) =E= sum(i, f_waste(i,JE) * eta_WtE(i,JE));
* Equation 4: Defining the total cost of the system.
Equation EQ4;
EQ4..cost_total =E= cost_processing + cost_transport;
* Equation 5: Defining the total GHG emissions.
Equation EQ5;
EQ5..ghg_total =E= ghg_processing + ghg_landfill;
* Equation 6: Calculating total processing cost.
Equation EQ6;
EQ6..cost_processing =E= sum((i,j), f_waste(i,j) * C_processing(j));
* Equation 7: Calculating total transportation cost, including landfill disposal.
Equation EQ7;
EQ7..cost_transport =E= sum((i,j), f_waste(i,j) * C_transport(j))
+ sum(i, f_landfill(i) * C_landfill);
* Equation 8: Calculating GHG emissions from processing.
Equation EQ8;
EQ8..ghg_processing =E= sum((i,j), f_waste(i,j) * EF_processing(j));
* Equation 9: Calculating GHG emissions from landfill disposal.
Equation EQ9;
EQ9..ghg_landfill =E= sum(i, f_landfill(i) * EF_landfill(i));
* Equation 10: Calculating the number of truck trips required.
Equation EQ10;
EQ10..v_number =E= sum((i,j), f_waste(i,j)) / CA_truck;
* Constraint 11: Ensuring processing capacity is not exceeded.
Equation EQ11(J);
EQ11(J)..sum(i, f_waste(i,j)) =L= CA_processing(j);
* -------------------------------------------------------
* Model declaration and solving
* -------------------------------------------------------
model waste_mgmt /all/;
* Solve the model to minimize total GHG emissions.
solve waste_mgmt using LP minimizing ghg_total;
****************************************************************
```

When running the previous code, the user will be able to view a GAMS report in.lst format.This report provides a detailed overview of the solution and the values of the variables, which clearly outlines how to manage MSW according to the specifications of the problem (see Fig. 6 where the waste i sent to each technology j can be observed).

In this particular example, by minimizing only ghg_total, the model calculates the following values of the main variables (see Table 5):

To calculate different points within the Pareto curve, you can add the following line of code to the equations section:

```
************************************************************************
Equation EPSILON C;
EPSILON C..ghg_total =L= 32.9500 ;
```

Table 5. Example 3.1 General results.

cost_processing [$]	70,591
cost_transport [$]	6,465
ghg_processing [tonCO2-eq]	32.95
ghg_landfill [tonCO2-eq]	0
v_number	22

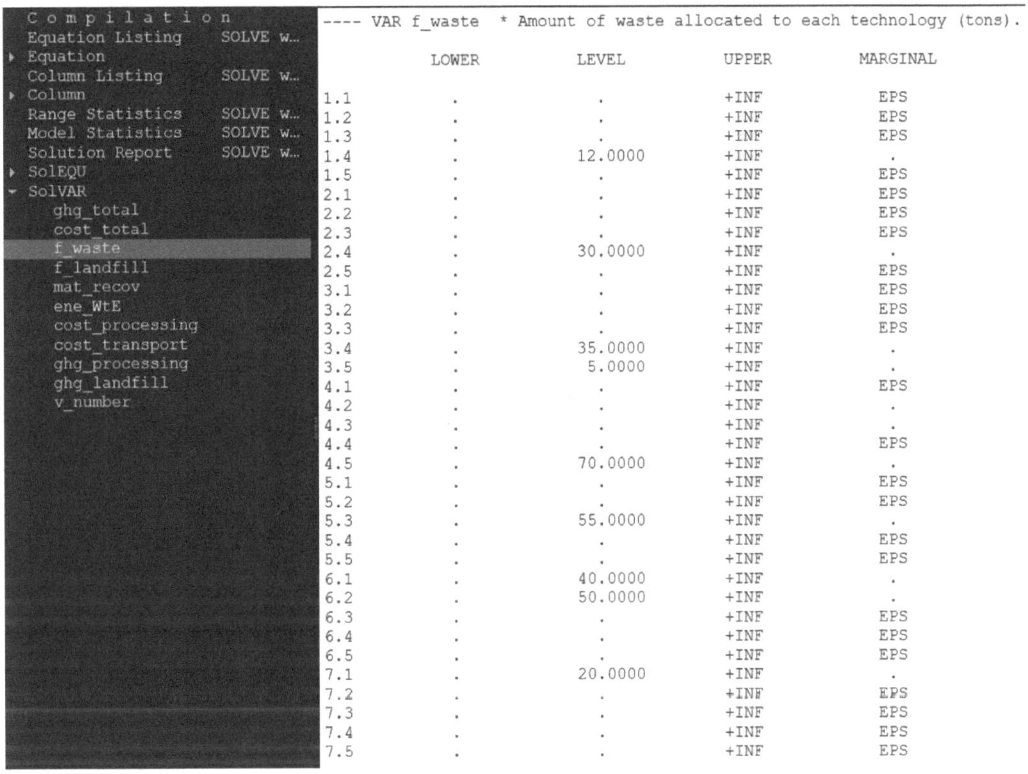

---- VAR f_waste * Amount of waste allocated to each technology (tons).

	LOWER	LEVEL	UPPER	MARGINAL
1.1	.	.	+INF	EPS
1.2	.	.	+INF	EPS
1.3	.	.	+INF	EPS
1.4	.	12.0000	+INF	.
1.5	.	.	+INF	EPS
2.1	.	.	+INF	EPS
2.2	.	.	+INF	EPS
2.3	.	.	+INF	EPS
2.4	.	30.0000	+INF	.
2.5	.	.	+INF	EPS
3.1	.	.	+INF	EPS
3.2	.	.	+INF	EPS
3.3	.	.	+INF	EPS
3.4	.	35.0000	+INF	.
3.5	.	5.0000	+INF	.
4.1	.	.	+INF	EPS
4.2	.	.	+INF	.
4.3	.	.	+INF	.
4.4	.	.	+INF	EPS
4.5	.	70.0000	+INF	.
5.1	.	.	+INF	EPS
5.2	.	.	+INF	EPS
5.3	.	55.0000	+INF	.
5.4	.	.	+INF	EPS
5.5	.	.	+INF	EPS
6.1	.	40.0000	+INF	.
6.2	.	50.0000	+INF	.
6.3	.	.	+INF	EPS
6.4	.	.	+INF	EPS
6.5	.	.	+INF	EPS
7.1	.	20.0000	+INF	.
7.2	.	.	+INF	EPS
7.3	.	.	+INF	EPS
7.4	.	.	+INF	EPS
7.5	.	.	+INF	EPS

Fig. 6. Solution report generated by GAMS.

```
* 32.95 is the value previously calculated with the previous code
***********************************************************************
```

Subsequently, the target variable to be minimized must be changed, as shown in the following lines of code:

```
***********************************************************************
* Solve the model to minimize total cost.
solve waste_mgmt using LP minimizing cost_total;
***********************************************************************
```

It's important to note that if the user wishes to plot the Pareto curve, they will need to utilize additional software, such as Excel.

3.3.4 Using Python for waste management optimization

For the implementation in Python, it is essential to first install the Pyomo library.You can do this using pip with the following command:

pip install pyomo

```
Variables:
  f_waste : Size=35, Index=I*J
      Key    : Lower : Value : Upper : Fixed : Stale : Domain
      (1, 1) :     0 :   0.0 :  None : False : False : NonNegativeReals
      (1, 2) :     0 :   0.0 :  None : False : False : NonNegativeReals
      (1, 3) :     0 :   0.0 :  None : False : False : NonNegativeReals
      (1, 4) :     0 :  12.0 :  None : False : False : NonNegativeReals
      (1, 5) :     0 :   0.0 :  None : False : False : NonNegativeReals
      (2, 1) :     0 :   0.0 :  None : False : False : NonNegativeReals
      (2, 2) :     0 :   0.0 :  None : False : False : NonNegativeReals
      (2, 3) :     0 :   0.0 :  None : False : False : NonNegativeReals
      (2, 4) :     0 :  25.0 :  None : False : False : NonNegativeReals
      (2, 5) :     0 :   5.0 :  None : False : False : NonNegativeReals
      (3, 1) :     0 :   0.0 :  None : False : False : NonNegativeReals
      (3, 2) :     0 :   0.0 :  None : False : False : NonNegativeReals
      (3, 3) :     0 :   0.0 :  None : False : False : NonNegativeReals
      (3, 4) :     0 :  40.0 :  None : False : False : NonNegativeReals
      (3, 5) :     0 :   0.0 :  None : False : False : NonNegativeReals
      (4, 1) :     0 :   0.0 :  None : False : False : NonNegativeReals
      (4, 2) :     0 :  50.0 :  None : False : False : NonNegativeReals
      (4, 3) :     0 :   5.0 :  None : False : False : NonNegativeReals
      (4, 4) :     0 :   0.0 :  None : False : False : NonNegativeReals
      (4, 5) :     0 :  15.0 :  None : False : False : NonNegativeReals
      (5, 1) :     0 :   0.0 :  None : False : False : NonNegativeReals
      (5, 2) :     0 :   0.0 :  None : False : False : NonNegativeReals
      (5, 3) :     0 :   0.0 :  None : False : False : NonNegativeReals
      (5, 4) :     0 :   0.0 :  None : False : False : NonNegativeReals
      (5, 5) :     0 :  55.0 :  None : False : False : NonNegativeReals
      (6, 1) :     0 :  40.0 :  None : False : False : NonNegativeReals
      (6, 2) :     0 :   0.0 :  None : False : False : NonNegativeReals
      (6, 3) :     0 :  50.0 :  None : False : False : NonNegativeReals
      (6, 4) :     0 :   0.0 :  None : False : False : NonNegativeReals
      (6, 5) :     0 :   0.0 :  None : False : False : NonNegativeReals
      (7, 1) :     0 :  20.0 :  None : False : False : NonNegativeReals
      (7, 2) :     0 :   0.0 :  None : False : False : NonNegativeReals
      (7, 3) :     0 :   0.0 :  None : False : False : NonNegativeReals
      (7, 4) :     0 :   0.0 :  None : False : False : NonNegativeReals
      (7, 5) :     0 :   0.0 :  None : False : False : NonNegativeReals
```

Fig. 7. Solution report generated with Pyomo.

Additionally, you will need to install a solver for the optimization model.For this case, the use of GLPK is suggested. To install GLPK, follow these steps:

Install GLPK using conda (recommended): If you have Anaconda or Miniconda installed, you can easily install GLPK with: *conda install-c conda-forge glpk*

You can also follow the instructions and install GLPK manually according to the guidelines provided on the official GLPK website for Windows. Here's the link to the GLPK for Windows installation page:

https://gnuwin32.sourceforge.net/packages/glpk.htm
****

import pyomo.environ as pyo

\# --

\# Model Initialization

\# --

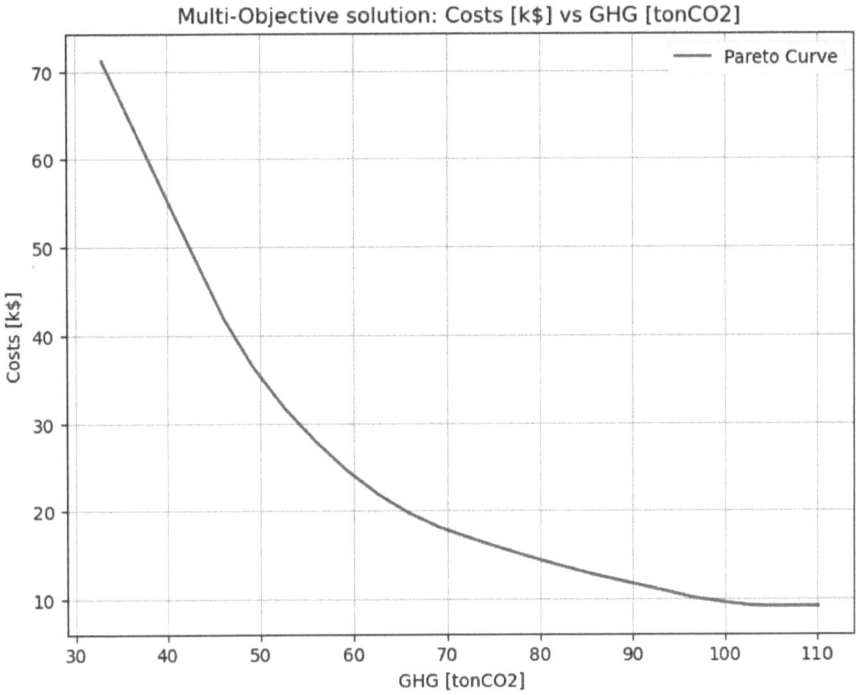

Fig. 8. Pareto Curve for multi-objective solution.

```
model = pyo.ConcreteModel()
# ---------------------------------------------------------
# Sets
# ---------------------------------------------------------
model.I = pyo.RangeSet(1, 7) # Types of subwaste
model.J = pyo.RangeSet(1, 5) # Technologies
model.JR = pyo.Set(initialize=[1, 2, 3]) # Material Recovery Technologies
model.JE = pyo.Set(initialize=[4, 5]) # Waste-to-Energy (WtE) Technologies
# ---------------------------------------------------------
# Parameters
# ---------------------------------------------------------
W = {1: 12, 2: 30, 3: 40, 4: 70, 5: 55, 6: 90, 7: 20} # (tons)
C_processing = {1: 112, 2: 2.82, 3: 144, 4: 380, 5: 354} # (USD/ton)
C_transport = {1: 15, 2: 20, 3: 18, 4: 25, 5: 22} # (USD/ton)
EF_processing = {1: 0.05, 2: 0.10, 3: 0.08, 4: 0.15, 5: 0.12} # (tons of CO2/ton processed)
EF_landfill = {1: 0.20, 2: 0.30, 3: 0.25, 4: 0.40, 5: 0.35, 6: 0.45, 7: 0.50} # (tons of CO2/ton disposed)
CA_processing = {1: 60, 2: 50, 3: 55, 4: 80, 5: 75} # (tons)
eta_recov = {
 (1, 1): 0.00,
 (1, 2): 0.60,
 (1, 3): 0.85,
 (2, 1): 0.95,
 (2, 2): 0.80,
 (2, 3): 0.00,
 (3, 1): 0.00,
```

```
 (3, 2): 0.50,
 (3, 3): 0.90,
 (4, 1): 0.75,
 (4, 2): 0.50,
 (4, 3): 0.00,
 (5, 1): 0.00,
 (5, 2): 0.65,
 (5, 3): 0.90,
 (6, 1): 0.00,
 (6, 2): 0.00,
 (6, 3): 0.00,
 (7, 1): 0.00,
 (7, 2): 0.70,
 (7, 3): 0.00
}
eta_WtE = {
 (1, 4): 750,
 (1, 5): 0,
 (2, 4): 0.00,
 (2, 5): 0.00,
 (3, 4): 0.00,
 (3, 5): 0.00,
 (4, 4): 0.00,
 (4, 5): 0.00,
 (5, 4): 0.00,
 (5, 5): 8000,
 (6, 4): 900,
 (6, 5): 0,
 (7, 4): 0,
 (7, 5): 300
} # (kWh/ton)
C_landfill = 30 # (USD/ton)
CA_truck = 15 # (tons)

# -------------------------------------------------------
# Variables
# -------------------------------------------------------
model.f_waste = pyo.Var(model.I, model.J, domain=pyo.NonNegativeReals)
model.f_landfill = pyo.Var(model.I, domain=pyo.NonNegativeReals)
model.mat_recov = pyo.Var(model.JR, domain=pyo.NonNegativeReals)
model.ene_WtE = pyo.Var(model.JE, domain=pyo.NonNegativeReals)
model.v_number = pyo.Var(domain=pyo.NonNegativeReals)
model.cost_total = pyo.Var(domain=pyo.NonNegativeReals)
model.ghg_total = pyo.Var(domain=pyo.NonNegativeReals)
model.cost_processing = pyo.Var(domain=pyo.NonNegativeReals)
model.cost_transport = pyo.Var(domain=pyo.NonNegativeReals)
model.ghg_processing = pyo.Var(domain=pyo.NonNegativeReals)
model.ghg_landfill = pyo.Var(domain=pyo.NonNegativeReals)
# -------------------------------------------------------
# Constraints
```

```
# --------------------------------------------------------
# Constraint 1: Waste allocation balance
model.EQ1 = pyo.ConstraintList()
for i in model.I:
 model.EQ1.add(W[i] == sum(model.f_waste[i, j] for j in model.J) + model.f_landfill[i])
# Constraint 2: Total recovered material
model.EQ2 = pyo.ConstraintList()
for jr in model.JR:
 model.EQ2.add(model.mat_recov[jr] == sum(model.f_waste[i, jr] * eta_recov.get((i, jr), 0)
for i in model.I))
# Constraint 3: Total energy generated by WtE technologies
model.EQ3 = pyo.ConstraintList()
for je in model.JE:
 model.EQ3.add(model.ene_WtE[je] == sum(model.f_waste[i, je] * eta_WtE.get((i, je), 0) for
i in model.I))
# Equation 4: Total system cost
model.EQ4 = pyo.Constraint(expr=model.cost_total == model.cost_processing + model.
cost_transport)
# Equation 5: Total GHG emissions
model.EQ5 = pyo.Constraint(expr=model.ghg_total == model.ghg_processing + model.
ghg_landfill)
# Equation 6: Total processing cost
model.EQ6 = pyo.Constraint(expr=model.cost_processing == sum(
 model.f_waste[i, j] * C_processing[j] for i in model.I for j in model.J))
# Equation 7: Total transportation cost including landfill
model.EQ7 = pyo.Constraint(expr=model.cost_transport == sum(
 model.f_waste[i, j] * C_transport[j] for i in model.I for j in model.J) +
 sum(model.f_landfill[i] * C_landfill for i in model.I))
# Equation 8: GHG emissions from processing
model.EQ8 = pyo.Constraint(expr=model.ghg_processing == sum(
 model.f_waste[i, j] * EF_processing[j] for i in model.I for j in model.J))
# Equation 9: GHG emissions from landfill disposal
model.EQ9 = pyo.Constraint(expr=model.ghg_landfill == sum(
 model.f_landfill[i] * EF_landfill[i] for i in model.I))
# Equation 10: Number of truck trips required
model.EQ10 = pyo.Constraint(expr=model.v_number == sum(
 model.f_waste[i, j] for i in model.I for j in model.J) / CA_truck)
# Constraint 11: Ensuring processing capacity is not exceeded
model.EQ11 = pyo.ConstraintList()
for j in model.J:
 model.EQ11.add(sum(model.f_waste[i, j] for i in model.I) <= CA_processing[j])
# --------------------------------------------------------
# Objective: Minimize GHG Emissions
# --------------------------------------------------------
model.Objective = pyo.Objective(expr=model.ghg_total, sense=pyo.minimize)
# --------------------------------------------------------
# Solver Execution
# --------------------------------------------------------
solver = pyo.SolverFactory('glpk')
result = solver.solve(model)
```

```
# -------------------------------------------------------
# Display Results
# -------------------------------------------------------
model.display()
**************************************************************
```

When executing the previous code, the user will see the explicit solution of the model printed in the console, as illustrated in Fig. 7. Additionally, by implementing a loop to plot the Pareto curve, the user will obtain the graph shown in Fig. 8. The complete code that integrates this graph can be found in the appendix of this book.

3.4 Challenges and Limitations

It is clear that implementing optimization models in MSW management allows for the optimization of the different MSW process routes, minimizing the environmental and economic impact that this generates. However, even today significant challenges related to MSW management persist. The majority of the models examined in this chapter rely entirely on readily available, trustworthy, and accurate data sources, which can be challenging in developing nations with limited infrastructure and resources. These nations are also the ones most impacted by this issue. Furthermore, social integration is a difficult problem to solve. To guarantee that an MSW management system is operating at its best, cooperation from all parties involved is required. While financial incentives appear to be a viable option, figuring out their actual values can be difficult. However, because predictive models have the potential to create more accurate models based on the historical background of each research region, they are also crucial to the management of MSW in the future. Therefore, for future work, the development of comprehensive models that allow the simultaneous interaction of different tools will be necessary to create sustainable and adaptable systems.

3.5 References

Alshaikh, R., and Abdelfatah, A. 2024. Optimization techniques in municipal solid waste management: A systematic review. Sustainability, *16*(15), 6585. https://doi.org/10.3390/su16156585

Ganguly, R. K., and Chakraborty, S. K. 2024.Plastic waste management during and post Covid19 pandemic: Challenges and strategies towards circular economy. Heliyon, *10*(4), e25613. https://doi.org/10.1016/j.heliyon.2024.e25613

Guerrero, L. A., Maas, G., and Hogland, W. 2013. Solid waste management challenges for cities in developing countries. Waste Management, *33*(1), 220–232. https://doi.org/10.1016/j.wasman.2012.09.008

Karimi, H., Wassan, N., Ehsani, B., Tavakkoli-Moghaddam, R., and Ghodratnama, A. 2024. Optimizing COVID-19 medical waste management using goal and robust possibilistic programming.Engineering Applications of Artificial Intelligence, *131*, 107838. https://doi.org/10.1016/j.engappai.2023.107838

Kumar, A., and Samadder, S. R. 2017. A review on technological options of waste to energy for effective management of municipal solid waste. Waste Management, *69*, 407–422.https://doi.org/10.1016/j.wasman.2017.08.046

Lam, S. S., and Chase, H. A. 2012. A review on waste to energy processes using microwave pyrolysis. Energies, *5*(10), 4209–4232. https://doi.org/10.3390/en5104209

Maity, Dr. S.K. 2018. Importance of municipal solid waste management. International Journal of Advanced Engineering Research and Science, *5*(5), 361–363. https://doi.org/10.22161/ijaers.5.5.49

Munguía-López, A. del C., Ochoa-Barragán, R., and Ponce-Ortega, J. M. 2022. Optimal waste management during the COVID-19 pandemic.Chemical Engineering and Processing - Process Intensification, *176*, 108942. https://doi.org/10.1016/j.cep.2022.108942

Nevrlý, V., Šomplák, R., Putna, O., and Pavlas, M. 2019. Location of mixed municipal waste treatment facilities: Cost of reducing greenhouse gas emissions. Journal of Cleaner Production, *239*, 118003. https://doi.org/10.1016/j.jclepro.2019.118003

Ponce-Ortega, J. M., Ochoa-Barragán, R., and Ramírez-Márquez, C. 2024. *Optimization of chemical processes: A sustainable perspective*. Springer Nature.

Šomplák, R., Kůdela, J., Smejkalová, V., Nevrlý, V., Pavlas, M., and Hrabec, D. 2019. Pricing and advertising strategies in conceptual waste management planning. Journal of Cleaner Production, *239*, 118068. https://doi.org/10.1016/j.jclepro.2019.118068

Yadav, V., Karmakar, S., Dikshit, A. K., and Vanjari, S. 2016. A feasibility study for the locations of waste transfer stations in urban centers: A case study on the city of Nashik, India. Journal of Cleaner Production, *126*, 191–205. https://doi.org/10.1016/j.jclepro.2016.03.017

Zaeimi, M. B., and Rassafi, A. A. 2021. Optimization model for integrated municipal solid waste system using stochastic chance-constraint programming under uncertainty: A case study in Qazvin, Iran. Journal of Advanced Transportation, *2021*, 1–16. https://doi.org/10.1155/2021/9994853

Zhao, S., Ren, T., Ma, L., and Li, Z. 2022. Multi-period planning of municipal solid waste management: A case study in Qingdao. Process Integration and Optimization for Sustainability, *7*(1–2), 107–126. https://doi.org/10.1007/s41660-022-00279-7

Section II

Sustainable Waste Utilization and Planning

4

Optimal Planning for Municipal Solid Waste Utilization

With inadequate landfill capacity and ineffective waste management systems, growing municipal solid waste generation is a significant issue for big cities. Several options associated with the supply chain for implementing a municipal solid waste management system are available; however, to determine the optimal solution, several technical, economic, environmental, and social aspects must be considered. Therefore, this chapter proposes a mathematical programming model for optimizing the supply chain associated with the MSW management system to maximize the economic benefit while accounting for technical and environmental issues. The optimization model simultaneously selects the processing technologies and their location, the distribution of wastes from cities as well as the distribution of products to markets. The problem was formulated as a multi-objective mixed-integer linear programming problem to maximize the profit of the supply chain and the amount of recycled wastes, where the results are shown through Pareto curves that tradeoff economic and environmental aspects. The proposed approach is applied to a case study for the west-central part of Mexico to consider the integration of wastes from several cities to yield useful products. The results show that an integrated utilization of MSW can provide economic, environmental, and social benefits.

4.1 Introduction

Waste is any residual material from industrial and human activities with no residual value (Ortiz et al. 2010). There is much interest in waste production, management, and disposal. Waste generation is a side effect of consumption and production, and it tends to increase with economic development (Beolchini et al. 2011). There is an increasing awareness of the environmental, health, economic, and social problems linked with waste disposal. Given its quantity and environmental significance, municipal solid waste treatment and management need special consideration. (Vasudevan et al. 2012).

The average amount of municipal solid waste generated in Malaysia is 0.5 kg/person/day, and it can be up to 1.7 kg/person/day in other cities of the world (Ramayah et al. 2012). In certain nations, this causes issues due to a shortage of landfills and a suitable municipal solid waste management system (MSW) management system. In several countries of Central and South America, waste management is not adequate, and it is improperly divided, which raises expenses, makes the treatment procedure more challenging and costly, and results in unusable conditions. Thus, the application of proper environmental, institutional, financial, economic, and social tools to guarantee sustainable waste management is required (Ghinea et al. 2012). There are also negative impacts on water, land, and air resulting from inadequate treatment and management of municipal solid waste (Krüger et al. 2012).

Several actions have been implemented to solve this problem, but most of them are focused only on one kind of waste, without considering the interaction between the waste composition and distribution, and without considering the entire supply chain optimization as well as the economies of scale. The distribution of wastes, and products and the design and selection of processing facilities are crucial to yield an integral solution. This chapter presents a general approach to the optimal planning of a municipal solid waste management system. This approach is based on a mathematical programming formulation for the optimal planning of the reuse of municipal solid waste to maximize the economic benefits while accounting for social and environmental aspects. The optimization model can select the processing technologies, consumers, cities producing wastes, amount of recycled waste, products, and location of processing facilities (Santibañez-Aguilar et al. 2013).

4.2 Problem Statement

Figure 1 shows a schematic representation of the addressed problem, which considers a set of *C* cities. These cities have a specific production rate of wastes that include different components such as plastics, metals, papers, and food. The wastes from cities can be separated into different subtypes (e.g., polyethylene and polyvinyl chloride for plastics, green and amber for glass). After the wastes are separated, they may be transported to processing plants located in the set of cities (*C*) to treat the waste and yield several products based on the used technologies and raw materials. A set of candidate technologies is considered for each class of waste. The products are transported to the consumers in the set of cities. Each city has a specific availability of generated waste and demand for products. The processing plants are in different geographical locations. Each processing plant can treat the waste using several technologies. The costs of raw materials and product transportation are also provided, as are the startup and ongoing expenses for processing technology.

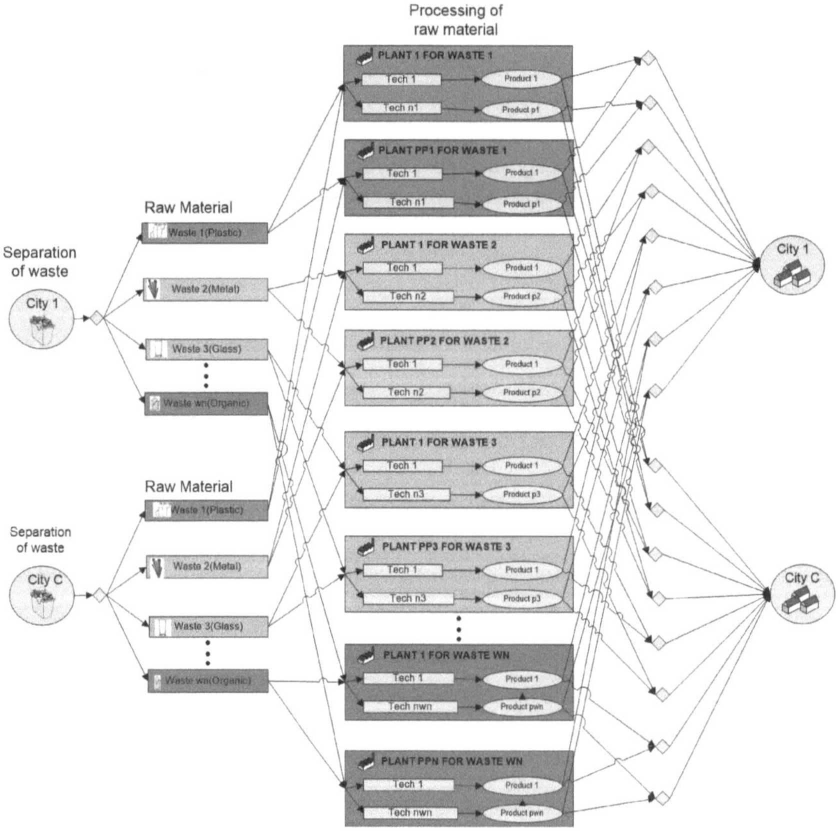

Fig. 1. Superstructure for optimizing a municipal solid waste management system. [Printed (adapted) with permission from Santibañez-Aguilar, J.E., Ponce-Ortega, J.M., González-Campos, J.B., Serna-González, M., and El-Halwagi, M. 2013. Optimal Planning for the sustainable utilization of municipal solid waste. Waste Management, *33*(12), 2607–2622].

The optimization task seeks to determine the optimal supply chain network for waste utilization including the transportation of wastes, the location of the processing facilities, the selection of technologies used in each place, and the distribution of products.

4.3 Mathematical Model Formulation

The index c represents the cities that generate waste and consume useful products. Since different wastes may be generated (e.g., plastics, metals, glass, organics, papers), these wastes are defined with the index wn. The model considers several processing facilities distributed in a specific region. There are several groups of processing facilities, and each one is specialized in a type of waste wn. To identify each group of processing facilities, it is necessary to use a coupled index: pp-wn, where pp represents a group of processing facilities and wn indicates that group of processing plants for the specific waste type. In each processing facility, there is a set of candidate processing technologies for each waste. The index to describe these sets is n-wn, where n is used for the technologies and wn is used to specify the set of processing technologies. The used nomenclature is presented in the Nomenclature section, and the model formulation is presented as follows.

4.3.1 Constraints for Waste Sources in the Cities

Each city has a specific production of waste, and it can be separated. Since not all the waste may be utilized, the useful waste is a portion of the total production of waste in each city. The used waste wn of the type p-wn in the city c ($F_{wn,c,p-wn}$) must be lower than the useful waste type p-wn in the city c ($\alpha^{sep}_{wn,c,p-wn} F^{total}_{c,p-wn}$), the useful waste wn is equal to the total waste type p-wn in the city c ($F^{total}_{wn,c,p-wn}$) multiplied by a separation factor (–), which is stated as follows:

$$F_{wn,c,p-wn} \leq \alpha^{sep}_{wn,c,p-wn} \mathrm{F}^{total}_{c,p-wn}, \quad \forall wn \in W, c \in C, p-wn \in S \tag{4.1}$$

4.3.2 Waste Distribution from Cities to Processing Plants

When waste is selected, it is distributed to different processing plants located in different regions. The processing plants specialize in a type of waste (plastic, metal, glass, or organic waste). Each processing facility has a set of processing routes depending on the treated waste. The proposed mass balance for a specific waste is stated as follows: The selected waste in the city c ($F_{wn,c,r}$) is equal to the delivered waste from the city c to the specific processing plant pp-wn (for example, $F^{plant}_{wn,c,pp-wn,p-wn}$ for plastic) for all the types of considered wastes:

$$F_{wn,c,p-wn} = \sum_{pp-wn \in PH} F^{plant}_{wn,c,pp-wn,p-wn}, \quad \forall wn \in W, c \in C, p-wn \in S \tag{4.2}$$

There are a lot of types of processing facilities, and each processing plant has a set of processing technologies depending on the waste. For example, each plastic plant has its set of technologies to treat plastic wastes; thus, for each waste wn, there is a set of technologies to treat all subtypes of wastes p-wn, these technologies are defined with the script n-wn.

4.3.3 Waste Distribution into the Processing Plants to be Treated with Different Technologies

The wastes in the processing plants are distributed to different processing technologies, and depending on the selected technology it is possible to produce several products. The sum of raw materials from the cities c ($F^{plant}_{wn,c,pp-wn,p-wn}$) is equal to the sum of the distributed raw materials to different technologies ($F^{tech}_{wn,pp-wn,n-wn,p-wn}$):

$$\sum_{c \in C} F^{plant}_{wn,c,pp-wn,p-wn} = \sum_{n-wn \in T} F^{tech}_{wn,pp-wn,n-wn,p-wn}, \quad \forall wn \in W, pp-wn \in P, p-wn \in S \tag{4.3}$$

There is no index for products because each technology can produce only one product.

4.3.4 Mass Balances in the Processing Plants

The model considers the transformation of raw materials to products through the processing plants. The conversion processes usually are modeled through nonlinear relationships; however, in this chapter the use of an efficiency factor for each technology and type of feedstock ($\alpha^{prod}_{wn,n-wn,p-wn}$) is proposed, this factor helps to obtain the amount of produced product ($P_{wn,pp-wn,n-wn,p-wn}$) from an amount of given feedstock ($F^{tech}_{wn,pp-wn,n-wn,p-wn}$):

$$P_{wn,pp-wn,n-wn,p-wn} = \sum_{c \in C} P^{city}_{wn,pp-wn,c,n-wn,p-wn}, \quad \forall wn \in W, pp-wn \in P, n-wn \in T, p-wn \in S \tag{4.4}$$

The efficiency factors are determined before the optimization process, and they are determined from simulation and/or previously reported data.

4.3.5 Balances in Splitters after the Processing Plants

When the raw materials are converted, the products are distributed from the processing plants to product consumers located in the different cities c. The sum of amounts of delivered products to all cities ($P^{city}_{wn,pp-wn,c,n-wn,p-wn}$) is equal to the produced product in each processing facility ($P_{wn,pp-wn,n-wn,p-wn}$):

$$P_{wn,pp-wn,n-wn,p-wn} = \sum_{c \in C} P^{city}_{wn,pp-wn,c,n-wn,p-wn}, \quad \forall wn \in W, pp-wn \in P, n-wn \in T, p-wn \in S \tag{4.5}$$

4.3.6 Balances in Product Distribution Points before the Cities

Before each city c, there is a product distribution point, where the amounts of the same product are mixed to obtain the total amount of product for a specific city. In this equation, the sum of the number of products that arrived in the city from the processing facilities is equal to the total product in the city:

$$\sum_{pp-wn \in PROCPLANT} P^{city}_{wn,pp-wn,c,n-wn,p-wn} = PT^{city}_{wn,c,n-wn,p-wn}, \quad \forall wn \in W, n-wn \in T, c \in C, p-wn \in S \tag{4.6}$$

4.3.7 Constraints for the Demand for Products in the Cities

The total product in a city ($PT^{city}_{wn,c,n-wn,p-wn}$) must be lower than the product demand ($PD^{city}_{wn,n-wn,c,p-wn}$):

$$PT^{city}_{wn,c,n-wn,p-wn} \leq \mathrm{PD}^{city}_{wn,n\text{-}wn,c,p\text{-}wn}, \quad \forall wn \in W, n-wn \in T, c \in C, p-wn \in S \tag{6.7}$$

4.3.8 Waste Separation Costs

The separation of wastes is one of the most expensive activities in this process; however, this activity is necessary because the MSW is a mixture of wastes. In countries like the USA and the European Union, the separation process is very efficient compared with countries like the ones of Central and South America, since in the USA the separation cost is linked to the collection stage. In Mexico, most people do not separate the waste and there is not an efficient municipal collection system. In this case, the separation cost must be considered, in addition to the transportation from houses to landfills. This approximation is given by a separation unitary cost ($C^{sep}_{wn,c,p-wn}$), which depends on the type of waste and the place.

The separation is given in two steps: first the separation of the generic wastes (plastics, metals, glasses, or organic compounds), and second the separation of different sub-wastes.

$$CostSep_{wn} = \sum_{c \in C} \sum_{p-wn \in S} C^{sep}_{wn,c,p-wn} F_{wn,c,p-wn}, \quad \forall wn \in W \tag{4.8}$$

The model considers a total separation cost, which considers all individual separation costs:

$$CostSep^{total} = \sum_{wn \in W} CostSep_{wn} \tag{4.9}$$

4.3.9 No Separation Cost Associated with the MSW

This separation cost is normally absorbed by the government through landfills. This cost is formed by a unitary cost to not separate multiplied by the amount of unseparated MSW. The unitary cost to not separate is lower than the unitary cost to separate; however, the amount of MSW that is not separated is normally greater than the amount of separated MSW.

$$CostNSep^{total} = \sum_{wn \in W} \sum_{c \in C} \sum_{p-wn \in S} C^{\text{no-sep}}_{wn,c,p-wn} \left(F^{\text{total}}_{wn,c,p-wn} - F_{wn,c,p-wn}\right) \tag{4.10}$$

4.3.10 Total Transportation Cost of Wastes from Cities to Processing Plants

An important activity is the transportation of wastes from cities to processing facilities; this activity has an associated cost, which considers the unitary transportation cost of any type of waste ($C^{unit-transp}_{wn,p-wn}$), the distance between cities and processing facilities ($d_{c,pp-wn}$), and the transported wastes from cities to processing plants ($F^{plant}_{wn,c,pp-wn,p-wn}$). The unitary transportation cost has units of money per mass and distance:

$$CostTransp^{\text{city}}_{\text{processing plant}} = \sum_{wn \in W} \sum_{c \in C} \sum_{pp-wn \in P} \sum_{p-wn \in S} C^{\text{unit-transp}}_{wn,p-wn} d_{c,pp-wn} F^{plant}_{wn,c,pp-wn,p-wn} \tag{4.11}$$

4.3.11 Total Transportation Cost of Products from Processing Plants to Consumer Cities

Another transportation activity is the transportation of products from the processing plants to markets or cities, in this case, the unitary transportation cost ($C^{unit-transp}_{prod\ wn,n-wn,p-wn}$) depends on the type of waste and the produced products. Furthermore, the total transportation cost considers the distance between the cities and the processing plants, as the amount of transported product ($P^{city}_{wn,pp-wn,c,n-wn,p-wn}$):

$$CostTransp^{\text{processing plant}}_{\text{city}} = \sum_{wn \in W} \sum_{n-wn \in T} \sum_{c \in C} \sum_{pp-wn \in P} \sum_{p-wn \in S} C^{\text{unit-transp}}_{\text{prod}\ wn,n-wn,p-wn} d_{c,pp-wn} P^{city}_{wn,pp-wn,c,n-wn,p-wn} \tag{4.12}$$

4.3.12 Processing cost in the Processing Facilities

There is a considered processing cost for each technology, which accounts for the operating cost that considers a unitary cost ($C^{\text{processing}}_{wn,pp-wn,n-wn,p-wn}$) that depends on the processing plant *pp-wn*, type of used technology *n-wn* and type of wastes *wn* and *p-wn*. The total operating cost also depends on the amount of processed waste in each processing facility ($F^{tech}_{wn,pp-wn,n-wn,p-wn}$):

$$CostOp^{\text{processing}} = \sum_{wn\in W}\sum_{pp-wn\in P}\sum_{n-wn\in T}\sum_{p-wn\in S} C^{\text{processing}}_{wn,pp-wn,n-wn,p-wn}F^{tech}_{wn,pp-wn,n-wn,p-wn} \tag{4.13}$$

Also, it is required to consider the capital costs for the processing facilities. The capital cost is a function of the type of technology: $A^{processing}_{wn,pp-wn,n-wn,p-wn} + B1^{cap}_{wn,pp-wn,n-wn,p-wn}\left(F^{tech}_{wn,pp-wn,n-wn,p-wn}\right)^{\alpha}$, the superscript *a* is used to consider the economies of scale; in this context, the function is nonlinear and can generate problems during the solution approach; thus, the function can be modified to yield a linear function as follows: $A^{processing}_{wn,pp-wn,n-wn,p-wn} + B^{cap}_{wn,pp-wn,n-wn,p-wn}$, where $B^{cap}_{wn,pp-wn,n-wn,p-wn}$ is the unit variable capital cost. The capital cost for each processing facility is calculated as follows:

$$CostCap^{\text{processing}}_{wn,pp-wn,n-wn,p-wn} = A^{processing}_{wn,pp-wn,n-wn,p-wn} + B^{cap}_{wn,pp-wn,n-wn,p-wn}, \quad \forall wn\in W, pp-wn\in P, n-wn\in T, p-wn\in S \tag{4.14}$$

Once the individual capital cost is obtained, it is possible to obtain the total annual capital cost, which is the sum of all individual capital costs ($CostCap^{\text{processing}}_{wn,pp-wn,n-wn,p-wn}$) for each type of waste, processing plant, and processing technology:

$$CostCap^{total} = \sum_{wn\in W}\sum_{pp-wn\in P}\sum_{n-wn\in T}\sum_{p-wn\in S} CostCap^{\text{processing}}_{wn,pp-wn,n-wn,p-wn} \tag{4.15}$$

The terms in the capital cost of the processing plants depend on the capacity of the plants; these are defined in the following disjunctions, which are activated when the processed waste is between the lower and upper limits ($\text{Fmin}^{processing}_{wn,pp-wn,n-wn,p-wn,q}$ and $\text{Fmax}^{processing}_{wn,pp-wn,n-wn,p-wn,q}$, respectively):

$$\bigvee_{q}\begin{bmatrix} Y^{processing}_{wn,pp-wn,n-wn,p-wn,} \\ A^{processing}_{wn,pp-wn,n-wn,p-wn} = \text{a}^{processing}_{wn,pp-wn,n-wn,p-wn,q} \\ B^{cap}_{wn,pp-wn,n-wn,p-wn} = \text{b}^{cap}_{wn,pp-wn,n-wn,p-wn,q}F^{tech}_{wn,pp-wn,n-wn,p-wn} \\ \text{Fmin}^{processing}_{wn,pp-wn,n-wn,p-wn,q} \le F^{tech}_{wn,pp-wn,n-wn,p-wn} \le \text{Fmax}^{processing}_{wn,pp-wn,n-wn,p-wn,q} \end{bmatrix}, \quad \forall wn\in W, pp-wn\in P, n-wn\in T, p\text{-}wn\in S, q\in I$$

The previous disjunction is algebraically reformulated as follows:

$$A^{processing}_{wn,pp-wn,n-wn,p-wn} = \sum_{q\in I}\text{a}^{processing}_{wn,pp-wn,n-wn,p-wn,q}y^{processing}_{wn,pp-wn,n-wn,p-wn,q}, \quad \forall wn\in W, pp-wn\in P, n-wn\in T, p-wn\in S \tag{4.16}$$

$$B^{cap}_{wn,pp-wn,n-wn,p-wn} = \sum_{q\in I}\text{b}^{cap}_{wn,pp-wn,n-wn,p-wn,q}dF^{tech}_{wn,pp-wn,n-wn,p-wn,q}, \quad \forall wn\in W, pp-wn\in P, n-wn\in T, p-wn\in S \tag{4.17}$$

$$F^{tech}_{wn,pp-wn,n-wn,p-wn} = \sum_{q\in I} dF^{tech}_{wn,pp-wn,n-wn,p-wn,q} \quad \forall wn\in W, pp-wn\in P, n-wn\in T, p-wn\in S \tag{4.18}$$

$$\text{Fmin}^{processing}_{wn,pp-wn,n-wn,p-wn,q}y^{processing}_{wn,pp-wn,n-wn,p-wn,q} \le dF^{tech}_{wn,pp-wn,n-wn,p-wn}, \quad \forall wn\in W, pp-wn\in P, n-wn\in T, p-wn\in S, q\in I \tag{4.19}$$

$$\text{Fmax}^{processing}_{wn,pp-wn,n-wn,p-wn,q}y^{processing}_{wn,pp-wn,n-wn,p-wn,q} \ge dF^{tech}_{wn,pp-wn,n-wn,p-wn}, \quad \forall wn\in W, pp-wn\in P, n-wn\in T, p-wn\in S, q\in I \tag{4.20}$$

$$\sum_{q\in I} y^{processing}_{wn,pp-wn,n-wn,p-wn,q} = 1, \quad \forall wn\in W, pp-wn\in P, n-wn\in T, p-wn\in S \tag{4.21}$$

Equation (4.17) is used to define the fixed capital cost ($A^{processing}_{wn,pp-wn,n-wn,p-wn}$), which depends on the processing plant, processing technology, and type of waste. The unit variable capital cost ($B^{cap}_{wn,pp-wn,n-wn,p-wn}$) is given in Equation (4.18). The amount of waste in the processing plants ($F^{tech}_{wn,pp-wn,n-wn,p-wn}$) is obtained through Equation (4.19) and these are limited by Equations (4.20) and (4.21). Equation (4.22) limits to one the number of defined intervals, which are based on the economies of scale.

4.3.13 Profit from Selling Value-added Products

The products can be sold to different markets, in this case, the markets are located around the cities; each product has a unit price, which depends on the type of product and the city where it is sold. The profit from the sold products (C^{sales}) is the sum of the amount of products distributed to the different markets ($P^{city}_{wn,pp-wn,c,n-wn,p-wn}$) multiplied by the corresponding unit cost ($C^{product}_{wn,n-wn,c}$):

$$C^{sales} = \sum_{wn\in W} \sum_{pp-wn\in P} \sum_{n-wn\in T} \sum_{c\in C} \sum_{p-wn\in S} P^{city}_{wn,pp-wn,c,n-wn,p-wn} C^{\text{product}}_{wn,n-wn,c} \quad (4.22)$$

4.3.14 Economic Objective Function

The economic objective function is the maximization of the total net annual profit (*NetProfit*), which takes into account the profit for the selling of products (C^{sales}) minus the total separation cost ($CostSep^{total}$), total capital cost ($CostCap^{total}$), total operating cost ($CostOp^{\text{processing}}$), transportation cost ($CostTransp^{\text{processing plant}}_{\text{city}}$ and ($CostTransp^{\text{city}}_{\text{processing plant}}$), and no separation cost ($CostNSep^{total}$):

$$\begin{aligned} \text{Max} \quad NetProfit = {} & C^{sales} - CostSep^{total} - CostCap^{total} - CostOp^{\text{processing}} \\ & - CostTransp^{\text{processing plant}}_{\text{city}} - CostTransp^{\text{city}}_{\text{processing plant}} - CostNSep^{total} \end{aligned} \quad (4.23)$$

4.3.15 Reused Wastes

Reused waste represents environmental and social concerns because, for greater reuse of the produced MSW, the waste discharged into landfills decreases. This way, the environmental problems associated with land and water pollution decrease. Similar to this, reusing MSW has a positive effect on society since it improves life quality and creates jobs that are needed for the various stages of the reusing process. Therefore, the percentage of the consumed waste ($Consume^{waste}$) is considered in this work as a measure of the social impact, and this objective function is stated as follows:

$$\text{Max} \quad Consume^{waste} = 100 \times \frac{\sum_{wn\in W} \sum_{c\in C} \sum_{p-wn\in S} F_{wn,c,p-wn}}{\sum_{wn\in W} \sum_{c\in C} \sum_{p-wn\in S} \acute{a}^{sep}_{wn,c,p-wn} F^{total}_{c,p-wn}} \quad (4.24)$$

4.4 Case Study

A case study is presented below to show the applicability of the proposed approach. This case study corresponds to the optimal management of the municipal solid waste for the region located in the central-west part of Mexico considering the profit as the economic objective and the percentage of utilized waste as the environmental objective, where several cities and markets have been considered as shown in Fig. 2. Five cities are considered as sources and consumer locations for the considered system; in addition, fifteen technologies to treat different wastes are considered.

The data for the considered conversion technologies (Table 1) were taken from EPA (2011), Alireza and Gordon, (2012), Plunkert, (2006), Hur et al. (2003), and Young (2010). Although each

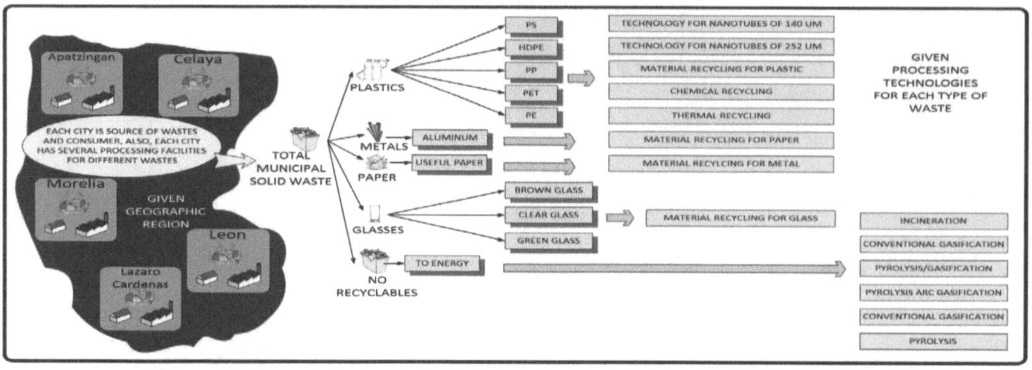

Fig. 2. General representation of the addressed case study in Chapter 4. [Printed (adapted) with permission from Santibañez-Aguilar, J.E., Ponce-Ortega, J.M., González-Campos, J.B., Serna-González, M., and El-Halwagi, M. 2013. Optimal Planning for the sustainable utilization of municipal solid waste. Waste Management, *33*(12), 2607–2622].

city has a given waste production rate, this production depends on the number of people living there and the economic activities developed in such a city; moreover, the composition (i.e. the types of MSW) is different for each considered city. The amount of the different wastes as well as the efficiency factors for recollection of wastes are presented in Table 2 (INE 2012, EPA 2011). Table 3 presents the data for the unit production costs, and the transportation costs for products and wastes as well as the distances between the considered places are shown in Table 4. Table 5 presents the separation costs for the different wastes and their assumptions.

4.5 Results

The mathematical model associated with the case study consists of 800 binary variables, 6,426 continuous variables, and 5,442 constraints. It was coded in the software GAMS taking an average of 0.08 seconds of CPU time in a processor i7-3720QM at 2.60GHz with 16GB of RAM for the solution of each point of the Pareto curve using the solver CPLEX, which is a GAMS solver that can efficiently solve large optimization problems (Brooke et al. 2013).

Pareto solutions that compensate for the two objective functions considered (net profit and percentage of reused MSW) are presented. Figure 3 demonstrates the Pareto curve that is produced when the net annual profit is the primary goal and is influenced by the percentage of trash that is consumed overall; in this instance, the net annual profit falls as the waste percentage rises.

It is important to note that a Pareto front is a graphical representation formed by optimal points subject to different constraints; normally this type of graphic is given in two dimensions which are two contradicting objectives. The points above the Pareto front (Fig. 3) are infeasible solutions, whereas the points below the Pareto front are sub-optimal solutions. Therefore, from this Pareto front, the decision maker can select the solution that best satisfies specific requirements.

In the Pareto curve of Fig. 3, several regions that are obtained for the selection of different configurations of the supply chain to satisfy all the constraints are identified. The annual profit is always negative because the separation and no separation costs are high, and a consumption of 100% of the total available waste is not equal to the 100% of waste reused, since each waste has a given percentage of separation and in most cases this is not greater than 10%.

Table 1. Efficiency factors for the processing technologies to treat municipal solid waste. [Printed (adapted) with permission from Santibañez-Aguilar, J.E., Ponce-Ortega, J.M., González-Campos, J.B., Serna-González, M., and El-Halwagi, M. 2013. Optimal Planning for the sustainable utilization of municipal solid waste. Waste Management, *33*(12), 2607–2622].

Waste	Sub-waste	Technology	Product	Production Factor kg Product/kg Waste
Plastic	PP	Technology 1 for nanotubes	Nanotubes of 140um diameter	0.11[1]
Plastic	PE	Technology 1 for nanotubes	Nanotubes of 140um diameter	0.16[1]
Plastic	PET	Technology 1 for nanotubes	Nanotubes of 140um diameter	0.19[1]
Plastic	PE	Technology 2 for nanotubes	Nanotubes of 252um diameter	0.32[1]
Plastic	PP	Technology 2 for nanotubes	Nanotubes of 252um diameter	0.51[1]
Plastic	PE	Material Recycling 1	Pellet	0.672[2]
Plastic	PS	Material Recycling 1	Pellet	0.672[2]
Plastic	PP	Material Recycling 1	Pellet	0.672[2]
Plastic	PE	Thermal Recycling	Gasoline	0.299[2]
Plastic	PE	Thermal Recycling	Diesel	0.089[2]
Plastic	PE	Thermal Recycling	Heavy Oil	0.068[2]
Plastic	PE	Chemical Recycling	Heat Energy	19.44*[2]
Plastic	HDPE	Pyrolysis	n-Paraffines	0.432[2]
Plastic	HDPE	Pyrolysis	n-Olefines	0.446[2]
Plastic	PP	Pyrolysis	Branched Paraffines	0.304[2]
Plastic	PP	Pyrolysis	Branched Paraffines	0.333[2]
Paper	Paper	Material Recycling 2	New Paper	0.84[3]
Metal	Aluminum	Material Recycling 3	Useful Aluminum	1[4]
Glass	Clear Glass	Material Recycling 4	New Glass	1[5]
Glass	Green Glass	Material Recycling 4	New Glass	1[5]
Glass	Brown Glass	Material Recycling 4	New Glass	1[5]
MSW	MSW	Mass Burn (Incineration)	Electricity	1.9584*[6]
MSW	MSW	Pyrolysis	Electricity	2.0556*[6]
MSW	MSW	Pyrolysis/Gasification	Electricity	2.466*[6]
MSW	MSW	Plasma Arc Gasification	Electricity	2.9376*[6]
MSW	MSW	Conventional gasification	Electricity	2.466*[6]

* This factor has units of MJ Energy/kg Waste.
[1]Alireza and Gordon (2012).
[2]Hur et al. (2003).
[3]Environmental Protection Agency Office of Resource Conservation and Recovery (2011).
[4]Plunkert (2006).
[5]http://elreciclaje.org/content/reciclaje-de-vidrio. Accessed (09/29/2012).
[6]Young (2010).

Figure 4.4 to 4.7 show the distribution of the wastes consumed at each point of the Pareto curve, these figures allow an understanding of the behavior of the Pareto curve, because for each point of the Pareto curve, the number, type, and distribution of the consumed wastes are different. In this context, metal is consumed at all the points of the Pareto curve, and it is totally used because this represents about 0.5% of total available consumed waste, plastic is consumed at just a few points of the Pareto curve because the operating and capital costs associated to the conversion technologies to treat plastics are very expensive. An important change in the behavior of the Pareto curve is

Table 2. Generated wastes and sub-wastes for each city and efficiency for recollection (kg). [Printed (adapted) with permission from Santibañez-Aguilar, J.E., Ponce-Ortega, J.M., González-Campos, J.B., Serna-González, M., and El-Halwagi, M. 2013. Optimal Planning for the sustainable utilization of municipal solid waste. Waste Management, *33*(12), 2607–2622].

		Morelia[1]	**Celaya**[1]	**Apatzingan** [1]	**Lazaro**[1] **Cardenas**	**Leon**[1]	**Efficiency for recollection**[2]
MSW	**Total production**	$4.35x10^{08}$	$2.00x10^{08}$	$5.46x10^{07}$	$9.72x10^{07}$	$5.23x10^{08}$	
	Non recyclables (70.3% MSW)	$3.06x10^{08}$	$1.41x10^{08}$	$3.84x10^{07}$	$6.83x10^{07}$	$3.68x10^{08}$	10%
Plastics	**Total production (5.8% MSW)**	$2.52x10^{07}$	$1.16x10^{07}$	$3.16x10^{06}$	$5.64x10^{06}$	$3.03x10^{07}$	
	PP (Tonnes)	$9.37x10^{06}$	$4.32x10^{06}$	$1.18x10^{06}$	$2.09x10^{06}$	$1.13x10^{07}$	6.40%
	PE (Tonnes)	$6.30x10^{06}$	$2.91x10^{06}$	$7.91x10^{05}$	$1.41x10^{06}$	$7.59x10^{06}$	6.40%
	PET (Tonnes)	$3.68x10^{05}$	$1.70x10^{05}$	$4.62x10^{04}$	$8.23x10^{04}$	$4.43x10^{05}$	6.40%
	PS (Tonnes)	$1.33x10^{06}$	$6.15x10^{05}$	$1.67x10^{05}$	$2.98x10^{05}$	$1.61x10^{06}$	6.40%
	HDPE (Tonnes)	$2.69x10^{06}$	$1.24x10^{06}$	$3.38x10^{05}$	$6.02x10^{05}$	$3.24x10^{06}$	6.40%
Metals	**Total production (3.1% MSW)**	$1.35x10^{07}$	$6.21x10^{06}$	$1.69x10^{06}$	$3.01x10^{06}$	$1.62x10^{07}$	
	Aluminum (Tonnes)	$8.24x10^{05}$	$3.80x10^{05}$	$1.03x10^{05}$	$1.84x10^{05}$	$9.91x10^{05}$	35.80%
Glasses	**Total production (6.6% MSW)**	$2.87x10^{07}$	$1.32x10^{07}$	$3.60x10^{06}$	$6.42x10^{06}$	$3.45x10^{07}$	
	Clear glass (Tonnes)	$5.21x10^{06}$	$2.40x10^{06}$	$6.54x10^{05}$	$1.17x10^{06}$	$6.27x10^{06}$	24.70%
	Green glass (Tonnes)	$6.10x10^{06}$	$2.81x10^{06}$	$7.66x10^{05}$	$1.36x10^{06}$	$7.34x10^{06}$	41.40%
	Brown glass (Tonnes)	$1.74x10^{07}$	$8.01x10^{06}$	$2.18x10^{06}$	$3.89x10^{06}$	$2.09x10^{07}$	18.10%
Paper	**Total production (14.2% MSW)**	$6.17x10^{07}$	$2.85x10^{07}$	$7.75x10^{06}$	$1.38x10^{07}$	$7.43x10^{07}$	
	Paper useful	$2.58x10^{07}$	$1.19x10^{07}$	$3.23x10^{06}$	$5.76x10^{06}$	$3.10x10^{07}$	71.60%

[1]Mexican National Institute of Ecology (2012).
[2]Environmental Protection Agency Office of Resource Conservation and Recovery (2011).

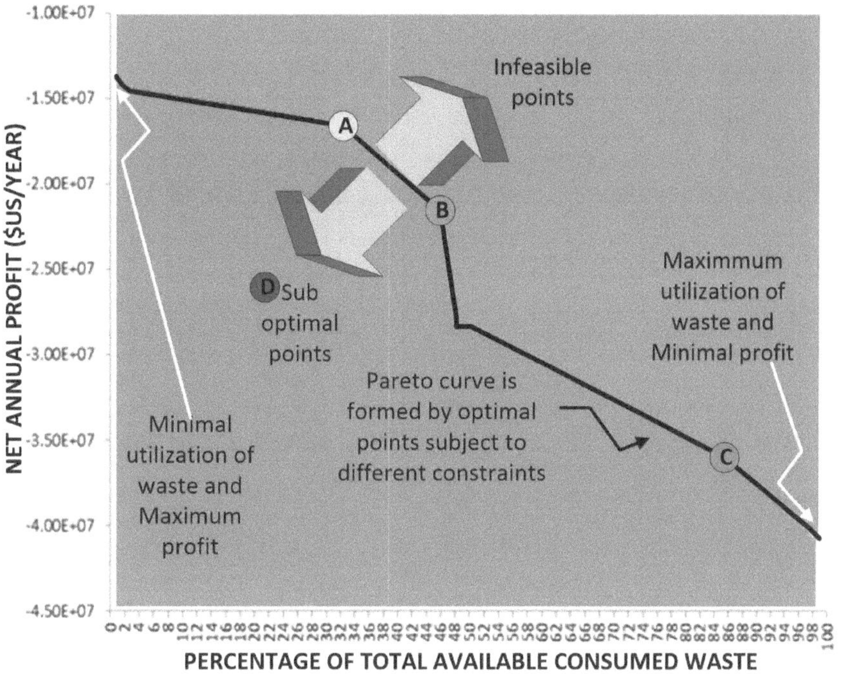

Fig. 3. Pareto curve for a system of processing facilities based on municipal solid waste. [Printed (adapted) with permission from Santibañez-Aguilar, J.E., Ponce-Ortega, J.M., González-Campos, J.B., Serna-González, M., and El-Halwagi, M. 2013. Optimal Planning for the sustainable utilization of municipal solid waste. Waste Management, 33 (12), 2607–2622]

Table 3. Unit price for the considered products in the cities. [Printed (adapted) with permission from Santibañez-Aguilar, J.E., Ponce-Ortega, J.M., González-Campos, J.B., Serna-González, M., and El-Halwagi, M. 2013. Optimal Planning for the sustainable utilization of municipal solid waste. Waste Management, *33*(12), 2607–2622].

Product	Price ($US/kg)
Nanotubes of 140 um	100[4]
Nanotubes of 252 um	70[4]
Pellet	0.3645[1]
Fuel	0.332[1]
Electricity*	0.052632[2]
Olefines	0.18[1]
Aluminum	1.3[3]
Clear glass	0.039057[4]
Green glass	0.01171[4]
Brown glass	0.02994[4]
Paper	0.2109[4]

*$US/MJ
[1]Hur et al. (2003).
[2]Young (2010).
[3]Plunkert (2006).
[4]Mexican National Institute of Ecology (2012).

observed in the point between 46% and 48% of the total available consumed waste, this behavior can be explained as follows: Figs, 6 and 7 show that when 46% of the total waste is consumed, the glass is totally consumed but no-recyclable wastes are not consumed at all; however, when the 48% of the total wastes are consumed, the glass is not consumed and around of 30% of the no-recyclable

Table 4. Distance from cities to processing facilities and unit transportation costs. [Printed (adapted) with permission from Santibañez-Aguilar, J.E., Ponce-Ortega, J.M., González-Campos, J.B., Serna-González, M., and El-Halwagi, M. 2013. Optimal Planning for the sustainable utilization of municipal solid waste. Waste Management, *33*(12), 2607–2622].

Cities	Distance to processing facilities located in cities (km)				
	Morelia	**Celaya**	**Apatzingan**	**Lazaro Cardenas**	**Leon**
Morelia	1	147	192	144	207
Celaya	147	1	326	447	135
Apatzingan	192	326	1	186	393
Lazaro Cardenas	144	447	186	1	514
Leon	207	135	393	514	1
Unit transportation cost ($US/kg km)	$1.65x10^{-5}$ *	$4x10^{-5}$ +			

* For Plastic, MSW, Paper
+ Metal and Glass

Table 5. Estimated separation cost in Mexico. [Printed (adapted) with permission from Santibañez-Aguilar, J.E., Ponce-Ortega, J.M., González-Campos, J.B., Serna-González, M., and El-Halwagi, M. 2013. Optimal Planning for the sustainable utilization of municipal solid waste. Waste Management, *33*(12), 2607–2622].

Type of wastes	Separation cost ($US/kg)	Reference and consideration
Plastic	0.235	Manual separation. Mexican National Institute of Ecology (2012)
Paper	0.235	
Glass	0.235	
Aluminum	0.235	
Mixture of wastes	0.235	

wastes are consumed, also the percentage used of no recyclables increases in this point to 86% of total available. Figure 5 represents the amount of consumed paper, in this case, paper is consumed at almost all the points of the Pareto curve and the percentage of recycled paper increases while the net profit increases, this suggests that the technologies to treat paper have a very positive effect in the net profit.

Figure 8 shows the distribution of the costs through the supply chain for the case study analyzed. In this context, the separation cost is the highest cost above the point where around 36% of total available waste is consumed, this is because the separation activity is the most important in the treatment of waste; furthermore, this activity is the one that more jobs generates, this way, this can promote tax credits or incentives from governments for the recycling process due to the social benefit obtained. The disposal cost is the highest cost from 0% to 34% of the total waste reused, this is because the amount of trash without treatment is huge; this cost decreases when the percentage of consumed waste increases, but this depletion is small because there is a high amount of wastes untreated. The transportation cost is not representative of the total cost because the cities are in a small geographic region with only five considered cities because of the optimization of the supply chain. Since the separation and no separation costs are considered, the total cost is always greater than the revenue from sales; however, this is an area of opportunity to make these processes more efficient.

In addition, three points were selected from Fig. 3 to show the different wastes consumed at each point. First, for point A, 32% of total consumed wastes are used, this represents the consumption of two wastes: aluminum and paper, five cities and two processing facilities (to treat aluminum and paper, respectively); both processing plants are in Morelia, it is important to note that the city of Morelia is in the center of the considered region. The products obtained from aluminum are distributed to

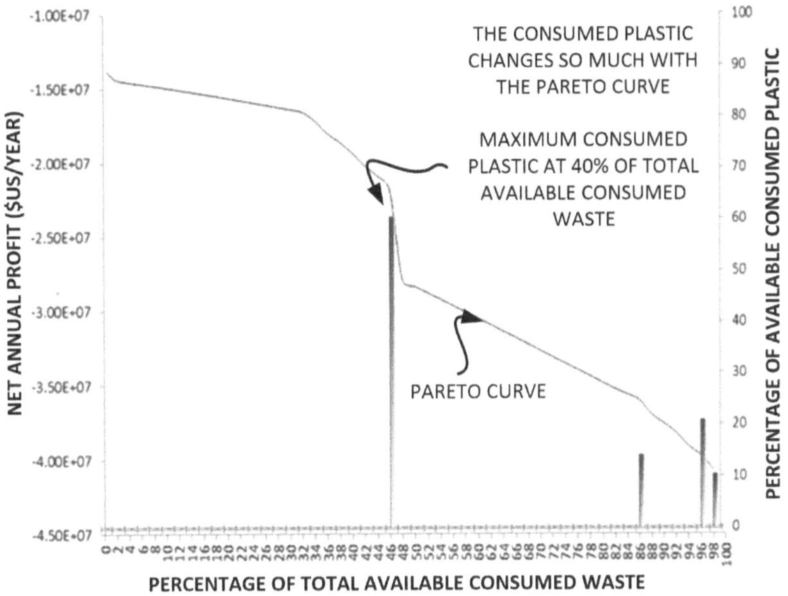

Fig. 4. Percentage of consumed plastic with respect to the Pareto curve. [Printed (adapted) with permission from Santibañez-Aguilar, J.E., Ponce-Ortega, J.M., González-Campos, J.B., Serna-González, M., and El-Halwagi, M. 2013. Optimal Planning for the sustainable utilization of municipal solid waste. Waste Management, *33*(12), 2607–2622].

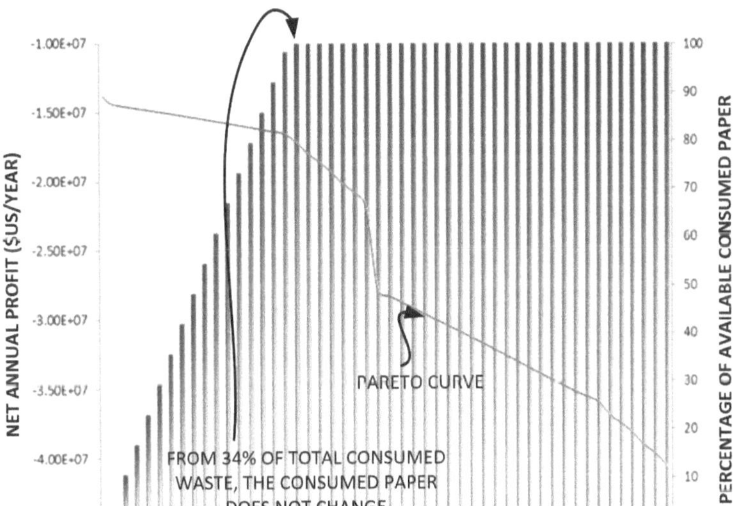

Fig. 5. Percentage of consumed paper concerning the Pareto curve. [Printed (adapted) with permission from Santibañez-Aguilar, J.E., Ponce-Ortega, J.M., González-Campos, J.B., Serna-González, M., and El-Halwagi, M. 2013. Optimal Planning for the sustainable utilization of municipal solid waste. Waste Management, *33*(12), 2607–2622].

the cities of Morelia and Lazaro Cardenas; on the other hand, the products obtained from paper are distributed to the considered cities except Leon, the selected technology in both cases was material recycling and a schematic representation for this solution is presented in Fig. 9.

Figure 10 shows the results for point B, which represents almost 46% of the total consumed waste, at this point seven subtypes of wastes were selected (i.e., aluminum, paper, clear glass, green

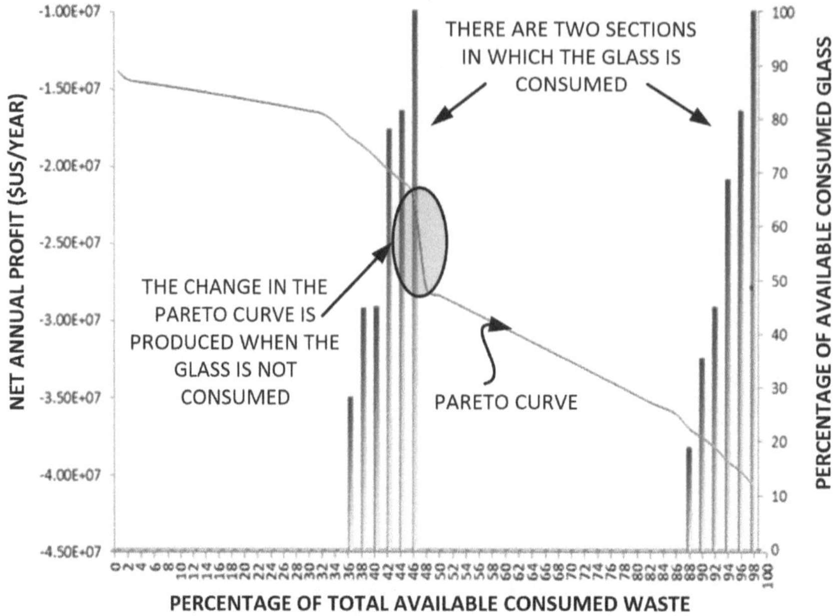

Fig. 6. Percentage of consumed glass concerning the Pareto curve. [Printed (adapted) with permission from Santibañez-Aguilar, J.E., Ponce-Ortega, J.M., González-Campos, J.B., Serna-González, M., and El-Halwagi, M. 2013. Optimal Planning for the sustainable utilization of municipal solid waste. Waste Management, 33(12), 2607–2622].

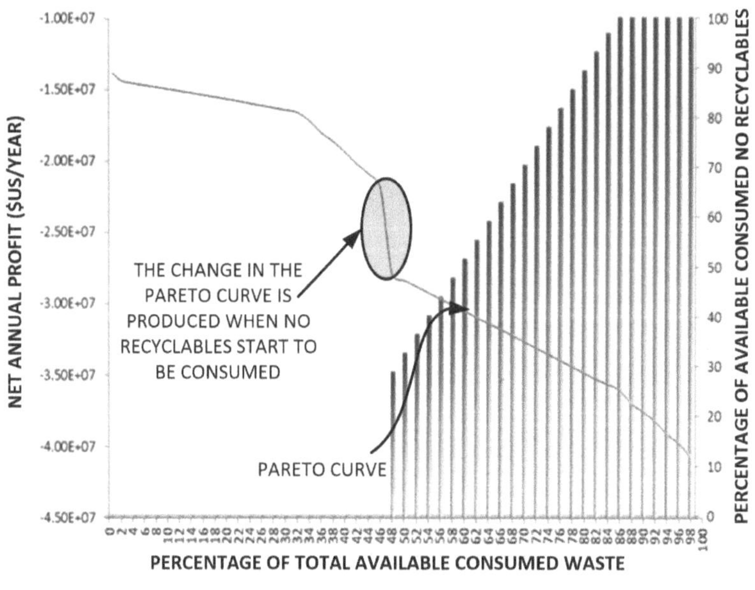

Fig. 7. Percentage of consumed no recyclable wastes concerning the Pareto curve. [Printed (adapted) with permission from Santibañez-Aguilar, J.E., Ponce-Ortega, J.M., González-Campos, J.B., Serna-González, M., and El-Halwagi, M. 2013. Optimal Planning for the sustainable utilization of municipal solid waste. Waste Management, 33(12), 2607–2622].

glass, brown glass, polyethylene, and polypropylene). The configuration for aluminum is the same as the one corresponding to point A; for paper, the difference is that the products are distributed to all considered cities. Two types of plastic are considered in this case, polypropylene is distributed from all cities to Morelia to treat this plastic by thermal recycling to produce a fuel; on the other hand,

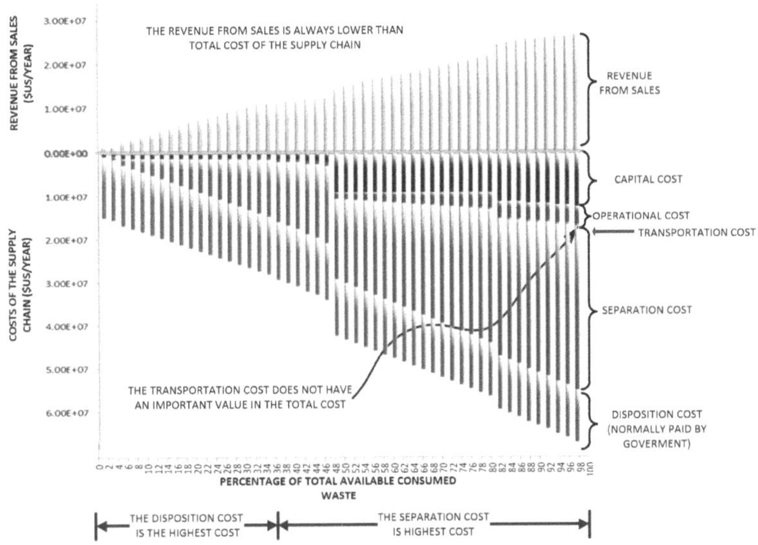

Fig. 8. Distribution for the costs of the supply chain based on solid waste. [Printed (adapted) with permission from Santibañez-Aguilar, J.E., Ponce-Ortega, J.M., González-Campos, J.B., Serna-González, M., and El-Halwagi, M. 2013. Optimal Planning for the sustainable utilization of municipal solid waste. Waste Management, *33*(12), 2607–2622]

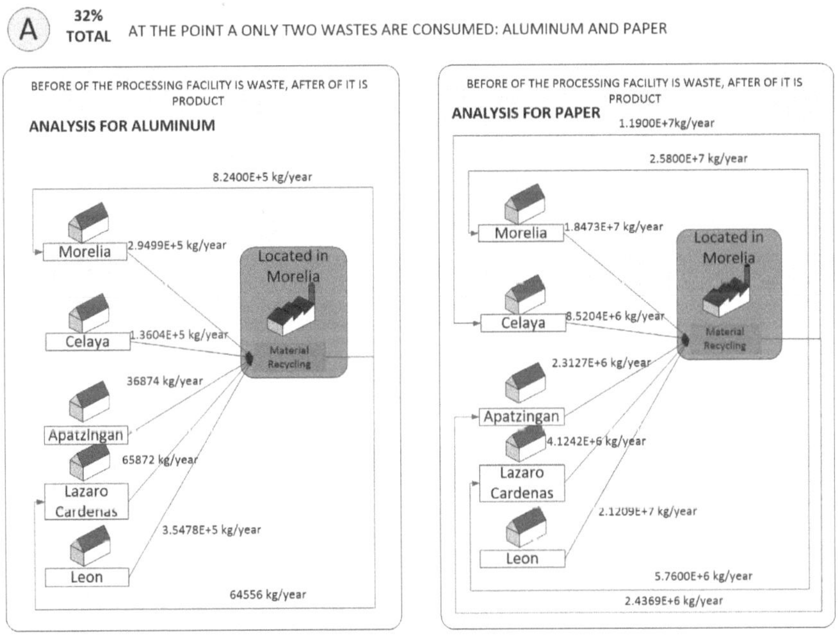

Fig. 9. Solution of the distribution in point A of Figure 4.3. [Printed (adapted) with permission from Santibañez-Aguilar, J.E., Ponce-Ortega, J.M., González-Campos, J.B., Serna-González, M., and El-Halwagi, M. 2013. Optimal Planning for the sustainable utilization of municipal solid waste. Waste Management, *33*(12), 2607–2622].

polyethylene is transported to Leon to obtain olefins through pyrolysis. Glass is delivered to two different plants, both located in Morelia; in the first one, clear and green glass are processed whereas in the second one, only brown glass is treated, in this case, all products are consumed in Morelia.

For Point C, plastic waste is obtained from two cities and this waste is processed by thermal recycling in a processing facility located in Leon. Also, no-recyclable wastes are processed in a

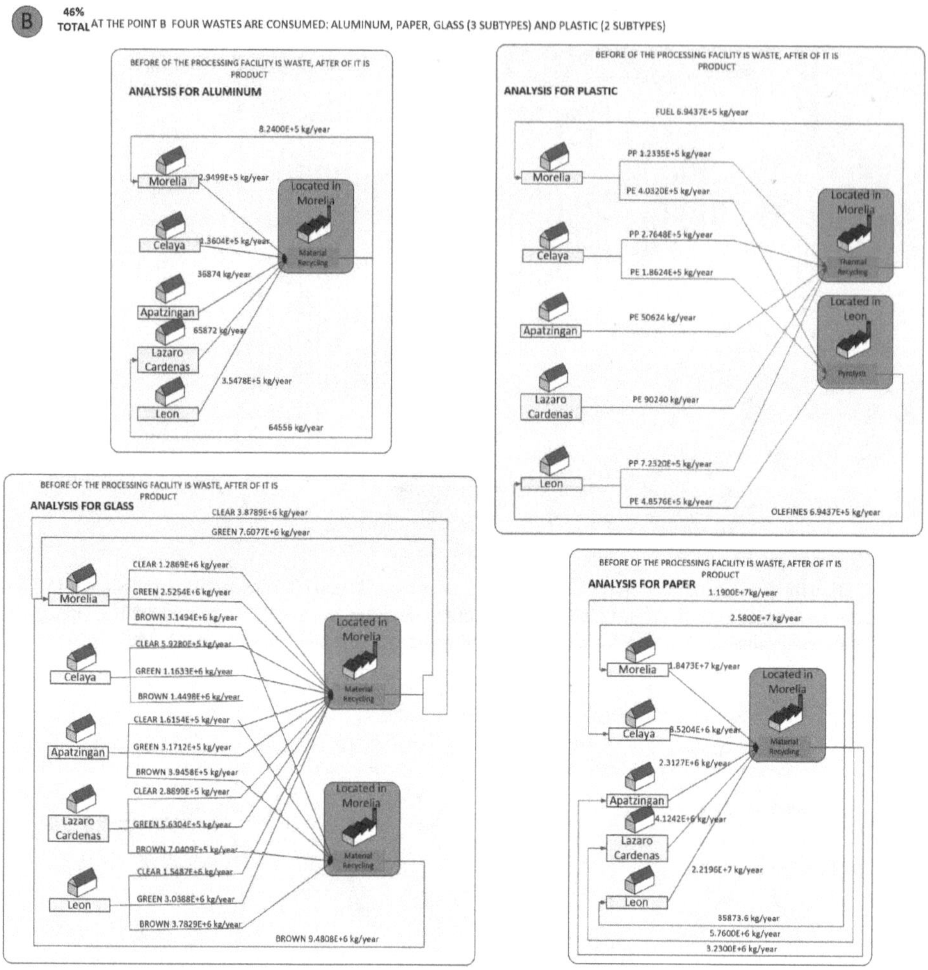

Fig. 10. Solution of the distribution in point B of Fig. 3. [Printed (adapted) with permission from Santibañez-Aguilar, J.E., Ponce-Ortega, J.M., González-Campos, J.B., Serna-González, M., and El-Halwagi, M. 2013. Optimal Planning for the sustainable utilization of municipal solid waste. Waste Management, *33*(12), 2607–2622].

processing plant located in Morelia through plasma arc gasification to obtain energy which is used in Morelia; this configuration is shown in Fig. 11.

4.6 Concluding Remarks

This chapter presented a mathematical model for the optimal planning of a distributed system to treat MSW while considering economic and environmental aspects. The issue has been mathematically formulated as a multi-objective mixed integer linear programming problem. The model considers simultaneously the supply chain optimization for the products obtained from the municipal solid waste. The application of the proposed methodology has been illustrated through a case study of a distributed system for treating MSW in the central-west region of Mexico. The results show that it is possible to implement a distributed processing system to reuse MSW while maximizing the net profit of the supply chain. In addition, the obtained Pareto solutions can provide valuable insights to decision-makers to select the solution that shows the best compromise among the considered objectives. The model can be useful to consider waste management in regions where waste management has not been developed or there is not an established way to control waste. This work can be used as a starting point to add

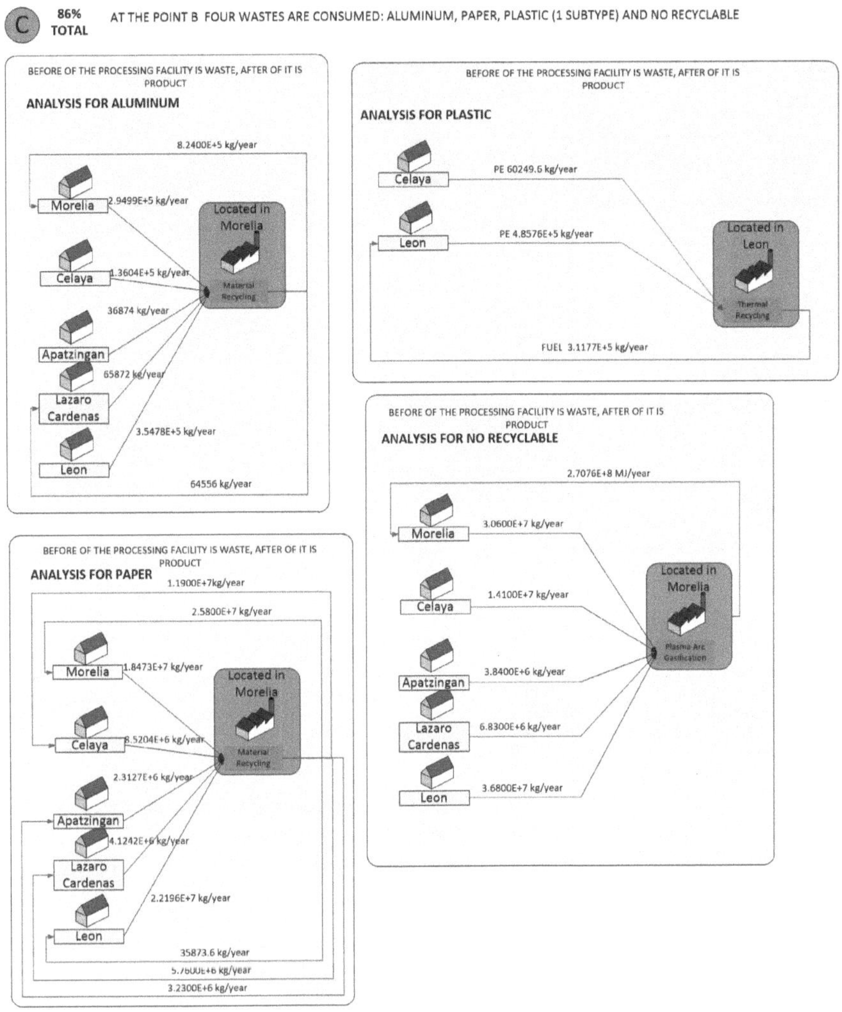

Fig. 11. Solution of the distribution in point C of Fig. 3. [Printed (adapted) with permission from Santibañez-Aguilar, J.E., Ponce-Ortega, J.M., González-Campos, J.B., Serna-González, M., and El-Halwagi, M. 2013. Optimal Planning for the sustainable utilization of municipal solid waste. Waste Management, *33*(12), 2607–2622].

more objectives for example social, safety, and health, between others as well as the scheduling of the waste management system. The model can also be extended to consider the associated uncertainty.

4.7 Nomenclature

The nomenclature used in the mathematical model has been separated as parameters, continuous and binary variables.

4.7.1 Parameters

$F_{c,p-wn}^{total}$	Maximum available waste type *p-wn* in the city *c*
$\alpha_{wn,c,p-wn}^{sep}$	Factor used to determine the amount of useful waste type *p-wn* in the city *c*

$\alpha^{prod}_{wn,n-wn,p-wn}$	Efficiency factor for each processing technology to obtain products from wastes
$PD^{city}_{wn,n\text{-}wn,c,p\text{-}wn}$	Demand for each product in each city
$C^{sep}_{wn,c,p\text{-}wn}$	Unitary separation cost of wastes
$C^{no\text{-}sep}_{wn,c,p\text{-}wn}$	Unitary no separation cost that is paid by the government for landfills
$C^{unit\text{-}transp}_{wn,p\text{-}wn}$	Unitary transportation cost of waste per amount of material and distance
$d_{c,pp-wn}$	Distance between the city *c* and the processing facility *pp-wn*
$C^{unit\text{-}transp}_{prod\ wn,n-wn,p\text{-}wn}$	Unitary transportation cost of products per amount of material and distance
$C^{product}_{wn,n-wn,c}$	Unitary cost for products

4.7.2 Continuous variables

$F_{wn,c,p-wn}$	Used waste *wn* type *p-wn* in the city *c*
$F^{plant}_{wn,c,pp-wn,p-wn}$	Distributed waste from city *c* to processing plants type *pp-wn*
$F^{tech}_{wn,pp-wn,n-wn,p-wn}$	Distributed waste to each processing technology into each processing facility
$P_{wn,pp-wn,n-wn,p-wn}$	Obtained product from each type of waste into processing facilities
$P^{city}_{wn,pp-wn,c,n-wn,p-wn}$	Delivered product to consumers
$PT^{city}_{wn,c,n-wn,p-wn}$	Total product from all processing facilities in each consumer
$CostSep_{wn}$	Individual separation cost for each type of waste *wn*
$CostSep^{total}$	Total separation cost
$CostNSep^{total}$	Total no separation cost
$CostTransp^{city}_{processing\ plant}$	Transportation costs from cities to processing facilities
$CostTransp^{processing\ plant}_{city}$	Transportation costs from processing facilities to cities
$C^{processing}_{wn,pp-wn,n-wn,p-wn}$	Unitary operational cost for processing plants
$CostOp^{processing}$	Total operational cost
$A^{processing}_{wn,pp-wn,n-wn,p-wn}$	Fixed part of the capital cost for processing facilities
$B^{cap}_{wn,pp-wn,n-wn,p-wn}$	Part of the capital cost that depends on the capacity of the processing facilities
$CostCap^{total}$	Total capital cost
C^{sales}	Profit from selling of products
NetProfit	Annual net profit
$Consume^{waste}$	Percentage of consumed waste

4.7.3 Binary variables

$y_{wn,pp-wn,n-wn,p-wn,q}^{processing}$	Binary variable used to define the existence of the processing plants and allows calculating the capital cost of the processing facility

4.8 References

Alireza, B., and Gordon, M. 2012. A review-synthesis of carbon nanotubes from plastic wastes. Chemical Engineering Journal, 195–196, 377–391.

Beolchini, F., Fonti, V., Dell'Anno, A., Rocchetti, L., and Vegliò, F. 2012. Assessment of biotechnological strategies for the valorization of metal bearing wastes. Waste Management, *32*(5), 949–956.

Brooke, A., Kendrick, D., Meeruas, A., and Raman, R. 2013. GAMS-Language guide. GAMS Development Corp. Washington, D.C.

Environmental Protection Agency Office of Resource Conservation and Recovery. 2011. Municipal Solid Waste Generation, Recycling, and Disposal in the United States Tables and Figures for 2010. U.S., USA.

Environmental Protection Agency Office of Resource Conservation and Recovery. 2011. Municipal Solid Waste Generation, Recycling, and Disposal in the United States. U.S. USA.

Hur, T., Lim, S., and Lee, H. 2003. A study on the eco-efficiencies for recycling methods of plastics wastes. InLCA. 22.

Krüger, S., Berger, A., and Krause, U. 2012. Chemical–analytical investigation of fire products in intermediate storages of recycling materials. Fire and Materials, *36*(3), 165–175.

Mexican National Institute of Ecology. 2012. Generaion and composition of the municipal solid wastes, Mexico City, Mexico. http://www2.ine.gob.mx/publicaciones/libros/128/cap3.html (Accessed 01/04/13).

Ortiz, O., Pasqualino, J. C., Diez, G., and Castells, F. 2010. The environmental impact of the construction phase: an application to composite walls from a life cycle perspective. Resources, Conservation and Recycling, *54*(11), 832–840.

Ramayah, T., Chow Lee, J. W., and Lim, S. 2012. Sustaining the environment through recycling: An empirical study. Journal of Environmental Management, *102*, 141–147.

Rogoff, M. J., and Screve, F. 2011. Waste to Energy, Technologies and Project Implementation. Elsevier, USA.

Santibañez-Aguilar, J. E., Ponce-Ortega, J. M., González-Campos, J. B., Serna-González, M., and El-Halwagi, M. 2013. Optimal Planning for the sustainable utilization of municipal solid waste. Waste Management, *33*(12), 2607–2622.

Santibañez-Aguilar, J. E., Ponce-Ortega, J. M., González-Campos, J. B., Serna-González, M., and El-Halwagi, M. M. 2013. Synthesis of distributed biorefining networks for the value-added processing of water hyacinth. ACS Sustainable Chemistry and Engineering, *1*(2), 284–305.

Vasudevan, R., Ramalinga Chandra Sekar, A., Sundarakannan, B., and Velkennedy, R. 2012. A technique to dispose waste plastics in an eco-friendly way Application in construction of flexible pavements. Construction and Building Materials, *28*(1), 311–320.

Wan-Alwi, S. R., Aripin, A., and Manan, Z. A. 2009. A generic graphical approach for simultaneous targeting and design of a gas network. Resources, Conservation and Recycling, *53*(10), 588–591.

You, F., and Wang, B. 2011. Life cycle optimization of biomass-to-liquids supply chains with distributed-centralized processing networks. Industrial and Engineering Chemistry Research, *50*(17), 10102–10127.

Young, G.C. 2010. Municipal solid waste to energy conversion processes: economic, technical, and renewable comparisons. Wiley, USA.

4.8.1 Web references

http://elreciclaje.org/content/reciclaje-de-vidrio. Accessed (09/29/2012)

Mexican Ministry of Transportation and Communications, 2012.

http://aplicaciones4.sct.gob.mx/sibuac_internet/ControllerUI?action=cmdEscogeRuta Accessed (12/20/2012)

5

Planning for Reuse of Municipal Solid Waste

This chapter presents a mathematical programming model for the optimal planning of the reuse of municipal solid waste to maximize economic benefit while simultaneously considering sustainability and safety criteria. The proposed approach considers several phases of the supply chain including waste separation, distribution to processing facilities, processing to obtain useful products, and distribution of products to consumers. Safety criteria are based on the potential fatalities associated with the supply chain for waste management. The proposed optimization model is formulated as a multi-objective optimization problem, which considers three different objectives including the maximization of the net annual profit, the maximization of the amount of reused municipal solid waste, and the minimization of the total potential fatalities associated with the supply chain. The proposed model is applied to a case study in the central-west region of Mexico. The results show that it is possible to implement a distributed processing system to reuse the municipal solid waste in an economically attractive way while minimizing the associated risk.

5.1 Introduction

Municipal solid waste management is an important practice that addresses critical societal needs. The enormous generation of municipal solid waste (MSW) is directly related to the current human lifestyle. Over the past decade, MSW disposal has increased drastically from 0.5 kg/(person day) to 1.7 kg/person day (Ramayah et al. 2013). This has been a significant issue, especially for nations with less developed effective MSW management strategies (Abarca-Guerrero et al. 2013). Inappropriate waste disposal may lead to serious side effects (Beolchini et al. 2012) including groundwater pollution, health risks, and serious safety issues such as fire and explosion that may lead to fatalities. As such, a sustainable and efficient waste management strategy is needed to balance the need for the development of the quality of human life and the protection of the environment (Tan et al. 2014). For the optimal planning of MSW management, it is necessary to consider the entire supply chain of the system including tasks like MSW recollection, transportation, treatment, production of value-added products, and distribution of products (Sabbas et al. 2003). In this context, Varbanov et al. (2012) introduced a new indicator, called waste energy potential utilization, to measure the impact of logistics and energy distribution from MSW. Hokkanen and Salminen (1997) reported a methodology to consider several criteria for the selection of an MSW management system. Hung et al. (2007) presented a review of models to support decision-making in MSW management. In addition, a classification for the models utilized in MSW management was reported by Morrisey and Brown (2004) and Karmperis et al. (2013). Santibañez-Aguilar et al. (2013) proposed a mathematical programming model for the optimal planning of a supply chain for MSW management considering economic and environmental aspects. Tan et al. (2014) reported an optimization model for synthesizing MSW processing networks to produce energy and value-added products achieving economic and environmental issues. Bowling et al. (2011) developed an approach to determine the optimal locations and sizes of biomass-management facilities. Minoglou and Komilis (2013) presented a simplified approach to optimize an integrated MSW management system considering the minimization of the total annual cost and the equivalent carbon dioxide emissions. Capón-García et al. (2013) presented a multi-period and multi-objective optimization formulation for MSW management considering economic and environmental issues, where treatment was modeled as a black-box model.

Although the economic and environmental issues are certainly critical in the management of MSW, the social aspects have been overlooked or considered as an afterthought following the design. Specifically, for the case of supply chains focused on biomass conversion, You et al. (2012) addressed the optimal design and planning of cellulosic ethanol supply chains under economic, environmental, and social objectives and Santibañez-Aguilar et al. (2014) proposed a three-objective optimization formulation for the optimal planning of supply chains involving economic, environmental and social objectives, where the social objective was measured through the number of generated jobs. For MSW management systems, Galan et al. (2013) introduced the consideration of economic, environmental, and social aspects in the design of a new MSW management network in the north of Spain. An additional way to consider the social aspect is through the risk assessment. In this regard, Sugiyama et al. (2008) presented a novel framework of chemical process design including four stages of decision-making considering economy, life-cycle environmental impacts, and people's health and safety aspects. Additionally, a structured methodology for risk identification for chemical plants was presented by Adhitya et al. (2009). Moreover, Ng et al. (2013) presented a systematic approach for the synthesis and optimization of a sustainable integrated biorefinery that considers economic, environmental, inherent safety, and inherent occupational health performances. Besides, Bernechea and Arnaldos (2013) proposed a methodology to minimize the risk associated with a storage facility by applying quantitative risk assessment to create a risk objective function; however, the risk assessment has not been considered simultaneously in the design phase of the supply chain. Fabiano et al. (2005) presented a site-oriented framework for risk assessment and developed a theoretical approach for emergency planning and optimization. Furthermore, El-Halwagi et al. (2013) proposed a new approach to simultaneously consider safety and economic factors for the optimal planning of supply chains of biorefineries. Also, van Raemdonck et al. (2013) carried out a study about risk analysis for the transportation of hazardous materials and introduced a methodology to calculate a local accidental risk. For the specific case of MSW, Das et al. (2012) proposed a risk assessment framework for the transportation of inflammable hazardous waste. In addition, Dadashzadeh et al. (2013) presented a risk-based approach for analyzing exposure to hazardous materials.

One of the most significant issues with the MSW supply chain is risk. Leaching of contaminants through the dissolution of soluble matter from the solid phase into a liquid phase is one of the primary concerns connected to the MSW supply chain. The characteristics of the leachate depend on the composition of the waste and water content because the leachate composition is the result of a reaction between various mineral phases in the waste and the leaching fluid. Leaching becomes an important risk when people live around landfills or dumps and drink water from shallow wells. In this context, pollutants are filtered through the land for an undefined period, causing serious health damage to the population. Gas emissions from accidents and the burning of the garbage that has been kept in the landfill pose major risks related to MSW management. Since most fires are uncontrolled, the combustion rate must be taken into account while assessing this danger. Additionally, the quantity of trash consumed and the fire's duration must be taken into account. Additionally, because carcinogenic and non-carcinogenic substances impact people differently, this risk mostly depends on the concentration of pollutants, the distribution and distance between the emissary and receptor, and the type of pollutant that is released.

None of the methodologies for the optimization of the MSW management system has simultaneously considered the risk associated with the system together with the economic, environmental, and social concerns. Particularly, safety is an issue of paramount importance in the

MSW supply chain that must be considered to provide sustainable processes. Therefore, this paper presents a mathematical programming model for the optimal planning of the reuse of MSW considering simultaneously economic, environmental, social, and safety issues. The optimization model can select the processing technologies, consumers, waste treated from different cities, recycled waste, products, and location of processing facilities (Santibañez-Aguilar et al. 2015).

5.2 Problem Statement

The addressed problem considers several cities with a given generation of MSW, which can be utilized to obtain useful products such as energy, recycled materials, olefins, fuels, etc. These residues are stored in several disposal centers (e.g., landfills, dumps). These disposal centers can pose risks to the population in different ways. One of the most significant risks is the consumption of polluted water since the pollutants in dumps may be transported through leaching from the surface to shallow wells that can serve as water sources for the population. Another risk is associated with the intoxication by gas emissions caused by burning in each dump and dispersed into the atmosphere (Fig. 1).

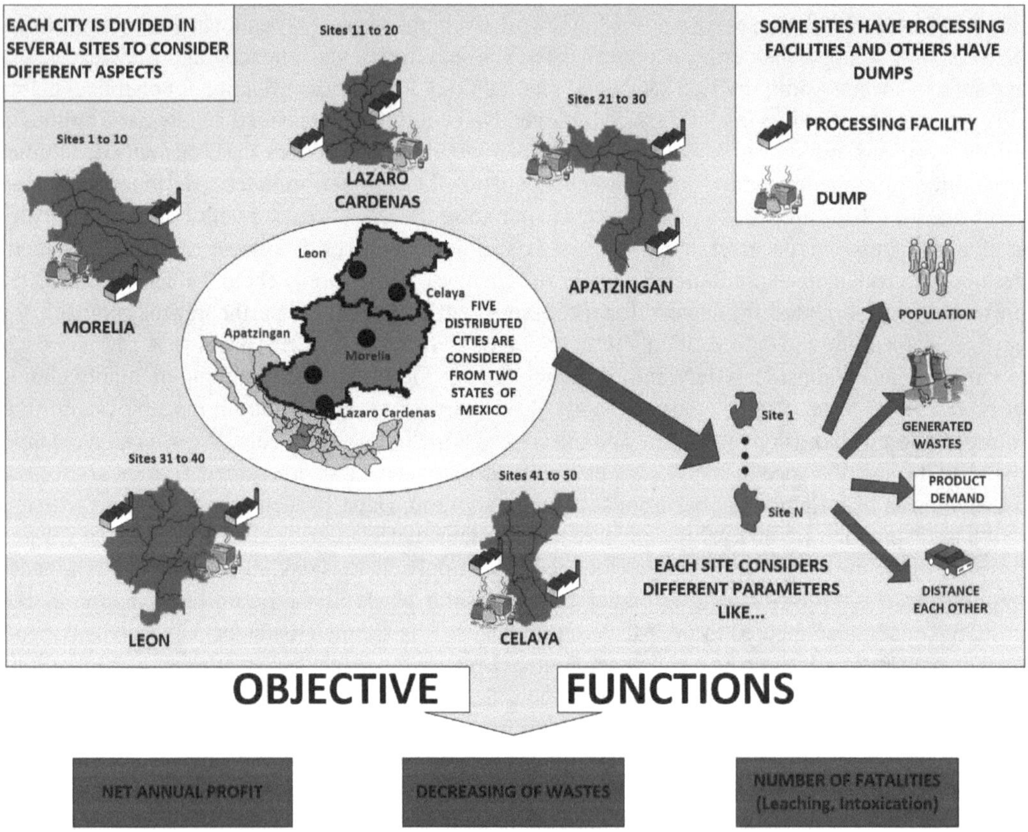

Fig. 1. General representation of the proposed case study. [Taken (adapted) from Santibañez-Aguilar, J.E., Martinez-Gomez, J., Ponce-Ortega, J.M., Nápoles-Rivera, F., Serna-González, M., González-Campos, J.B., and El-Hawlagi, M. 2015. Optimal planning for the reuse of municipal solid waste considering economic, environmental, and safety objectives. *AIChE Journal*, *61*(6), 10881–1899].

Each city is divided into several grids, called sites. Each site has a certain population, waste production, processing plants, and dumps. Figure 2 shows the proposed superstructure for solving the addressed problem, which is the basis for the optimization formulation presented in the next section.

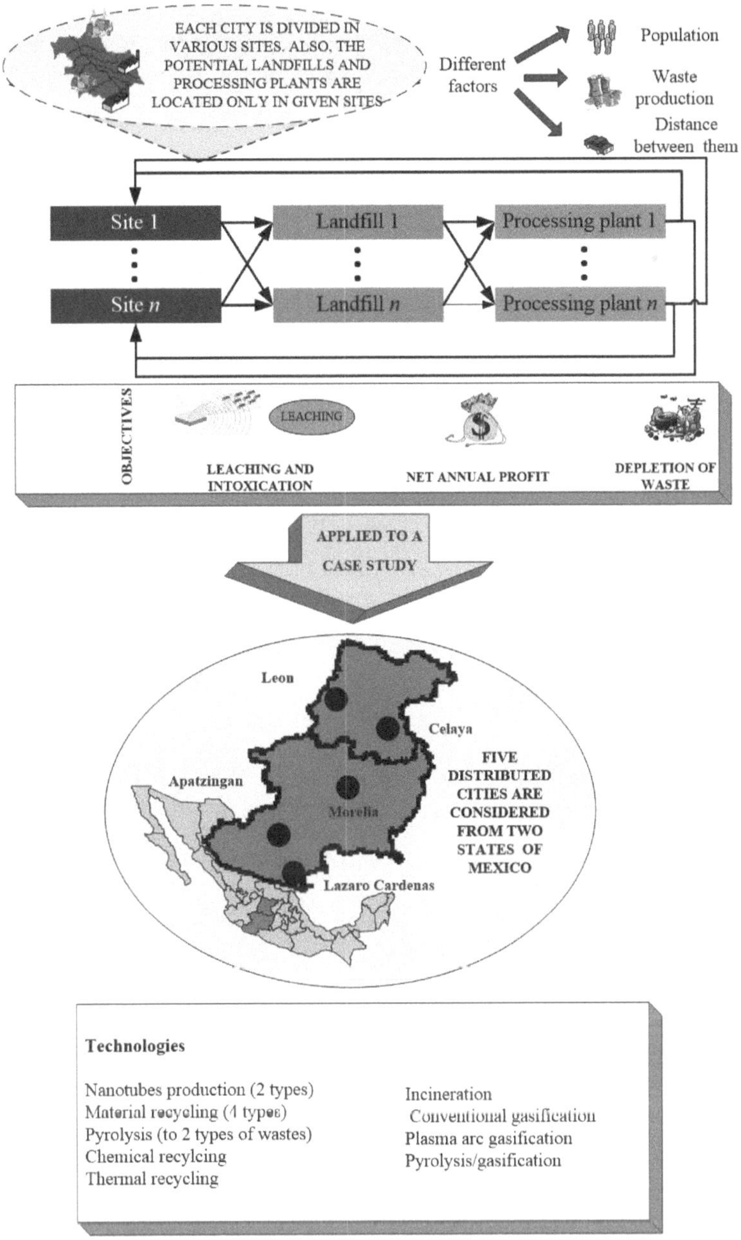

Fig. 2. Proposed superstructure for optimizing the MSW management system considering risk assessment. [Taken (adapted) from Santibañez-Aguilar, J.E., Martinez-Gomez, J., Ponce-Ortega, J.M., Nápoles-Rivera, F., Serna-González, M., González-Campos, J.B., and El-Hawlagi, M. 2015. Optimal planning for the reuse of municipal solid waste considering economic, environmental, and safety objectives. *AIChE Journal*, 61(6), 10881–1899].

5.3 Model Formulation

The indexes to be used in the optimization formulation are first defined. These indexes are used to define materials, locations, processing routes, pollutants, and intervals. For the case of materials, the index *wn* represents the considered waste in the supply chain and *p* is utilized for products. There are several considered locations, the sites are represented through the index *s*, the processing plants are defined by the index *pk*, the landfills are given by the index, and the affected sites by the gas emissions are given by the index a. Furthermore, the index g represents the pollutant in the gas emissions for burning of dumps and the pollutants in the leaching are given by the index k. On the other hand, the processing routes for the waste are defined through the index r and finally, the indexes i, j and n represent the different intervals used for the disjunctions. Then the model formulation is presented as follows.

5.3.1 Waste Transportation from Sites in Cities to Dumps

The sites in each city produce different MSW ($W^{\text{produced}}_{\text{wn,s}}$) and this is distributed to the different dumps ($\overset{distributed}{W}{}^{dump}_{wn,s,d}$); in addition, the waste at the inlet of the dumps ($\overset{in}{W}{}^{dump}_{wn,d}$) is the sum of the distributed waste from sites to dumps ($\overset{distributed}{W}{}^{dump}_{wn,s,d}$).

$$W^{\text{produced}}_{\text{wn,s}} = \sum_{d \in DUMPS} \overset{distributed}{W}{}^{dump}_{wn,s,d}, \quad \forall wn \in WASTE, s \in SITES \tag{5.1}$$

$$\overset{in}{W}{}^{dump}_{wn,d} = \sum_{s \in SITES} \overset{distributed}{W}{}^{dump}_{wn,s,d}, \quad \forall wn \in WASTE, d \in DUMPS \tag{5.2}$$

5.3.2 Waste Balances in Dumps

The MSW stored in dumps ($W^{total}_{wn,d}$) is equal to the MSW inlet to the dump ($\overset{in}{W}{}^{dump}_{wn,d}$) minus the MSW outlet from the dump ($\overset{out}{W}{}^{dump}_{wn,d}$). The outlet MSW ($\overset{out}{W}{}^{dump}_{wn,d}$) is limited by the maximum possible percentage of waste separation. Each site s has several dumps d to store MSW and the amount of available waste ($\overset{separation}{\alpha}{}^{factor}_{wn,d} \; W^{total}_{wn,d}$) should be greater than the used waste ($\overset{out}{W}{}^{dump}_{wn,d}$) to produce the considered products, where the available waste is equal to a separation factor ($\overset{separation}{\alpha}{}^{factor}_{wn,d}$) multiplied by the total stored waste in the dump d ($W^{total}_{wn,s,d}$).

$$W^{total}_{wn,d} = \overset{in}{W}{}^{dump}_{wn,d} - \overset{out}{W}{}^{dump}_{wn,d}, \quad \forall wn \in WASTE, d \in DUMPS \tag{5.3}$$

$$\overset{out}{W}{}^{dump}_{wn,d} \leq \overset{separation}{\acute{a}}{}^{factor}_{wn,d} \; W^{total}_{wn,d}, \quad \forall wn \in WASTE, d \in DUMPS \tag{5.4}$$

5.3.3 Waste Distribution from Dumps to Processing Plants

To consider the distribution of waste from dumps to processing plants it is necessary to remark that there are several processing plants in each city; then, a different index for processing plants is considered, in this case, the transportation is considered from dump d to processing facility pk. In this way, the transported waste from each dump ($\overset{out}{W}{}^{dump}_{wn,d}$) is equal to the sum of the distributed waste in each processing facility ($\overset{distributed}{W}{}^{plant}_{wn,d,pk}$).

$$\overset{out}{W}{}^{dump}_{wn,d} = \sum_{pk \in FACILITIES} \overset{distibuted}{W}{}^{plant}_{wn,d,pk}, \quad \forall wn \in WASTE, d \in DUMPS \tag{5.5}$$

The waste in each processing plant ($W_{wn,pk}^{\overset{in}{plant}}$) must be equal to the sum of the transported waste from each dump ($W_{wn,d,pk}^{\overset{distributed}{plant}}$).

$$W_{wn,pk}^{\overset{in}{plant}} = \sum_{d \in DUMPS} W_{wn,d,pk}^{\overset{distributed}{plant}}, \quad \forall wn \in WASTE, pk \in FACILITIES \tag{5.6}$$

There is required the inclusion of upper and lower limits for transportation from dumps to processing facilities.

$$W_{wn,d,pk}^{\overset{distributed}{plant}} \leq \mathrm{uppW}_{\mathrm{wn,d,pk}}^{\overset{\mathrm{distributed}}{\mathrm{plant}}}, \quad \forall wn \in WASTE, d \in DUMPS, pk \in FACILITIES \tag{5.7}$$

$$W_{wn,d,pk}^{\overset{distributed}{plant}} \geq \mathrm{lowW}_{\mathrm{wn,d,pk}}^{\overset{\mathrm{distributed}}{\mathrm{plant}}}, \quad \forall wn \in WASTE, d \in DUMPS, pk \in FACILITIES \tag{5.8}$$

5.3.4 MSW Processing

The MSW is processed to obtain value-added products. The wastes in the plant ($W_{wn,pk}^{\overset{in}{plant}}$) are distributed to various processing routes ($W_{wn,pk,r}^{\overset{distributed}{routes}}$) to obtain products using a conversion factor ($_{\mathrm{wn,p,r}}^{\mathrm{conversion}}$) as follows:

$$W_{wn,pk}^{\overset{in}{plant}} = \sum_{r \in ROUTES} W_{wn,pk,r}^{\overset{distributed}{routes}}, \quad \forall wn \in WASTE, pk \in FACILITIES \tag{5.9}$$

$$P_{p,pk}^{\overset{processing}{plant}} = \sum_{r \in ROUTES} \sum_{wn \in WASTE} \alpha_{\mathrm{wn,p,r}}^{\mathrm{conversion}} \alpha_{wn,pk,r}^{\overset{distributed}{routes}}, \tag{5.10}$$

$\forall p \in PRODUCTS, pk \in FACILITIES$

5.3.5 Distribution of Products from Processing Facilities to Consumers

Products can be distributed from processing plants to consumers; these consumers are in the cities. The products in processing plants ($P_{p,pk}^{\overset{processing}{plant}}$) can be distributed to the consumers ($P_{p,pk,s}^{\overset{distributed}{product}}$) as follows:

$$P_{p,pk}^{\overset{processing}{plant}} = \sum_{s \in SITES} P_{p,pk,s}^{\overset{distributed}{product}}, \quad \forall p \in PRODUCTS, pk \in FACILITIES \tag{5.11}$$

The sum of distributed products ($P_{p,pk,s}^{\overset{distributed}{product}}$) is equal to the products for a given consumer ($P_{p,s}^{\overset{product}{consumer}}$):

$$\overset{product}{P}{}^{consumer}_{p,s} = \sum_{pk \in FACILITIES} \overset{distributed}{P}{}^{product}_{p,pk,s}, \quad \forall p \in PRODUCTS, s \in SITES \tag{5.12}$$

5.3.6 Demand for Products in Cities

Each site has a specific demand for each product ($\overset{\text{product}}{\text{PD}}{}^{\text{consumer}}_{\text{p,s}}$), which must be greater than the products sent to consumers ($\overset{product}{P}{}^{consumer}_{p,s}$) from the recycled MSW.

$$\overset{product}{P}{}^{consumer}_{p,s} \leq \overset{\text{product}}{\text{PD}}{}^{\text{consumer}}_{\text{p,s}} \tag{5.13}$$

5.3.7 Revenue from Sales

The revenue from selling products ($REVENUE$) is the sum of the unitary product cost ($\overset{\text{unitary}}{\text{C}}{}^{\text{product}}_{\text{p}}$) multiplied by the total consumed product ($\overset{product}{P}{}^{consumer}_{p,s}$).

$$REVENUE = \sum_{p \in PRODUCTS} \sum_{s \in SITES} \overset{\text{unitary}}{\text{C}}{}^{\text{product}}_{\text{p}} \overset{product}{P}{}^{consumer}_{p,s} \tag{5.14}$$

5.3.8 Operating Costs

The operating cost for processing facilities ($OPCOST$) is obtained by multiplying the unitary processing cost for each processing technology in each processing plant ($\overset{\text{unitary}}{\text{C}}{}^{\text{processing}}_{\text{wn,p,pk,r}}$) times the amount of MSW that is processed in the processing facility for the given processing route ($\overset{distributed}{W}{}^{routes}_{wn,pk,r}$).

$$OPCOST = \sum_{wn \in WASTE} \sum_{p \in PRODUCTS} \sum_{pk \in FACILITIES} \sum_{r \in ROUTES} \overset{\text{unitary}}{\text{C}}{}^{\text{processing}}_{\text{wn,p,pk,r}} \overset{distributed}{W}{}^{routes}_{wn,pk,r} \tag{5.15}$$

5.3.9 Capital Cost

The capital cost for processing technologies usually has a nonlinear behavior concerning the amount of processed waste; however, this dependence can be linearized through a disjunctive mathematical programming formulation, this linearization is like the one used for risk assessment. The disjunction considers that if the processed waste is between given limits, the binary variable associated with the disjunction is activated and so is the corresponding cost.

$$\vee_q \begin{bmatrix} Y^{economies}_{wn,pk,r,q} \\ \overset{distributed}{W}{}^{routes}_{wn,pk,r} \geq \text{low}\overset{\text{distributed}}{\text{W}}{}^{\text{routes}}_{\text{wn,pk,r,q}} \\ \overset{distributed}{W}{}^{routes}_{wn,pk,r} \leq \text{upp}\overset{\text{distributed}}{\text{W}}{}^{\text{routes}}_{\text{wn,pk,r,q}} \\ CAPCOST^{economies}_{wn,pk,r} = \overset{\text{cap}}{\text{B}}{}^{\text{economies}}_{\text{wn,pk,r,q}} \overset{distributed}{W}{}^{routes}_{wn,pk,r} + \overset{\text{cap}}{\text{A}}{}^{\text{economies}}_{\text{wn,pk,r,q}} \end{bmatrix}, \quad \begin{matrix} \forall wn \in WASTE, \\ pk \in FACILITIES, \\ r \in ROUTES, \\ q \in ECON \end{matrix}$$

The previous disjunction is reformulated as follows. First, just one segment can be selected:

$$\sum_{q \in ECON} y^{economies}_{wn,pk,r,q} = 1, \quad \forall wn \in WASTE, pk \in FACILITIES, r \in ROUTES \tag{5.16}$$

Then, the continuous variables are disaggregated for each segment:

$$W^{\substack{distributed \\ routes}}_{wn,pk,r} = \sum_{q \in ECON} Wd^{\substack{distributed \\ routes}}_{wn,pk,r,q}, \quad \forall wn \in WASTE, pk \in FACILITIES, r \in ROUTES \tag{5.17}$$

$$CAPCOST^{economies}_{wn,pk,r} = \sum_{q \in ECON} dCAPCOST^{economies}_{wn,pk,r,q}, \quad \forall wn \in WASTE, pk \in FACILITIES, r \in ROUTES \tag{5.18}$$

$$CAPCOST^{total} = \sum_{wn \in WASTES} \sum_{pk \in FACILITIES} \sum_{r \in ROUTES} CAPCOST^{economies}_{wn,pk,r} \tag{5.19}$$

The constraints are stated in terms of the disaggregated variables:

$$Wd^{\substack{distributed \\ routes}}_{wn,pk,r,q} \geq \mathrm{lowW}^{\substack{distributed \\ routes}}_{wn,pk,r,q}\, y^{economies}_{wn,pk,r,q}, \quad \forall wn \in WASTE, pk \in FACILITIES, r \in ROUTES, q \in ECON \tag{5.20}$$

$$Wd^{\substack{distributed \\ routes}}_{wn,pk,r,q} \leq \mathrm{uppW}^{\substack{distributed \\ routes}}_{wn,pk,r,q}\, y^{economies}_{wn,pk,r,q}, \quad \forall wn \in WASTE, pk \in FACILITIES, r \in ROUTES, q \in ECON \tag{5.21}$$

$$dCAPCOST^{economies}_{wn,pk,r} = \mathrm{B}^{\substack{cap \\ economies}}_{wn,pk,r,q} Wd^{\substack{distributed \\ routes}}_{wn,pk,r,q} + \mathrm{A}^{\substack{cap \\ economies}}_{wn,pk,r,q}\, y^{economies}_{wn,pk,r,q},$$

$$\forall wn \in WASTE, pk \in FACILITIES, r \in ROUTES, q \in ECON \tag{5.22}$$

5.3.10 Transportation Costs

The transportation cost is important in any supply chain as this cost depends on the distance between the facilities as well as the transported amount of materials; furthermore, this cost allows for determining the amount that should be transported from one place to another. One of the most used ways to model the cost of this activity is given as follows: the transportation cost is equal to the distance between the facilities multiplied by the transported amount of material and multiplied by a unitary transportation cost that depends on the specific material. Equation (5.23) shows the total transportation cost, which is the sum of the transportation cost of sites-landfills (Equation (5.24)), landfills-processing facilities (Equation (5.25)), and processing facilities-consumers (Equation (5.26)):

$$TRANSPCOST = TRANSPCOST^{\substack{SITES- \\ LANDFILLS}} + TRANSPCOST^{\substack{LANDFILLS- \\ FACILITIES}} + TRANSPCOST^{\substack{FACILITIES- \\ CONSUMERS}} \tag{5.23}$$

$$TRANSPCOST^{\substack{SITES- \\ LANDFILLS}} = \sum_{wn \in WASTE} \sum_{d \in DUMPS} \sum_{s \in SITES} D1^{transp}_{s,d} W^{\substack{distributed \\ dump}}_{wn,s,d} C1^{transp}_{wn} \tag{5.24}$$

$$TRANSPCOST^{\substack{LANDFILLS- \\ FACILITIES}} = \sum_{wn \in WASTE} \sum_{d \in DUMPS} \sum_{s \in FACILITIES} D2^{transp}_{d,pk} W^{\substack{distibuted \\ plant}}_{wn,d,pk} C2^{transp}_{wn} \tag{5.25}$$

$$TRANSPCOST^{\substack{FACILITIES \\ COSNUMERS}} = \sum_{p \in PRODUCTS} \sum_{pk \in FACILITES} \sum_{s \in SITES} D3^{transp}_{s,pk} P^{\substack{distributed \\ product}}_{p,pk,s} C3^{transp}_{p} \tag{5.26}$$

5.3.11 Disposal Cost

The disposal cost considers all activities needed for the disposal of MSW, this cost is given through a linear function, and this cost considers unused waste. The disposal cost is equal to the unused waste multiplied by a unitary cost for disposal.

$$DISPCOST = \sum_{d \in DUMPS} \left(C^{disp}_{d} \sum_{wn \in WASTE} W^{total}_{wn,d} \right) \tag{5.27}$$

5.3.12 Separation Cost

The separation cost considers the amount of waste that is separated and distributed to the processing facilities, it is important to note that the separated waste must be processed. This cost is equal to the separated waste multiplied by a unitary separation cost.

$$SEPCOST = \sum_{wn \in WASTE} \sum_{d \in DUMPS} C_{wn}^{sep} W_{wn,d}^{\overset{out}{dump}} \tag{5.28}$$

5.3.13 Economic Objective Function

The economic objective function is the net annual profit and considers all the costs of the entire supply chain; in this context, the net annual profit is the revenue from sales minus the operating, capital, transportation, disposal, and separation costs.

$$\begin{aligned} NETPROFIT = {} & REVENUE - OPCOST - CAPCOST^{total} \\ & - TRANSPCOST - DISPCOST - SEPCOST \end{aligned} \tag{5.29}$$

5.3.14 Total Consumed Waste

The measure for the total consumed waste is a way to consider the environmental impact. Although there are many other forms to assess the environmental impact, this is an adequate way for the case of the MSW because one of the main impacts is the huge amount of waste that is disposed of in landfills. In this way, the total consumed waste is equal to the total waste that is sent to the processing facilities from dumps divided by the total waste that is produced in each site of the different cities.

$$WASTECONS = \sum_{wn \in WASTE} \sum_{d \in DUMPS} W_{wn,d}^{\overset{out}{dump}} \Bigg/ \sum_{wn \in WASTES} \sum_{s \in SITES} \mathrm{W}_{\mathrm{wn,s}}^{\mathrm{produced}} \tag{5.30}$$

5.3.15 Risk Assessment

The risk for the entire process accounts for the potential fatalities associated with leaching and intoxication for the burning of dumps in the entire supply chain. These potential fatalities are determined as follows.

5.3.15.1 Fatalities for leaching

The fatalities for leaching depend on each shallow well (each site is considered a shallow well) as well as each dump; in this case, each shallow well is associated with a given population. Then, the fatalities ($F_{k,s}^{\overset{leaching}{fatalaties}}$) are obtained by multiplying the exposed population ($\mathrm{P}_{\mathrm{s}}^{\mathrm{population}}$) times the probability of fatalities ($\alpha_{k,s}^{\overset{leaching}{fatalities}}$). It is important to note that one dump can pollute several shallow wells.

$$F_{k,s}^{\overset{leaching}{fatalaties}} = \alpha_{k,s}^{\overset{leaching}{fatalities}} \mathrm{P}_{\mathrm{s}}^{\mathrm{population}}, \qquad \forall s \in SITES, k \in POLLUTANT \tag{5.31}$$

The probability of fatalities ($\alpha_{k,s}^{\overset{leaching}{fatalities}}$) is obtained from the concentration of pollutants ($C_{k,s}^{\overset{phreatic}{concentration}}$) in the shallow well because the concentration of pollutants in the shallow well is associated with the intoxication of the population. The function to obtain the fatalities is nonlinear; however, the function can be limited to a given concentration to consider a linear function. The linearization is modeled with a disjunction to approximate the Probit curve with several linear equations.

$$\vee_i \begin{bmatrix} Y_{k,s,i}^{\substack{leaching\\fatalities}} \\ C_{k,s}^{\substack{phreatic\\concentration}} \geq \mathrm{lowC}_{k,i}^{\substack{phreatic\\concentration}} \\ C_{k,s}^{\substack{phreatic\\concentration}} \leq \mathrm{uppC}_{k,i}^{\substack{phreatic\\concentration}} \\ \alpha_{k,s}^{\substack{leaching\\fatalities}} = \mathrm{M}_{k,i}^{\substack{probit\\leaching}} C_{k,s}^{\substack{phreatic\\concentration}} + \mathrm{B}_{k,i}^{\substack{probit\\leaching}} \end{bmatrix}, \quad \begin{matrix} s \in SITES \\ \forall\ k \in POLLUTANT \\ i \in INTERVALS \end{matrix}$$

The previous formulation is reformulated through the convex hull technique, where Equations (5.33) and (5.39) are used to activate the binary variables in each valid interval ($y_{k,s,i}^{\substack{leaching\\fatalities}}$), which are 1 when the concentration is between the corresponding lower ($\mathrm{lowC}_{k,i}^{\substack{phreatic\\concentration}}$) and upper ($\mathrm{uppC}_{k,i}^{\substack{phreatic\\concentration}}$) limits.

$$Cd_{k,s.i}^{\substack{phreatic\\concentration}} \geq \mathrm{lowC}_{k,i}^{\substack{phreatic\\concentration}} y_{k,s,i}^{\substack{leaching\\fatalities}}, \quad \forall k \in POLLUTANT, d \in DUMPS, s \in SITES, i \in INTERVALS \tag{5.32}$$

$$Cd_{k,s.i}^{\substack{phreatic\\concentration}} \leq \mathrm{uppC}_{k,i}^{\substack{phreatic\\concentration}} y_{k,s,i}^{\substack{leaching\\fatalities}}, \quad \forall k \in POLLUTANT, d \in DUMPS, s \in SITES, i \in INTERVALS \tag{5.33}$$

Also, the number of selected intervals should be equal to 1. This way, the sum of the binary variables concerning the intervals must be equal to 1 to ensure that only one interval is chosen:

$$\sum_{i \in INTERVALS} y_{k,s,i}^{\substack{leaching\\fatalities}} = 1, \quad \forall k \in POLLUTANT, d \in DUMPS, s \in SITES \tag{5.34}$$

Furthermore, the continuous variables need to be disaggregated, where the total variable is the sum of each interval of the discretized variables.

$$C_{k,s}^{\substack{phreatic\\concentration}} = \sum_{i \in INTERVALS} Cd_{k,s.i}^{\substack{phreatic\\concentration}}, \quad \forall k \in POLLUTANT, d \in DUMPS, s \in SITES \tag{5.35}$$

$$\alpha_{k,s}^{\substack{leaching\\fatalities}} = \sum_{i \in INTERVALS} \alpha d_{k,s,i}^{\substack{leaching\\fatalities}}, \quad \forall k \in POLLUTANT, d \in DUMPS, s \in SITES \tag{5.36}$$

And finally, the constraints inside the disjunction are stated in terms of the disaggregated variables:

$$\alpha d_{k,s,i}^{\substack{leaching\\fatalities}} = \mathrm{M}_{k,i}^{\substack{probit\\leaching}} Cd_{k,s.i}^{\substack{phreatic\\concentration}} + \mathrm{B}_{k,i}^{\substack{probit\\leaching}} y_{k,s,i}^{\substack{leaching\\fatalities}}, \quad \forall k \in POLLUTANT, d \in DUMPS, s \in SITES, i \in INTERVALS \tag{5.37}$$

The pollutant concentration depends on a lot of factors, some of them are the quantity of pollutants in the surface of a dump ($C_{k,d}^{\substack{dump\\quantity}}$) and the level of the water. This is important because the land is a natural filter, which can eliminate some pollutants and can be modeled as a black box model associated with an efficiency factor ($\alpha_{k,d,s}^{\substack{filtration\\factor}}$) that depends on the type of land, water level, and type of pollutant among others factors, and this should be obtained experimentally and modeled as follows:

$$C^{\substack{phreatic \\ concentration}}_{k,s} \, \delta\left(\substack{\text{quantity-} \\ \text{concentration} \\ \text{s}}\right) \sum_{d\in DUMPS} \alpha^{\substack{\text{filtration} \\ \text{factor}}}_{\text{k,d,s}} \, C^{\substack{dump \\ quantity}}_{k,d}, \quad \forall k \in POLLUTANT \;\; d \in DUMPS \;\; s \in SITES \tag{5.38}$$

The quantity of pollutants in each dump ($C^{\substack{dump \\ quantity}}_{k,d}$) is calculated through a factor ($\beta^{\substack{\text{pollutant} \\ \text{surface}}}_{\text{k,d}}$) multiplied by the total amount of waste in each dump.

$$C^{\substack{dump \\ quantity}}_{k,d} = \beta^{\substack{\text{pollutant} \\ \text{surface}}}_{\text{k,d}} \left(\sum_{wn\in WASTE} W^{total}_{wn,d} \right), \quad \forall k \in POLLUTANT, d \in DUMPS \tag{5.39}$$

5.3.15.2 Fatalities for Intoxication for Burning of Dumps

Due to the ambient conditions in the dumps, it is common to consider the risk of fire, therefore, to consider the fatalities for intoxication when the dump is burning, it is required to use a dispersion gas model. In this way, the Pasquill Gifford model is an adequate way to consider the dispersion of gasses. According to the equation of Pasquill Gifford, the concentration of pollutants for the dispersion of gasses ($C^{\substack{toxic \\ concentration}}_{g,s,d}$) depends on the emission rate of the source ($Q_{g,d}$) multiplied by a factor that depends on the position of the receptor concerning the emitter:

$$C^{\substack{toxic \\ concentration}}_{g,d,a} = Q_{g,d} \, \alpha^{\text{dispersion}}_{\text{d,a}}, \quad \forall g \in GAS\ POLLUTANT, d \in DUMPS, a \in AFFECTED\ POINTS \tag{5.40}$$

where the factor $\alpha^{\text{dispersion}}_{\text{d,a}}$ depends on the position obtained from the Pasquill Gifford equation, which is given by Equation (5.25). The position is taken into account by the parameters $\sigma^{y}_{d,a}$ and $\sigma^{z}_{d,a}$, which are calculated separately from the model as a function of the distance ($\text{x}^{\text{distance}}_{\text{d,a}}$) between the dumps and the affected regions as follows:

$$\alpha^{\text{dispersion}}_{\text{d,a}} = \frac{1}{2\pi\,\sigma^{y}_{d,a}\,\sigma^{z}_{d,a}\,\text{u}} \exp\left[-\frac{1}{2}\left(\frac{\text{y}_{d,a}}{\sigma^{z}_{d,a}}\right)^2\right]\left\{\exp\left[-\frac{1}{2}\left(\frac{\text{z}_{d,a}\text{-Hr}_{d,a}}{\sigma^{z}_{d,a}}\right)^2\right] + \exp\left[-\frac{1}{2}\left(\frac{\text{z}_{d,a}\text{-Hr}_{d,a}}{\sigma^{z}_{d,a}}\right)^2\right]\right\}, \quad \forall d \in DUMPS, a \in AFFECTED\ POINTS$$

It should be noted that other parameters that determine the dispersion factor ($\alpha^{\text{dispersion}}_{\text{d,a}}$) are the height of the receptor ($\text{z}_{d,a}$), the perpendicular distance of the receptor ($\text{y}_{d,a}$), the wind velocity (u), the effective height of the emitter ($\text{H}_{\text{r,a}}$), and all these parameters are fixed for any given case study. This way, this chapter considers the worst-case scenario to define the parameters to obtain the dispersion factor, the worst-case data are predefined and for this case, these are shown in Table 1. The dependence of the distance ($\text{x}^{\text{distance}}_{\text{d,a}}$) for the parameters ($\sigma^{y}_{d,a}$) and ($\sigma^{z}_{d,a}$) are defined in the next relationships:

$$\sigma^{y}_{d,a} = 0.04\text{x}^{\text{distance}}_{\text{d,a}}\left(1 + 0.0001\text{x}^{\text{distance}}_{\text{d,a}}\right)^{-1/2} \quad \forall d \in DUMPS, a \in AFFECTED\ POINTS \tag{5.41}$$

$$\sigma^{z}_{d,a} = 0.016\text{x}^{\text{distance}}_{\text{d,a}}\left(1 + 0.0003\text{x}^{\text{distance}}_{\text{d,a}}\right)^{-1} \quad \forall d \in DUMPS, a \in AFFECTED\ POINTS \tag{5.42}$$

On the other hand, the emission rate of the source ($Q_{g\,d}$) is obtained from the total waste in a dump ($W^{total}_{wn,d}$) multiplied by a conversion factor ($\beta^{\substack{\text{pollutant} \\ \text{emission}}}_{\text{g,d}}$); this conversion factor is a function of the amount of waste emitted and burned in case of fire.

$$Q_{g,d} = \beta_{\mathrm{g,d}}^{\mathrm{pollutant\ emission}} \left(\sum_{wn \in WASTE} W_{wn,d}^{total} \right), \quad \forall d \in DUMPS, s \in SITES \tag{5.43}$$

The concentration for intoxication is useful to calculate the Probit value for the fatalities for intoxication when a dump is burned, which is proposed for a specific pollutant and the function is of the form $Y = a + b \ln\left(\mathrm{C}^{\mathrm{n}}\ \mathrm{t}\right)$, where each pollutant has its own Probit function; this is, each pollutant has given values for the parameters for the Probit function. This function can be linearized in a similar way to the fatalities for leaching. It is crucial to note that the Probit function is only specified for pure compounds, meaning that the pollutant under consideration is distributed across the gas emissions. The data of the Probit function for the pollutant considered is reported in Table 1.

Table 1. Considered data to obtain the emission factors for the gas dispersion model. [Taken (adapted) from Santibañez-Aguilar, J.E., Martinez-Gomez, J., Ponce-Ortega, J.M., Nápoles-Rivera, F., Serna-González, M., González-Campos, J.B., and El-Hawlagi, M. 2015. Optimal planning for the reuse of municipal solid waste considering economic, environmental, and safety objectives. *AIChE Journal*, *61*(6), 10881–1899].

Data for the gas dispersion model			
Symbol	**Description**	**Value**	**Consideration**
$z_{d,a}$	Height of receptor	1.63 m	Average height of population
$y_{d,a}$	Perpendicular distance of receptor	0 m	Line segment from emissary to receptor
u	Wind velocity	1.5 m	Defined as the worst-case
$H_{r,a}$	Effective height of emissary	0 m	The emissions are from the surface of the landfill
Data for the Probit function for the pollutant considered			
Symbol	**Description**	**Value**	**Pollutant**
a	Independent term in Probit function	–37.98	CO (Carbon monoxide)
b	Coefficient of the logarithm	2.1	
n	Exponent of the concentration	1	
t	Exposition time	28 800 s	

$$\vee_j \left[\begin{array}{c} Y_{g,d,a,j}^{intoxication\ linealized} \\ C_{g,d,a}^{toxic\ concentration} \geq \mathrm{lowC}_{\mathrm{g,d,a,j}}^{\mathrm{toxic\ concentration}} \\ C_{g,d,a}^{toxic\ concentration} \leq \mathrm{uppC}_{\mathrm{g,d,a,j}}^{\mathrm{toxic\ concentration}} \\ F_{g,d,a}^{toxic\ probit} = \mathrm{M}_{\mathrm{g,j}}^{\mathrm{toxic\ probit}} C_{g,d,a}^{toxic\ concentration} + \mathrm{B}_{\mathrm{g,j}}^{\mathrm{toxic\ probit}} \end{array} \right], \quad \forall \begin{array}{c} g \in GAS\ POLLUTANT \\ d \in DUMPS \\ a \in AFFECTED\ POINTS \\ j \in INTERVALS2 \end{array}$$

The previous disjunction is reformulated as follows. First, only one segment can be selected:

$$\sum_{j \in INTERVALS2} y_{g,d,a,j}^{intoxication\ linealized} = 1, \quad \forall g \in GAS\ POLLUTANT, d \in DUMPS, a \in AFFECTED\ POINTS \tag{5.44}$$

Then, the continuous variables are disaggregated:

$$\overset{toxic}{F}{}^{probit}_{g,d,a} = \sum_{j \in INTERVALS2} \overset{toxic}{Fd}{}^{probit}_{g,d,a,j}, \quad \forall g \in GAS\ POLLUTANT, d \in DUMPS, a \in AFFECTED\ POINTS \tag{5.45}$$

$$\overset{toxic}{C}{}^{concentration}_{g,d,a} = \sum_{j \in INTERVALS2} \overset{toxic}{Cd}{}^{concentration}_{g,d,a,j}, \quad \forall g \in GAS\ POLLUTANT, d \in DUMPS, a \in AFFECTED\ POINTS \tag{5.46}$$

Finally, the constraints are stated as a function of the disaggregated variables:

$$\overset{toxic}{Cd}{}^{concentration}_{g,d,a,j} \geq \overset{toxic}{\text{lowC}}{}^{\text{concentration}}_{\text{g,d,a,j}} \overset{intoxication}{y}{}^{linealized}_{g,d,a,j}, \quad \forall g \in GAS\ POLLUTANT, d \in DUMPS, a \in AFFECTED\ POINTS, j \in INTERVALS2 \tag{5.47}$$

$$\overset{toxic}{Cd}{}^{concentration}_{g,d,a,j} \leq \overset{toxic}{\text{uppC}}{}^{\text{concentration}}_{\text{g,d,a,j}} \overset{intoxication}{y}{}^{linealized}_{g,d,a,j}, \quad \forall g \in GAS\ POLLUTANT, d \in DUMPS, a \in AFFECTED\ POINTS, j \in INTERVALS2 \tag{5.48}$$

$$\overset{toxic}{Fd}{}^{probit}_{g,d,a,j} = \overset{toxic}{\text{M}}{}^{\text{probit}}_{\text{g,j}} \overset{toxic}{Cd}{}^{concentration}_{g,d,a,j} + \overset{toxic}{\text{B}}{}^{\text{probit}}_{\text{g,j}} \overset{intoxication}{y}{}^{linealized}_{g,d,a,j}, \quad \forall g \in GAS\ POLLUTANT, d \in DUMPS, a \in AFFECTED\ POINTS, j \in INTERVALS2 \tag{5.49}$$

Notice that the exposure time for the risk analysis is 8 hours since the considered toxicity is acute. In addition, it is needed to propose other equations to obtain the probability of fatalities ($\overset{toxic}{P}{}^{probit}_{g,d,a}$) as a function of the Probit value ($\overset{toxic}{F}{}^{probit}_{g,d,a}$) for the pollutant in the considered risk. The dependence is nonlinear and has the following form $P = 50\left[1 + \frac{Y-5}{|Y-5|} erf\left(\frac{|Y-5|}{\sqrt{2}}\right)\right]$. But this function is easily modeled through the following disjunctive formulation.

$$\vee_n \begin{bmatrix} \overset{intoxication}{Y}{}^{fatalities}_{g,d,a,n} \\ \overset{toxic}{F}{}^{probit}_{g,d,a} \geq \overset{toxic}{\text{lowF}}{}^{\text{probit}}_{\text{g,d,a,n}} \\ \overset{toxic}{F}{}^{probit}_{g,d,a} \leq \overset{toxic}{\text{lowF}}{}^{\text{probit}}_{\text{g,d,a,n}} \\ \overset{toxic}{P}{}^{probit}_{g,d,a} = \overset{\text{fatalities}}{\text{M}}{}^{\text{probit}}_{\text{g,n}} \overset{toxic}{F}{}^{probit}_{g,d,a} + \overset{\text{fatalities}}{\text{B}}{}^{\text{probit}}_{\text{g,n}} \end{bmatrix}, \quad \forall \begin{matrix} g \in GAS\ POLLUTANT \\ d \in DUMPS \\ a \in AFFECTED\ POINTS \\ n \in INTERVALS3 \end{matrix}$$

Which is algebraically reformulated as follows. First, only one segment can be selected:

$$\sum_{n \in INTERVALS3} \overset{intoxication}{y}{}^{fatalities}_{g,d,a,n} = 1, \quad \forall g \in GAS\ POLLUTANT, d \in DUMPS, a \in AFFECTED\ POINTS \tag{5.50}$$

Then, the continuous variables are disaggregated:

$$\overset{toxic}{F}{}^{probit}_{g,d,a} = \sum_{n \in INTERVALS3} \overset{toxic}{Fd2}{}^{probit}_{g,d,a,n}, \quad \forall g \in GAS\ POLLUTANT, d \in DUMPS, a \in AFFECTED\ POINTS \tag{5.51}$$

$$\overset{toxic}{P}{}^{probit}_{g,d,a} = \sum_{n \in INTERVALS3} \overset{toxic}{Pd}{}^{probit}_{g,d,a,n}, \quad \forall g \in GAS\ POLLUTANT, d \in DUMPS, a \in AFFECTED\ POINTS \tag{5.52}$$

The relationships are stated in terms of the disaggregated variables as follows:

$$\overset{toxic}{Fd2}{}^{probit}_{g,d,a,n} \geq \text{low}\overset{\text{toxic}}{\text{F}}{}^{\text{probit}}_{\text{g,d,a,n}} \overset{intoxication}{y}{}^{fatalities}_{g,d,a,n}, \quad \forall g \in GAS\ POLLUTANT, d \in DUMPS, a \in AFFECTED\ POINTS, n \in INTERVALS3 \tag{5.53}$$

$$\overset{toxic}{Fd2}{}^{probit}_{g,d,a,n} \leq \text{upp}\overset{\text{toxic}}{\text{F}}{}^{\text{probit}}_{\text{g,d,a,n}} \overset{intoxication}{y}{}^{fatalities}_{g,d,a,n}, \quad \forall g \in GAS\ POLLUTANT, d \in DUMPS, a \in AFFECTED\ POINTS, n \in INTERVALS3 \tag{5.54}$$

$$\overset{toxic}{Pd}{}^{probit}_{g,d,a,n} = \overset{\text{fatalities}}{\text{M}}{}^{\text{probit}}_{\text{g,n}} \overset{toxic}{F}{}^{probit}_{g,d,a} + \overset{\text{fatalities}}{\text{B}}{}^{\text{probit}}_{\text{g,n}} \overset{intoxication}{y}{}^{fatalities}_{g,d,a,n}, \quad \forall g \in GAS\ POLLUTANT, d \in DUMPS, a \in AFFECTED\ POINTS \tag{5.55}$$

In this context, the total fatalities for intoxication when a dump is burned are obtained by multiplying the number of exposed people ($\overset{\text{toxic}}{\text{P}}{}^{\text{population}}_{\text{d,a}}$) and the affectation fraction ($\overset{toxic}{P}{}^{probit}_{g,d,a}$).

$$\overset{intoxication}{F}{}^{fatalities}_{g,d,a} = \left(\frac{\overset{toxic}{P}{}^{probit}_{g,d,a}}{100} \right) \overset{\text{toxic}}{\text{P}}{}^{\text{population}}_{\text{d,a}}, \quad \forall g \in GAS\ POLLUTANT, d \in DUMPS, a \in AFFECTED\ POINTS \tag{5.56}$$

5.3.16 Safety Objective Function

The safety objective function is based on the total potential fatalities that can be generated by intoxication and leaching for the implementation of the system. For the fatalities for intoxication, there is considered a frequency factor ($\overset{frequency}{f}{}^{factor}$) to obtain the number of times that the landfill burns yearly (this information can be obtained from historical registered information). The total fatalities are equal to the sum of the total fatalities ($\overset{leaching}{F}{}^{fatalities}_{k,s}$) for leaching plus the total fatalities for intoxication ($\overset{intoxication}{F}{}^{fatalities}_{g,d,a}$).

$$\overset{total}{F}{}^{fatalities} = \overset{frequency}{f}{}^{factor} \sum_{g}\sum_{d}\sum_{a} \overset{intoxication}{F}{}^{fatalities}_{g,d,a} + \sum_{k}\sum_{s} \overset{leaching}{F}{}^{fatalities}_{k,s} \tag{5.57}$$

5.4 Case Study

The considered case study corresponds to the west region of Mexico, where five distributed cities are considered (Morelia, Lazaro Cardenas, Apatzingan, Leon, and Celaya). Each city is divided into ten grids or sites, and each site has its own population density and waste generation (Table 2). The Mexican Institute of Statistics, Geography and Informatics (INEGI 2013) reported the population and surface of each city, and data employed were taken from its database to generate the data for each grid. Furthermore, the generated waste is distributed in twenty dumps and can be processed in two processing plants per city, notice that each city has four dumps, and these dumps are in specific sites, and the distance between each dump and the processing facilities is different, the location of the different dumps and the distance between dumps and processing plants are given in Table 3. To obtain the relevant data, the distance between cities was considered (SCT 2013). Additionally, Table 4 shows the considered MSW, products, and processing routes; the different data for the processing technologies were taken from Alireza and Gordon (2012), Hur et al. (2003), Environmental Protection Agency

Table 2. Data of waste generation and population for each site of each city. [Taken (adapted) from Santibañez-Aguilar, J.E., Martinez-Gomez, J., Ponce-Ortega, J.M., Nápoles-Rivera, F., Serna-González, M., González-Campos, J.B., and El-Hawlagi, M. 2015 . Optimal planning for the reuse of municipal solid waste considering economic, environmental, and safety objectives. *AIChE Journal*, 61(6), 10881–1899].

			WASTE PRODUCTION												
	SITES	POPULATION (PEOPLE)	PP	PE	PET	PS	HDPE	ALUMINIUM	CLEAR GLASS	GREEN GLASS	BROWN GLASS	PAPER	NON-RECICLABLES	SURFACE	POPULATION DENSITY
		$x10^{-4}$	(kg) $x10^{-4}$	(kg) $x10^{-4}$	(kg) $x10^{-4}$	(kg) $x10^{-4}$	kg) $x10^{-4}$	(kg) $x10^{-4}$	(kg) $x10^{-4}$	(kg) $x10^{-4}$	(kg) $x10^{-4}$	(kg) $x10^{-4}$	(kg) $x10^{-4}$	(km^2)	Peopl/km^2
MORELIA	**1**	3.57	45.85	30.83	1.80	6.51	13.16	4.03	25.49	29.85	85.14	126.24	1497.2	193.03	185
	2	7.53	96.79	65.08	3.80	13.74	27.79	8.51	53.82	63.01	179.73	266.50	3160.8	58.38	1290
	3	12.28	157.76	106.07	6.20	22.39	45.29	13.87	87.72	102.70	292.96	434.38	5151.9	181.31	677
	4	11.36	145.98	98.15	5.73	20.72	41.91	12.84	81.17	95.03	271.08	401.95	4767.2	194.31	585
	5	5.70	73.23	49.24	2.88	10.39	21.02	6.44	40.72	47.67	135.98	201.63	2391.4	122.51	465
	6	11.08	142.32	95.69	5.59	20.20	40.86	12.52	79.13	92.65	264.28	391.87	4647.7	120.16	922
	7	8.75	112.39	75.57	4.41	15.95	32.27	9.88	62.49	73.17	208.71	309.46	3670.3	68.82	1271
	8	7.78	99.97	67.22	3.93	14.19	28.70	8.79	55.59	65.08	185.65	275.27	3264.8	191.11	407
	9	4.51	57.95	38.96	2.28	8.22	16.64	5.10	32.22	37.72	107.60	159.55	1892.3	11.29	3994
	10	0.37	4.78	3.21	0.19	0.68	1.37	0.42	2.66	3.11	8.87	13.15	155.96	56.03	66
LAZARO CARDENAS	**11**	1.32	31.90	21.49	1.26	4.54	9.16	2.81	17.72	20.75	59.15	87.88	1041.2	130.14	101
	12	3.19	77.03	51.89	3.03	10.97	22.11	6.78	42.80	50.11	142.83	212.19	2514.2	89.47	356
	13	1.75	42.30	28.49	1.66	6.02	12.14	3.72	23.50	27.52	78.43	116.52	1380.6	218.13	80
	14	1.53	36.95	24.89	1.45	5.26	10.61	3.25	20.53	24.04	68.51	101.79	1206.0	67.04	228
	15	0.67	16.15	10.88	0.64	2.30	4.64	1.42	8.97	10.51	29.95	44.50	527.23	108.45	62
	16	0.43	10.50	7.07	0.41	1.49	3.01	0.92	5.83	6.83	19.47	28.92	342.70	21.44	203
	17	0.06	1.41	0.95	0.06	0.20	0.41	0.12	0.79	0.92	2.62	3.89	46.13	13.80	42
	18	2.87	69.36	46.72	2.73	9.87	19.91	6.10	38.53	45.11	128.60	191.06	2263.7	205.32	140
	19	2.96	71.48	48.15	2.81	10.18	20.52	6.29	39.71	46.49	132.53	196.90	2332.9	115.60	256
	20	3.10	74.91	50.46	2.95	10.66	21.50	6.59	41.62	48.73	138.90	206.35	2445.0	182.64	170

APATZINGAN	**21**	2.20	21.01	14.08	0.82	2.97	6.02	1.83	11.65	13.64	38.82	57.51	683.76	126.29	174
	22	0.75	7.20	4.82	0.28	1.02	2.06	0.63	3.99	4.67	13.29	19.70	234.16	90.20	84
	23	0.90	8.62	5.78	0.34	1.22	2.47	0.75	4.78	5.60	15.93	23.60	280.54	156.51	58
	24	1.42	13.56	9.09	0.53	1.92	3.89	1.18	7.52	8.80	25.06	37.13	441.38	76.67	185
	25	1.68	16.07	10.77	0.63	2.27	4.60	1.40	8.91	10.43	29.69	43.99	522.92	35.63	473
	26	0.54	5.17	3.47	0.20	0.73	1.48	0.45	2.87	3.36	9.56	14.16	168.32	142.97	38
	27	0.94	8.97	6.01	0.35	1.27	2.57	0.78	4.97	5.82	16.56	24.54	291.76	212.43	44
	28	1.90	13.14	12.16	0.71	2.57	5.20	1.58	10.05	11.77	33.51	49.65	590.25	195.74	97
	29	1.74	16.60	11.13	0.65	2.35	4.75	1.45	9.20	10.77	30.66	45.43	540.13	439.30	40
	30	0.28	2.67	1.79	0.10	0.38	0.76	0.23	1.48	1.73	4.93	7.30	86.78	164.17	17
		WASTE PRODUCTION													
LEON	**31**	18.33	26.66	17.99	1.05	3.80	7.68	2.35	14.93	17.35	49.63	73.49	871.39	28.30	6476
	32	3.93	5.72	3.86	0.23	0.82	1.65	0.50	3.20	3.73	10.65	15.78	187.08	33.76	1166
	33	13.22	19.23	12.98	0.76	2.74	5.54	1.69	10.77	12.52	35.80	53.01	628.52	141.51	934
	34	18.71	27.22	18.36	1.07	3.88	7.84	2.40	15.24	17.71	50.66	75.01	889.44	62.06	3014
	35	20.98	30.53	20.60	1.20	4.35	8.79	2.69	17.09	19.87	56.83	84.14	997.75	185.50	1131
	36	2.49	3.62	2.44	0.14	0.52	1.04	0.32	2.02	2.35	6.73	9.96	118.15	91.04	273
	37	23.06	33.54	22.63	1.32	4.78	9.66	2.95	18.78	21.83	62.43	92.45	1096.2	237.32	971
	38	28.96	42.13	28.42	1.66	6.01	12.14	3.71	23.59	27.42	78.42	116.11	1376.8	194.36	1490
	39	7.21	10.50	7.08	0.41	1.50	3.02	0.92	5.88	6.83	19.53	28.92	342.98	189.93	380
	40	6.76	9.84	6.64	0.39	1.40	2.83	0.87	5.51	6.40	18.32	27.13	321.64	56.60	1195

Table 2. contd. ...

	SITES	POPULATION (PEOPLE)	PP	PE	PET	PS	HDPE	ALUMINIUM	CLEAR GLASS	GREEN GLASS	BROWN	PAPER	NON-RECICLABLES	SURFACE	POPULATION DENSITY
		x10^{-4}	(kg) x10^{-4}	(kg) x10^{-4}	(kg) x10^{-4}	(kg) x10^{-4}	kg) x10^{-4}	(kg) x10^{-4}	(kg) x10^{-4}	(kg) x10^{-4}	(kg) x10^{-4}	(kg) x10^{-4}	(kg) x10^{-4}	(km^2)	Peopl/km^2
CELAYA	**41**	5.90	142.23	95.53	5.58	20.26	40.78	12.47	78.92	92.38	263.06	390.18	4631.8	41.82	1410
	42	2.90	69.93	46.97	2.74	9.96	20.05	6.13	38.80	45.43	129.35	191.85	2277.4	12.91	2245
	43	4.66	112.42	75.51	4.41	16.02	32.24	9.86	62.38	73.03	207.94	308.42	3661.2	124.90	373
	44	3.52	84.83	56.98	3.33	12.09	24.32	7.44	47.07	55.11	156.91	232.73	2762.7	73.40	479
	45	3.77	90.88	61.05	3.56	12.95	26.06	7.97	50.43	59.03	168.10	249.33	2959.7	8.98	4195
	46	5.76	138.98	93.35	5.45	19.80	39.85	12.19	77.12	90.28	257.05	381.28	4526.1	29.33	1964
	47	6.08	146.65	98.50	5.75	20.89	42.05	12.86	81.37	95.26	271.24	402.32	4775.9	28.35	2145
	48	5.39	130.13	87.41	5.10	18.54	37.31	11.41	72.20	84.53	240.68	356.99	4237.8	107.64	501
	49	5.52	133.08	89.39	5.22	18.96	38.16	11.67	73.84	86.44	246.14	365.09	4333.9	54.73	1008
	50	3.35	80.85	54.31	3.17	11.52	23.18	7.09	44.86	52.52	149.54	221.80	2633.0	71.15	471

Table 3. Distance and location for each dump with respect to the processing facilities. [Taken (adapted) from Santibañez-Aguilar, J.E., Martinez-Gomez, J., Ponce-Ortega, J.M., Nápoles-Rivera, F., Serna-González, M., González-Campos, J.B., andEl-Hawlagi, M. 2015 . Optimal planning for the reuse of municipal solid waste considering economic, environmental, and safety objectives. *AIChE Journal*, *61*(6), 10881–1899].

DUMP		DISTANCE BETWEEN DUMPS AND PROCESSING PLANTS (km)									
		MORELIA		LAZARO CARDENAS		APATZINGAN		LEON		CELAYA	
	SITE IN WHICH IS LOCATED	PLANT 1	PLANT 2	PLANT 3	PLANT 4	PLANT 5	PLANT 6	PLANT 7	PLANT 8	PLANT 9	PLANT 10
1	1	3.4	7.7	151.7	159.6	195.4	213.7	222.6	234.1	166.0	168.7
2	3	14.5	1.0	145.0	152.9	206.5	207.0	215.9	227.4	159.4	162.0
3	5	22.4	26.7	170.7	145.0	214.4	199.1	208.0	219.5	151.5	154.1
4	7	28.5	32.7	176.7	184.6	220.5	193.0	247.6	213.4	191.1	148.0
5	11	148.0	152.7	8.7	14.5	190.0	203.2	528.5	539.1	463.0	464.2
6	14	161.8	166.5	22.5	4.3	203.8	192.9	518.3	528.9	452.8	453.9
7	16	166.6	171.3	27.3	33.2	208.6	188.2	547.2	524.1	448.0	449.2
8	20	130.0	184.7	40.7	46.5	222.0	235.2	560.5	571.1	495.0	496.2
9	21	196.0	200.0	587.0	591.6	4.0	21.0	405.6	425.1	342.4	347.0
10	24	209.3	213.2	207.2	188.9	17.3	11.3	395.9	415.3	332.7	337.3
11	27	219.6	223.5	217.5	222.2	27.6	1.0	429.2	405.1	366.0	327.0
12	30	234.7	238.6	232.6	237.3	42.7	59.7	444.3	463.7	381.1	385.7
13	33	211.5	208.0	515.0	521.8	397.5	408.7	7.8	24.5	145.8	150.7
14	35	222.9	218.6	500.0	518.0	408.9	401.9	1.0	17.7	139.0	143.9
15	37	230.0	226.4	533.4	540.2	416.0	394.0	26.2	9.8	164.3	136.0
16	38	234.4	230.9	537.9	544.7	420.4	431.5	30.7	5.4	168.7	173.5
17	41	149.1	152.7	452.7	456.3	328.1	338.7	144.3	153.3	11.0	12.7
18	42	148.0	151.5	451.5	455.2	327.0	337.6	143.2	152.2	9.9	11.6
19	45	159.5	163.0	463.0	448.0	338.5	330.4	136.0	145.0	2.7	4.4
20	49	168.5	172.1	472.1	475.7	347.5	358.1	163.7	136.0	30.4	32.1

Office of Resource Conservation and Recovery (2011), Plunkert, (2006) and Young (2010). Table 5 shows the unit capital and operating costs associated with the considered processing technologies; these costs were estimated from the provided information by the same references in Table 4.

5.5 Results

The solution approach applied is based on generating several Pareto curves considering only two objectives for each Pareto curve. The mathematical model is a Mixed Integer Linear Programming (MILP) problem, and it was coded in the software GAMS (Brooke et al. 2014) and consists of 208,681

Table 4. Efficiency factors for the processing technologies to treat municipal solid waste. [Taken (adapted) from Santibañez-Aguilar, J.E., Martinez-Gomez, J., Ponce-Ortega, J.M., Nápoles-Rivera, F., Serna-González, M., González-Campos, J.B., and El-Hawlagi, M. 2015 . Optimal planning for the reuse of municipal solid waste considering economic, environmental, and safety objectives. *AIChE Journal*, *61*(6), 10881–1899].

Waste	Technology	Product	Production Factor (kg Product/kg Waste)
PP	Technology 1 for nanotubes	Nanotubes of 140um diameter	0.11[1]
PE	Technology 1 for nanotubes	Nanotubes of 140um diameter	0.16[1]
PET	Technology 1 for nanotubes	Nanotubes of 140um diameter	0.19[1]
PE	Technology 2 for nanotubes	Nanotubes of 252um diameter	0.32[1]
PP	Technology 2 for nanotubes	Nanotubes of 252um diameter	0.51[1]
PE	Material Recycling 1	Pellet	0.672[2]
PS	Material Recycling 1	Pellet	0.672[2]
PP	Material Recycling 1	Pellet	0.672[2]
PE	Thermal Recycling	Gasoline	0.299[2]
PE	Chemical Recycling	Heat Energy	19.44*[2]
HDPE	Pyrolysis	n-Paraffins	0.432[2]
HDPE	Pyrolysis	n-Olefines	0.446[2]
PP	Pyrolysis	Branched Paraffins	0.304[2]
PP	Pyrolysis	Branched Paraffins	0.333[2]
Paper	Material Recycling 2	New Paper	0.84[3]
Aluminum	Material Recycling 3	Useful Aluminum	1[4]
Clear Glass	Material Recycling 4	New Glass	1[5]
Green Glass	Material Recycling 4	New Glass	1[5]
Brown Glass	Material Recycling 4	New Glass	1[5]
MSW	Mass Burn (Incineration)	Electricity	1.9584*[6]
MSW	Pyrolysis	Electricity	2.0556*[6]
MSW	Pyrolysis/Gasification	Electricity	2.466*[6]
MSW	Plasma Arc Gasification	Electricity	2.9376*[6]
MSW	Conventional gasification	Electricity	2.466*[6]

[1]Alireza and Gordon (2012)
[2]Hur et al. (2003)
[3]Environmental Protection Agency Office of Resource Conservation and Recovery (2011)
[4]Plunkert (2006)
[5] http://elreciclaje.org/content/reciclaje-de-vidrio. Accessed (09/29/2013)
[6]Young (2010)

Table 5. Processing cost for the different technologies and wastes for the considered case study. [Taken (adapted) from Santibañez-Aguilar, J.E., Martinez-Gomez, J., Ponce-Ortega, J.M., Nápoles-Rivera, F., Serna-González, M., González-Campos, J.B., and El-Hawlagi, M. 2015. Optimal planning for the reuse of municipal solid waste considering economic, environmental, and safety objectives. *AIChE Journal*, *61*(6), 10881–1899].

Waste	Technology	Fixed Capital Cost ($US)	Variable Capital Cost ($US/kg processed)	Operational Cost ($US/kg processed)
PP	Technology 1 for nanotubes	26038000	0.052076	2.6038
PE	Technology 1 for nanotubes	26038000	0.052076	2.6038
PET	Technology 1 for nanotubes	26038000	0.052076	2.6038
PP	Technology 2 for nanotubes	26038000	0.052076	2.6038
PE	Technology 2 for nanotubes	26038000	0.052076	2.6038
PE	Material recycling	3880000	0.0051165	0.388
PS	Material recycling	3880000	0.0051165	0.388
PP	Material recycling	3880000	0.0051165	0.388
PE	Thermal recycling	2057000	0.0059503	0.2057
PE	Chemical recycling	6705000	0.0026819	0.6705
HDPE	Pyrolysis 1	2057000	0.0026819	0.2057
PP	Pyrolysis 1	2057000	0.0026819	0.2057
Aluminum	Material recycling 2	5000000	0.0009	0.09
Glass different types	Material recycling 3	4000000	0.00025	0.01
Paper	Material recycling 4	5000000	0.00044	0.0044
MSW	Mass burn (incineration)	115997700	0.004	0.04
MSW	Pyrolysis 1	86936900	0.004	0.04
MSW	Pyrolysis/Gasification	102593400	0.004	0.04
MSW	Plasma arc gasification	101583800	0.004	0.04
MSW	Conventional gasification	80337800	0.004	0.04

constraints, 134,710 continuous variables and 60,240 binary variables taking from 5 to 120 seconds of CPU time in a processor i7-3720QM at 2.60 GHz with 16 GB of RAM for the solution of each Pareto curve using the solver CPLEX.

Figure 3 shows the tradeoffs between the number of fatalities and the net annual profit; in this way, the number of fatalities decreases up to 21% from the maximum number of fatalities while the change observed in the profit is 19%. It is important to note that the fatalities for leaching are the most vital in the proposed case study. Furthermore, Figure 3 shows that the consumed waste is almost constant for the Pareto curve of the net annual profit versus the number of fatalities, determining that the number of fatalities is not a function of the consumed waste.

Figure 4 shows the distribution of the stored MSW in landfills for each point of the Pareto curve of Fig. 3; notice that this waste distribution is very similar in all points. In this context, the main changes are observed in the cities of Apatzingan and Leon, since the waste that is sent to dump 14 in Leon is distributed to the dumps 9 and 12 from Apatzingan city.

On the other hand, the net annual profit and the consumed waste are considered in another Pareto curve shown in Fig. 5, which has a similar behavior that the one reported by Santibañez-Aguilar et al. (2013) obtained without considering the safety objective. The differences are for the consideration of several sites in each city that influence the transportation costs. Figure 5 shows that the number of fatalities is almost constant when the net annual profit and the consumed waste are considered as main objectives; in this way, the number of fatalities changes only 2.5% along the Pareto curve. Moreover, the minimum value for potential fatalities is 2,087 for the case of Fig. 5 and 1,660 for the case of the minimization of fatalities and the maximization of the net annual profit (Fig. 3). This change represents an increment of 12.8% in the fatalities for the case of Fig. 5 with respect to the one in Fig. 3. Figure 6 shows that the amount of the total waste is distributed uniformly in the cities for the points of the Pareto curve of Fig. 5; however, the number of fatalities is not affected for the distribution of the total waste in the landfills, this is because there are two types of considered fatalities, where the fatalities for intoxication are affected by the amount of waste in each dump in major proportion than the fatalities for leaching, and for this specific case the fatalities for leaching keep the total number of fatalities almost constant.

Additionally, Fig. 7 presents the tradeoffs between the number of fatalities and the consumed waste; in this context, the number of fatalities increases from 1,250 to 1,860, and the consumed waste from 5.8% to 12%. Moreover, the net annual profit changes considerably along the Pareto curve. It is important to note that the number of fatalities obtained in Fig. 7 is lower than the ones in Figs. 5 and 5.3; however, the net annual profit has the worst values, and these values fluctuate significantly. This means that Fig. 7 can be used to obtain the best configuration considering the safety of the supply chain involving deterioration in the economic objective function.

Figure 8 shows the distribution of total waste for the points of the Pareto curve of Fig. 7; where the number of fatalities is directly affected by the distribution of the total waste, and the minimum number of fatalities is obtained when the waste is stored mainly in the cities of Morelia, Apatzingan and Leon. Also, the maximum number of fatalities is obtained when the distribution has a similar value in all cities.

Finally, Table 6 presents a summary of the results of the considered case study.

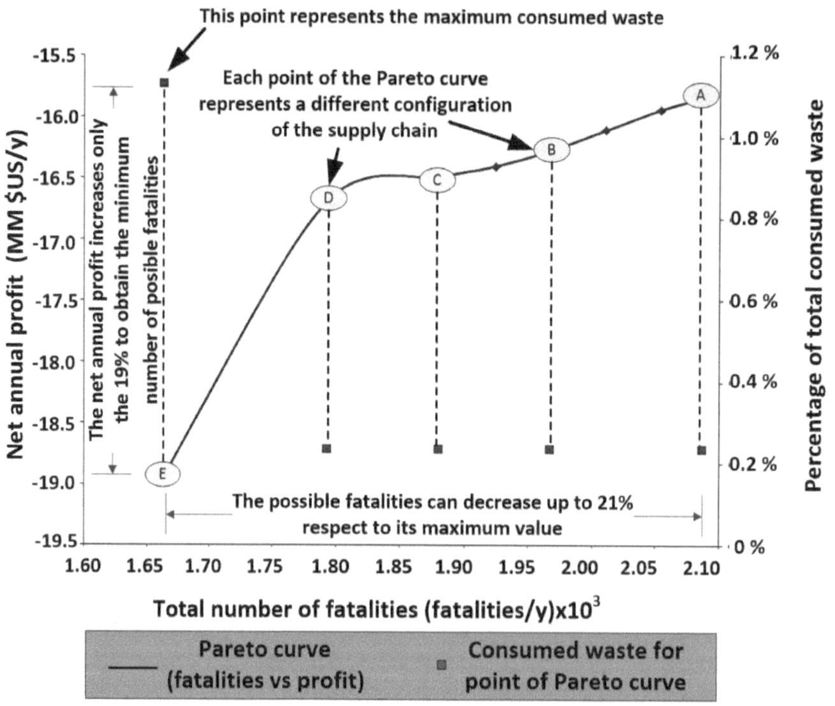

Fig. 3. Pareto curve for the net annual profit and the number of fatalities. [Taken (adapted) from Santibañez-Aguilar, J.E., Martinez-Gomez, J., Ponce-Ortega, J.M., Nápoles-Rivera, F., Serna-González, M., González-Campos, J.B., and El-Hawlagi, M. 2015 . Optimal planning for the reuse of municipal solid waste considering economic, environmental, and safety objectives. *AIChE Journal*, 61(6), 10881–1899].

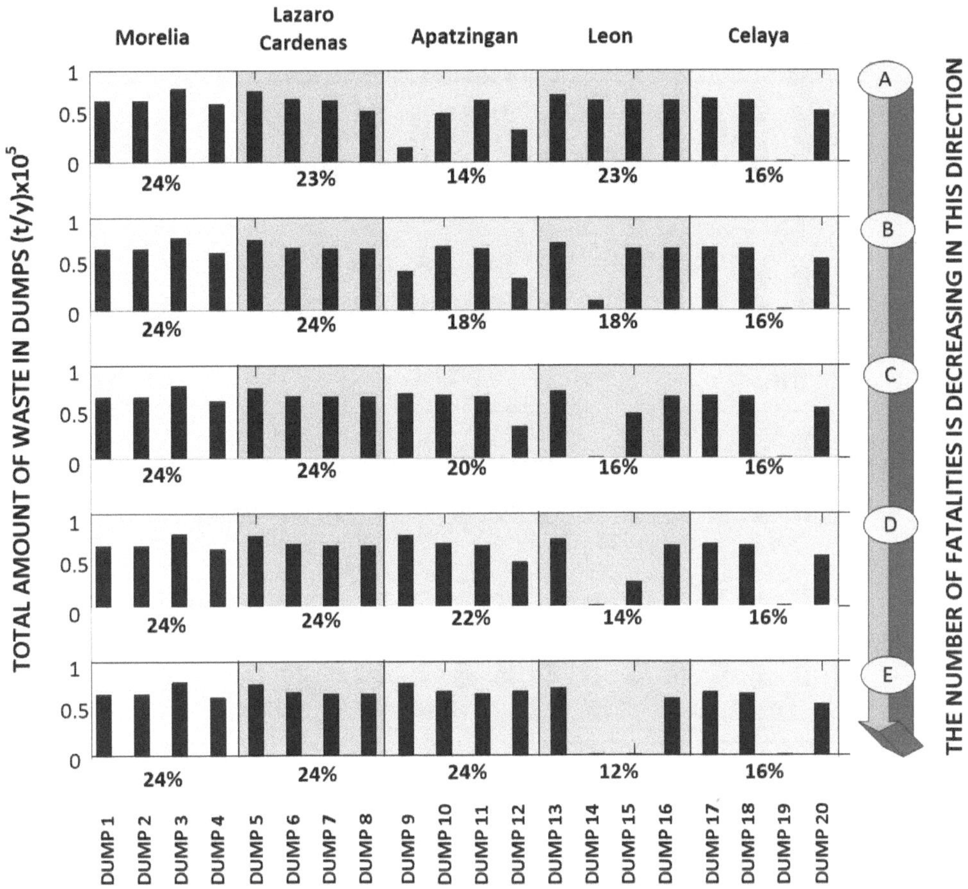

Fig. 4. Distribution of the amount of stored waste in each dump for selected points of the Pareto curve of **Fig. 3**. [Taken (adapted) from Santibañez-Aguilar, J.E., Martinez-Gomez, J., Ponce-Ortega, J.M., Nápoles-Rivera, F., Serna-González, M., González-Campos, J.B., and El-Hawlagi, M. 2015. Optimal planning for the reuse of municipal solid waste considering economic, environmental, and safety objectives. *AIChE Journal*, *61*(6), 10881–1899].

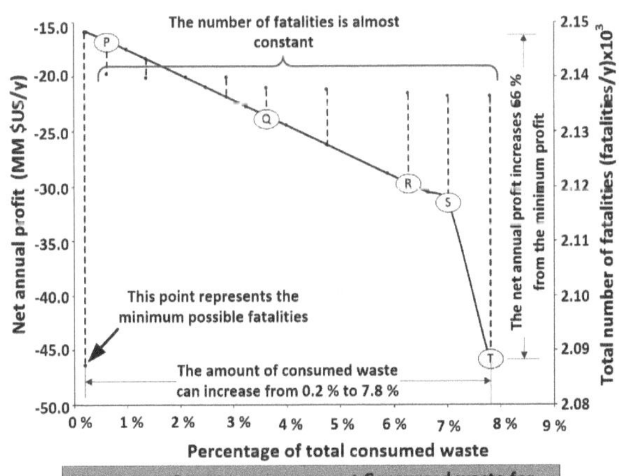

Fig. 5. Pareto curve for the net annual profit and consumed waste. [Taken (adapted) from Santibañez-Aguilar, J.E., Martinez-Gomez, J., Ponce-Ortega, J.M., Nápoles-Rivera, F., Serna-González, M., González-Campos, J.B., and El-Hawlagi, M. 2015 . Optimal planning for the reuse of municipal solid waste considering economic, environmental, and safety objectives. *AIChE Journal*, 61(6), 10881–1899].

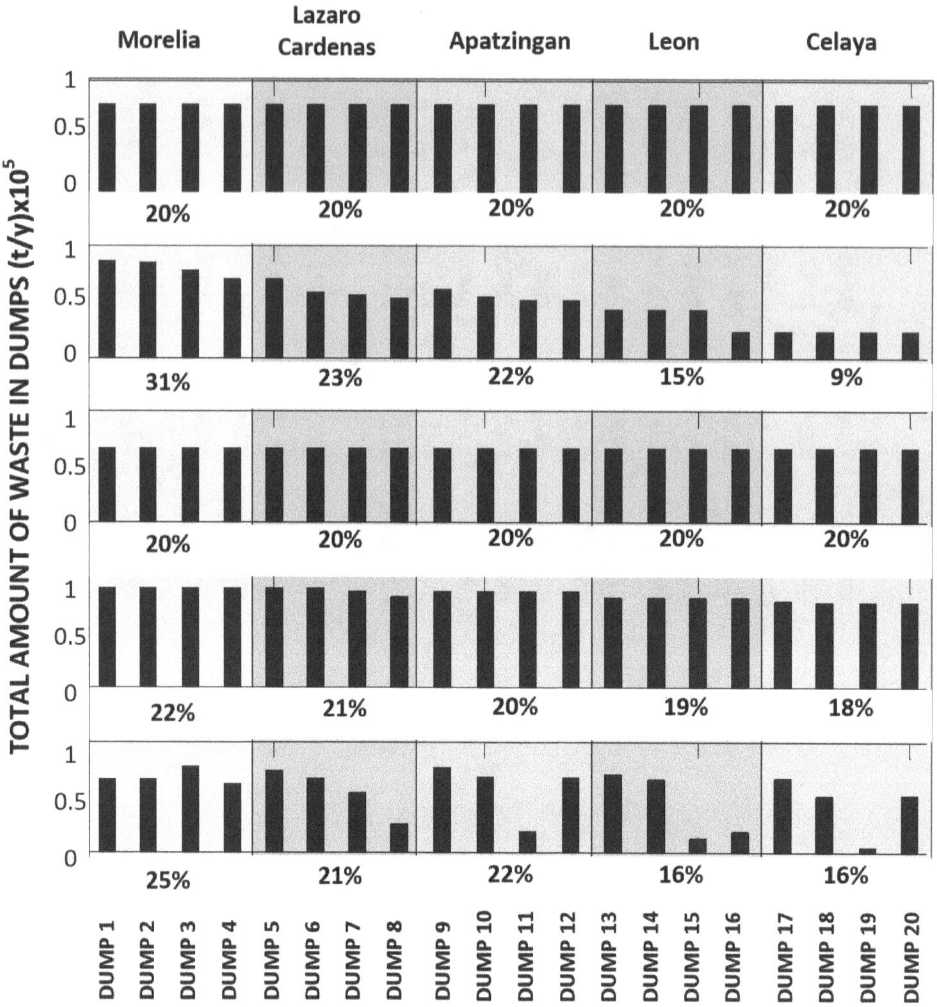

Fig. 6. Distribution of the amount of stored waste in each dump for selected points of the Pareto curve of **Fig. 5**. [Taken (adapted) from Santibañez-Aguilar, J.E., Martinez-Gomez, J., Ponce-Ortega, J.M., Nápoles-Rivera, F., Serna-González, M., González-Campos, J.B., and El-Hawlagi, M. 2015 . Optimal planning for the reuse of municipal solid waste considering economic, environmental, and safety objectives. *AIChE Journal*, 61(6), 10881–1899].

5.6 Concluding Remarks

This chapter presented a mathematical programming formulation for the optimal planning of a distributed system to treat MSW while simultaneously accounting for economic, environmental, and safety objectives. Multi-objective mixed-integer linear programming is the mathematical formulation of the issue being addressed. The model takes into account both the supply chain optimization for items derived from MSW and the risk assessment for actions that could result in deaths when a specific waste management system configuration is put into place. A case study of a distributed waste management system in central-western Mexico has been used to demonstrate the use of the suggested methodology. The obtained Pareto solutions can provide valuable insights to decision-makers to select the solution that shows the best compromise among the considered objectives. Other case studies can benefit from the general methodology that has been suggested. Additionally, the suggested optimization model's solution showed no numerical issues.

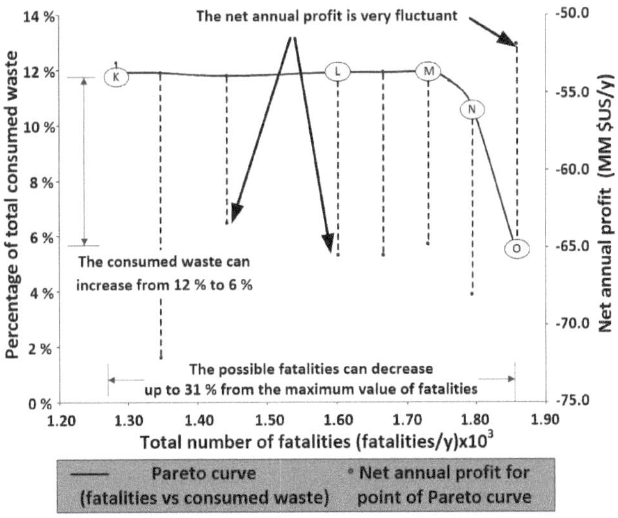

Fig. 7. Pareto curve for the net annual profit and consumed waste. [Taken (adapted) from Santibañez-Aguilar, J.E., Martinez-Gomez, J., Ponce-Ortega, J.M., Nápoles-Rivera, F., Serna-González, M., González-Campos, J.B., and El-Hawlagi, M. 2015. Optimal planning for the reuse of municipal solid waste considering economic, environmental, and safety objectives. *AIChE Journal*, 61(6), 10881–1899].

Table 6. Summary of the results from the Pareto curves. [Taken (adapted) from Santibañez-Aguilar, J.E., Martinez-Gomez, J., Ponce-Ortega, J.M., Nápoles-Rivera, F., Serna-González, M., González-Campos, J.B., and El-Hawlagi, M. 2015. Optimal planning for the reuse of municipal solid waste considering economic, environmental, and safety objectives. *AIChE Journal*, 61(6), 10881–1899].

Case	Minimum Value			Maximum value		
	Total Fatalities	Annual Profit ($US)	Consumed Waste	Total Fatalities	Annual Profit ($US)	Consumed Waste
Figure 3	1660	-19×10^6	0.24%	2100	-15.7×10^6	1.15%
Figure 5	2087	-46×10^6	0.24%	2140	-15.7×10^6	8%
Figure 7	1250	-73×10^6	5.8%	1860	-52×10^6	12%
		Worst Value	**Best Value**	**Analyzed case for best case**		
Total Fatalities		2140	1660	Minimizing the fatalities vs Maximizing the annual profit		
Annual Profit ($US)		-73×10^6	-15.7×10^6	Maximizing the annual profit vs Minimizing the fatalities Maximizing the annual profit vs Maximizing the consumed waste		
Consumed Waste		0.24 %	12 %	Maximizing the consumed waste vs Minimizing the fatalities		

5.7 Nomenclature

The symbols used in the model formulation have been segregated into indexes, parameters, and continuous and binary variables as follows.

5.7.1 Indexes

a Index that defines the affected points for gas emissions of pollutants

d Index used for dumps

g Index for the considered pollutant in the gas emissions

i, j, n	Indexes used to define the intervals in disjunctions
k	Index for the considered pollutants in leaching
p	Index used for products
pk	Index for processing facilities for waste processing
r	Index used for waste processing technologies
s	Index that represents the sites in the cities
wn	Index for the considered waste in the supply chain

5.7.2 Parameters

$A^{cap\ economies}_{wn,pk,r,q}$	Unitary fixed cost associated with the capital of processing technology r within the plant pk
$B^{cap\ economies}_{wn,pk,r,q}$	Unitary variable cost associated with the capital of processing technology r within the plant pk
$B^{probit\ leaching}_{k,i}$	Intercept to obtain the probability of fatalities for leaching from the Probit curve for a specific interval
$B^{toxic\ probit}_{g,j}$	Intercept to obtain the value of the Probit curve for the fatalities for intoxication
$B^{toxic\ probit}_{g,j}$	Intercept to obtain the value of the probability for the fatalities for intoxication
$C^{unitary\ product}_{p}$	Unitary price of product p at site s
$C^{unitary\ processing}_{wn,p,pk,r}$	Unitary operating cost for processing plants
C^{disp}_{d}	Unitary cost for waste disposal
C^{sep}_{wn}	Unitary separation cost for useful waste
$C1^{transp}_{wn}$	Unitary transportation cost from sites to dumps
$C2^{transp}_{wn}$	Unitary transportation cost from dumps to processing facilities
$C3^{transp}_{wn}$	Unitary transportation cost from processing facilities to consumers
$D1^{transp}_{s,d}$	Distance from site s to dump d
$D2^{transp}_{d,pk}$	Distance from dump d to processing plant pk
$D3^{transp}_{s,pk}$	Distance from processing facility pk to consumer s
$f^{frequency\ factor}$	Frequency factor to obtain the number of times that the landfill is burn yearly
$H_{r,a}$	Effective height of emissary
$lowW^{distributed\ plant}_{wn,d,pk}$	Lower limit for transportation from dumps to processing facilities
$lowW^{distributed\ routes}_{wn,pk,r,q}$	Lower limit for processing depending on the economies of scale
$lowC^{phreatic\ concentration}_{k,i}$	Lower limit for the concentration in the Probit curve to obtain the fatalities for leaching
$lowC^{toxic\ concentration}_{g,d,a,j}$	Lower limit for the intervals for the Probit curve for the gas pollutant g

$lowF^{toxic\ probit}_{g,d,a,n}$	Lower limit for the Probit value for each interval to obtain the final fatalities for intoxication
$M^{probit\ leaching}_{k,i}$	Pendent to obtain the probability of fatalities for leaching from the Probit curve for a specific interval
$M^{toxic\ probit}_{g,j}$	Pendent to obtain the value of the Probit curve for the fatalities for intoxication
$M^{fatalities\ probit}_{g,n}$	Pendent to obtain the value of the probability for the fatalities for intoxication
$P^{population}_{s}$	Population that is affected by leaching in each site
$PD^{product\ consumer}_{p,s}$	Product demand in the site *s*
$P^{toxic\ population}_{d,a}$	Exposed people to the gas pollutant
u	Wind velocity
$uppW^{distributed\ plant}_{wn,d,pk}$	Upper limit for transportation from dumps to processing facilities
$uppW^{distributed\ routes}_{wn,pk,r,q}$	Upper limit for processing depending on the economies of scale
$uppC^{phreatic\ concentration}_{k,i}$	Upper limit for the concentration in the Probit curve to obtain the fatalities for leaching
$uppC^{toxic\ concentration}_{g,d,a,j}$	Upper limit for the intervals for the Probit curve for the gas pollutant *g*
$uppF^{toxic\ probit}_{g,d,a,n}$	Upper limit for the Probit value for each interval to obtain the final fatalities for intoxication
$W^{produced}_{wn,s}$	Waste production of waste *wn* in the site *s*
$y_{d,a}$	Perpendicular distance of receptor
$x^{distance}_{d,a}$	Distance from the emissary to the affected area or population
$z_{d,a}$	Height of the receptor
$\alpha^{conversion}_{wn,p,r}$	Conversion factor to determine the amount of obtained product respect to the utilized waste
$\alpha^{separation\ factor}_{wn,d}$	Waste separation factor to obtain the available waste
$\alpha^{filtration\ factor}_{k,d,s}$	Filtration factor to model the natural sieve of the environment
$\alpha^{dispersion}_{d,a}$	Dispersion factor that depends of the position for the gas dispersion model
$\beta^{pollutant\ surface}_{k,d}$	Factor to obtain the quantity of the pollutant *k* in the dump from the total amount of waste
$\beta^{pollutant\ emission}_{g,d}$	Relationship between the amount of waste emitted and burned in case of fire
$\delta^{quantity\text{-}concentration}_{s}$	Conversion factor that is used to obtain the concentration within the water table from the quantity of waste
$\sigma^{y}_{d,a}$, $\sigma^{z}_{d,a}$	Parameters for the Pasquill Gifford model that depends on the position of the affected surface and dump

5.7.3 Continuous variables

$CAPCOST_{wn,pk,r}^{economies}$	Capital cost for the processing route r within the plant pk
$CAPCOST^{total}$	Total capital cost for the processing plants
$C_{k,s}^{phreatic\ concentration}$	Concentration of the pollutant k in the site s
$C_{k,d}^{dump\ quantity}$	Quantity of pollutant in the surface of dump
$C_{g,s,d}^{toxic\ concentration}$	Concentration of the gas pollutant g that affects the site s from the dump d
$Cd_{k,s.i}^{phreatic\ concentration}$	Discretized concentration of the pollutant k in the site s
$Cd_{g,d,a,j}^{toxic\ concentration}$	Discretized concentration of the gas pollutant g that affects the site s from the dump d
$dCAPCOST_{wn,pk,r}^{economies}$	Discretized capital cost for the processing route r within the plant pk
$F_{k,s}^{leaching\ fatalaties}$	Fatalities for leaching
$F_{g,d,a}^{toxic\ probit}$	Value for the Probit curve to obtain the fatalities for intoxication
$Fd_{g,d,a,j}^{toxic\ probit}$	Discretized value for the Probit curve to obtain the probability of fatalities for intoxication
$Fd2_{g,d,a,n}^{toxic\ probit}$	Discretized value for the Probit curve to obtain the fatalities for intoxication
$F_{g,d,a}^{intoxication\ fatalities}$	Fatalities for the intoxication by the gas pollutant
$NETPROFIT$	Net annual profit for the entire supply chain
$OPCOST$	Total operational cost for the processing facilities
$P_{p,pk,s}^{distributed\ product}$	Distributed product from processing facilities to consumers
$P_{p,s}^{product\ consumer}$	Total product in the market s
$P_{p,pk}^{processing\ plant}$	Obtained product from the utilized waste
$P_{g,d,a}^{toxic\ probit}$	Affectation fraction of fatalities for intoxication by a gas pollutant
$Pd_{g,d,a,n}^{toxic\ probit}$	Discretized affectation fraction of fatalities for intoxication by a gas pollutant
$Q_{g,d}$	Emission rate when the landfill is burned
$REVENUE$	Obtained revenue for the supply chain based on waste management
$SEPCOST$	Total separation cost for the waste that is processed
$TRANSPCOST$	Total transportation cost for the supply chain
$TRANSPCOST^{SITES-LANDFILLS}$	Transportation cost from sites to landfills
$TRANSPCOST^{LANDFILLS-FACILITIES}$	Transportation cost from landfills to facilities
$TRANSPCOST^{FACILITIES-COSNUMERS}$	Transportation cost from facilities to consumers

$W^{distributed\ dump}_{wn,s,d}$	Distributed waste from site *s* to dump *d*
$W^{in\ dump}_{wn,d}$	Amount of waste inlet to dump *d*
$W^{total}_{wn,d}$	Amount of waste stored in dumps
$W^{out\ dump}_{wn,d}$	Amount of waste outlet from dump *d*
$W^{distributed\ plant}_{wn,d,pk}$	Distributed waste in each processing facility
$W^{in\ plant}_{wn,pk}$	Amount of waste inlet to processing plant *pk*
$W^{distributed\ routes}_{wn,pk,r}$	Distributed waste to processing routes within the processing plants
$Wd^{distributed\ routes}_{wn,pk,r,q}$	Distributed waste to the processing routes within the processing plants for each interval of the economies of scale
$WASTECONS$	Ratio for consumed waste with respect to the total produced waste in cities
$\alpha^{leaching\ fatalities}_{k,s}$	Probability of fatalities for leaching
$\alpha d^{leaching\ fatalities}_{k,s,i}$	Discretized probability of fatalities for leaching

5.7.4 *Binary variables*

$y^{economies}_{wn,pk,r,q}$	Binary variable to consider the economies of scale in the processing facilities
$y^{leaching\ fatalities}_{k,s,i}$	Binary variable to consider the intervals for the Probit curve to obtain the fatalities for leaching
$y^{intoxication\ linealized}_{g,d,a,j}$	Binary variable to consider the intervals for the Probit curve to obtain the Probit value for burning
$y^{intoxication\ fatalities}_{g,d,a,n}$	Binary variable to consider the intervals to obtain the fatalities for burning

5.8 References

Abarca-Guerrero, L., Maas, G., and Hogland, W. 2013. Solid waste management challenges for cities in developing countries. Waste Management, *33*(1), 220–232.

Adhitya, A., Srinivasan, R., and Karimi, I. A. 2009. Supply chain risk identification using a HAZOP-based approach. *AIChE* Journal, *55*(6), 1447–1463.

Beolchini, F., Fonti, V., Dell'Anno, A., Rocchetti, L., and Vegliò, F. 2012. Assessment of biotechnological strategies for the valorization of metal bearing wastes. Waste Management, *32*(5), 949–956.

Bernechea E. J., and Arnaldos J. 2013. Optimizing the design of storage facilities through the application of ISD and QRA. Process Safety and Environmental Protection, 92, *6*, 598–615. DOI: 10.1016/j.psep.2013.06.002

Bowling, I. M., Ponce-Ortega, J. M., and El-Halwagi, M.M. 2011. Facility location and supply chain optimization for a biorefinery. Industrial and Engineering Chemistry Research, *50*(10), 6276–6286.

Brooke, A., Kendrick, D., Meeruas, A., and Raman, R. 2014. GAMS-Language Guide. GAMS Development Corp. Washington, DC.

Capón-García, E., Papadokonstantakism, S., and Hungerbühler, K. 2013. Efficient waste management based on scheduling optimization of waste treatment plants. Computer Aided Chemical Engineering, *32*, 949–954.

Das A., Gupta A.K., and Mazumder T.N. 2012. A comprehensive risk assessment framework for offsite transportation of inflammable hazardous waste. Journal of Hazardous Materials, *227*, 88–96.

Dadashzadeh, M., Khan, F., Abbassi, R., and Hawboldt, K. 2014. Combustion products toxicity risk assessment in an offshore installation. *Process Safety and Environmental Protection*, 92, 6 616–624, DOI: 10.1016/j.psep.2013.07.001

El-Halwagi, A.M., Rosas C., Ponce-Ortega, J.M., Jiménez-Gutiérrez, A., Mannan, M.S., and El-Halwagi, M.M. 2013. Multiobjective optimization of biorefineries with economic and safety objectives. AIChE Journal, *59*(7), 2427–2434.

EPA. 2011). Environmental Protection Agency Office of Resource Conservation and Recovery. Municipal Solid Waste Generation, Recycling, and Disposal in the United States Tables and Figures for 2010. U.S., USA.

Fabiano, B., Currò, F., Reverberi, A.P., and Pastorino, R. 2005. Dangerous good transportation by road: from risk analysis to emergency planning. Journal of Loss Prevention in the Process Industries, *18*(4), 403–413.

Galan, B., Dosal, E., Andrés, A., and Viguri, J. 2013. Optimisation of the construction and demolition waste management facilities location in Cantabria (Spain) under economical and environmental criteria. Waste and Biomass Valorization, *4*(4), 797–808.

Hokkanen, J., and Salminen, P. 1997. Choosing a solid waste management system using multicriteria decision analysis. European Journal of Operational Research, *98*(1), 19–36.

Hung, M. L., Ma, H. W., and Yang, W. F. 2007. A novel sustainable decision making model for municipal solid waste management. Waste Management, *27*(2), 209–219.

Hur, T., Lim, S., and Lee, H. 2003. A study on the eco-efficiencies for recycling methods of plastics waste. *InLCA*, 22.

INEGI. 2013. Mexican Information for States and Municipalities. http://www3.inegi.org.mx/sistemas/mexicocifras/default.aspx?e=16 (Accessed 20/03/2014)

Karmperis, A.,C., Aravossis, K., Tatsiopoulos, I. P., and Sotirchos, A. 2013. Decision support models for solid waste management: Review and game-theoretic approaches. Waste Management, *33*(5), 1290–1301.

Minoglou, M., and Komilis, D. 2013. Optimizing the treatment and disposal of municipal solid wastes using mathematical programming—A case study in a Greek region. Resources, Conservation and Recycling, *80*, 46–57.

Morrissey A. J., and Browne J. 2004. Waste management models and their application to sustainable waste management. Waste Management, *24*(3), 297–308.

Ng, R.,T., Hassim, M.,H., and Ng, D. K. 2013. Process synthesis and optimization of a sustainable integrated biorefinery via fuzzy optimization. AIChE Journal, *59*(11), 4212–4227.

Plunkert, P. A. 2006. Aluminium recycling in the United States in 2000. US Department of the Interior, US Geological Survey, USA.

Sabbas, T., Polettini, A., Pomi, R., Astrup, T., Hjelmar, O., Mostbauer, P., and Lechner, P. 2003. Management of municipal solid waste incineration residues. Waste Management, *23*(1), 61–88.

Santibañez-Aguilar, J. E., Ponce-Ortega, J. M., González-Campos, J. B., Serna-González, M., and El-Halwagi, M. M. 2013. Optimal planning for the sustainable utilization of municipal solid waste. Waste Management, *33*(12), 2607–2622.

Santibañez-Aguilar, J. E., González-Campos, J. B., Ponce-Ortega, J. M., Serna-González, M., and El-Halwagi, M. M. 2014. Optimal planning and site selection for distributed multiproduct biorefineries involving economic, environmental and social objectives. Journal of Cleaner Production, *65*, 270–294.

Santibañez-Aguilar, J. E., Martinez-Gomez, J., Ponce-Ortega, J. M., Nápoles-Rivera, F., Serna-González, M., González-Campos, J. B., and El-Hawlagi, M. 2015. Optimal planning for the reuse of municipal solid waste considering economic, environmental, and safety objectives. AIChE Journal, *61*(6), 10881–1899.

SCT. 2013. Mexican Ministry of Transportation and Communications. http://aplicaciones4.sct.gob.mx/sibuac_internet/ControllerUI?action=cmdEscogeRuta (Accessed 20/03/2014)

Sugiyama, H., Fischer, U., Hungerbühler, K., and Hirao, M. 2008. Decision framework for chemical process design including different stages of environmental, health, and safety assessment. AIChE Journal, *54*(4), 1037–1053.

Ramayah, T., Lee, J. W. C., and Lim, S. 2012. Sustaining the environment through recycling: An empirical study. Journal of Environmental Management, *102*, 141–147.

Tan, S. T., Lee, C. T., Hashim, H., Ho, W. S., and Lim, J. S. 2014. Optimal process network for municipal solid waste management in Iskandar Malaysia. Journal of Cleaner Production, *71*, 48–58.

Van Raemdonck, K., Macharis, C., and Mairesse, O. 2013. Risk analysis system for the transport of hazardous materials. Journal of Safety Research, *45*, 55–63.

Varbanov, P. S., Lam, H. L., Friedler, F., and Klemes, J. J. 2012. Energy generation and carbon footprint of waste to energy: Centralised vs. distributed processing. Computer Aided Chemical Engineering, *31*, 1402–1406.

You, F., Tao L., Graziano D. J., and Snyder, S.W. 2012. Optimal design of sustainable cellulosic biofuel supply chains: Multiobjective optimization coupled with life cycle assessment and input–output analysis. AIChE Journal, *58*(4), 1157–1180.

Young, G. C. 2010. Municipal solid waste to energy conversion processes: economic, technical, and renewable comparisons, Wiley, USA.

6

Strategic Planning for Municipal Solid Waste Management

Management of municipal solid wastes (MSW) involves multiple stakeholders such as government agencies, suppliers, consumers, providers of treatment/recycling services, and transporters. An optimal management strategy should be based on creating synergistic opportunities that benefit the multiple participants. This chapter presents a multi-objective optimization approach for the strategic planning of a municipal solid waste management system. The formulation considers the involved tasks such as recycling, reuse, transportation, separation, and distribution. The presented approach also accounts for multiple stakeholders to maximize the benefit to all participants. The Latin Hypercube sampling technique is adopted to systematically generate weights for the different stakeholders. A case study from Mexico is analyzed and three scenarios are considered. The first one considers that the separation cost is absorbed by each recycling company. The second one assumes that the government is responsible for the separation cost. The third one requires household inhabitants to carry out waste sorting. The optimization approach is used to analyze the results of the various scenarios and to deduce valuable insights into the interaction among the various stakeholders and the building blocks of the supply chain of MSW management systems.

6.1 Introduction

The generation of municipal solid waste (MSW) has increased drastically because of population growth and economic development. The average MSW generation is 1.2 kg per person per day with an estimated increase of 1.4 kg/capita/day by the year 2025 (Hoornweg and Bhada-Rata 2012). Landfilling represents the main option for waste disposal in several parts of the world (Leme et al. 2014). Nonetheless, landfilling is an ineffective way to manage the residues because it causes degradation of valuable land resources, increases land cost, and creates long-term environmental and human health problems (Gracida-Alvarez et al. 2016). Therefore, sustainable, and efficient waste management systems are needed to improve MSW management (Tan et al. 2014).

Developing an optimal MSW management system is a complex task that involves the optimal selection of waste collection routes, transfer station facilities, treatment strategies, treatment plant locations, and energy recovery (Dewi et al. 2014). Recent research efforts have endeavored to address MSW management systems. Warren and El-Halwagi (1996) studied the conversion of MSW to hydrogen and liquid transportation fuels. Chen and Chang (2010) developed a model to assess the performance of an MSW recycling system in Taiwan. Tan et al. (2014) reported an optimization model to integrate the four most used recycling technologies to obtain a cost-effective processing network in Malaysia. Vadenbo et al. (2014) presented a multi-objective mixed-integer linear programming (MILP) model for making decisions about waste and resource management in industrial networks. Rada et al. (2014) reported an approach to consider the interactions between tourism and the selective collection of MSW.

Ng et al. (2014) incorporated waste energy in the MSW management analyzing a case study from Malaysia. Sukholthaman and Shirahada (2015) presented approaches to reduce the impacts of ineffective waste management services in Bangkok. Santibañez-Aguilar et al. (2013) considered economic and environmental issues in the design of MSW reuse systems, and then included safety objectives (Santibañez-Aguilar et al. 2015). Carvalho et al. (2015) presented an approach that considers the economies of scale in the MSW management system.

Mirmajlesi and Shafei (2016) reported a multi-period, multi-product, and multi-echelon model for designing supply chains. Chifari et al. (2016) developed a holistic framework to organize and integrate quantitative information for characterizing the performance of MSW management systems. Sukholthaman and Sharp (2016) analyzed the dynamics of an MSW management system. Ho et al. (2017) incorporated the pinch analysis in the MSW management systems. Santibañez-Aguilar et al. (2017) reported an optimal control approach for strategic planning in an MSW supply chain. Korai et al. (2017) evaluated the feasibility of using MSW to generate energy in Pakistan. Harijani et al. (2017) developed a systematic approach to build an integrated recycling and disposal network for MSW by considering sustainability to maximize the total profit under a budget constraint. Struck (2017) analyzed the incentives needed for an MSW management system in the Czech Republic. Gilardino et al. (2017) reported a study to locate collection sites for MSW. A review of the MSW management was reported by Lee et al. (2016), and Yadav et al. (2017) developed an algorithm, based on interval analysis, to deal with the involved uncertainties and to identify the optimal location of a facility.

In many cases, the MSW management systems involve multiple stakeholders with different objectives and roles (Achillas et al., 2017). It is important to note that previously reported approaches for the strategic planning of the MSW management have not considered the interactions and conflicting interests of the different involved stakeholders (Morrissey and Browne, 2004). The previous methods have considered the overall profit associated with the entire MSW management system without accounting for the individual profit of the participants in the supply chain. This may produce impractical solutions where the overall profits are maximized at the expense of suboptimal solutions or even economic losses for some of the participants. Therefore, it is important to develop an optimization approach for the strategic planning for the MSW management system that accounts for the objectives of the different participants in the supply chain and to obtain solutions where all the participants benefit. This chapter proposes an optimization approach for the strategic planning of MSW management systems through a multi-objective scheme that accounts for the maximum benefit for the different participants in the supply chain. The proposed model incorporates a new optimization scheme for the multi-stakeholder problem which involves all the activities considered in the MSW management system such as recollection, transportation, recycling, and distribution of products. To demonstrate the significance of taking into account the various stakeholders engaged in the entire supply chain for the MSW management system, a case study from Mexico is provided and several scenarios are examined (Diaz.Barriga-Fernandez et al. 2017).

6.2 Problem Statement

The addressed problem aims to determine the optimal supply chain for the MSW management system while endeavoring to attain the maximum benefit for the different stakeholders (Fig. 1). The domain includes several neighboring cities which generate different quantities and qualities of MSW. The intent is to determine the optimal tasks for recollection, transportation, separation, and treatment for the MSW and the location of the processing facilities. The considered stakeholders include the government, suppliers, consumers, and the service providers for treating/recycling the MSW.

An illustration of the problem statement is shown in Fig. 1, which includes five adjacent cities. These cities produce different amounts of MSW with different compositions. A government-run system is responsible for collecting MSW. The recoverable part of the MSW is separated into five main materials: plastics, metals, glass, paper, and un-recyclables. The un-recyclables are landfilled. Depending on the analyzed scenario, the separation is carried out by each processing industry,

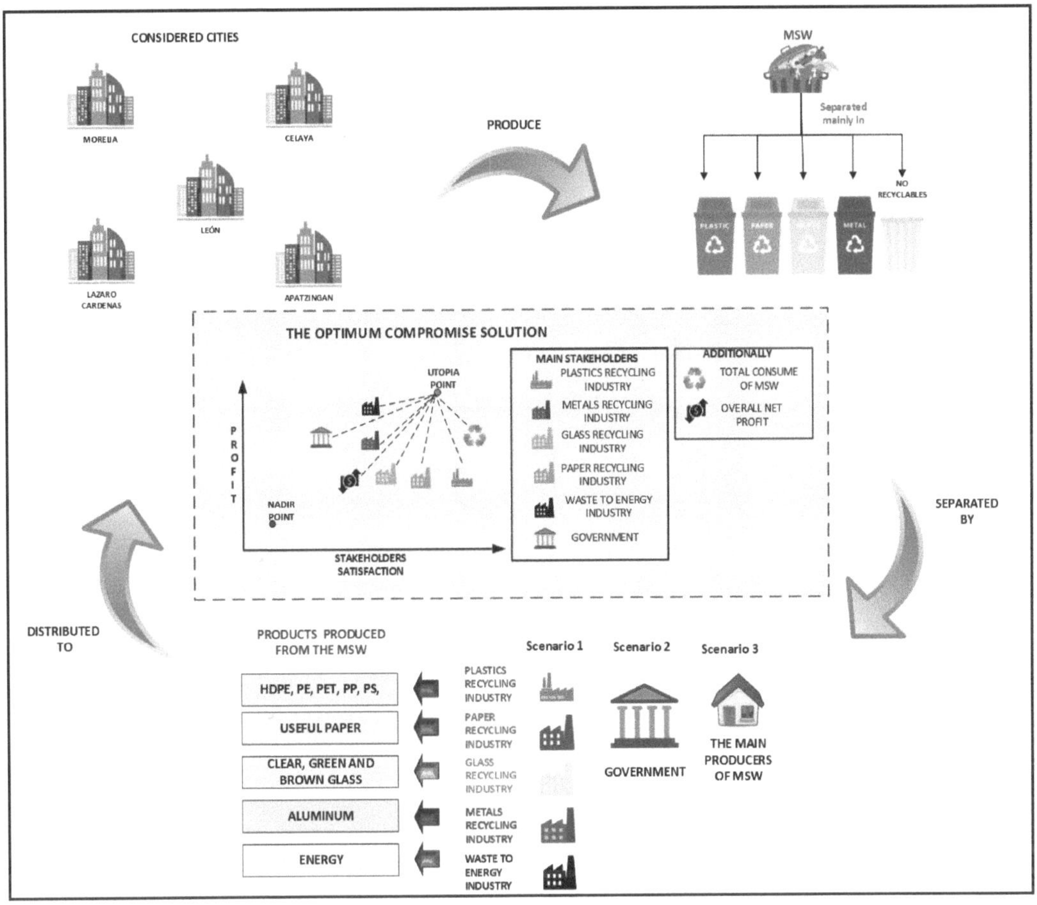

Fig. 1. Problem statement for the problem addressed in Chapter 6. [Printed (adapted) with permission from Diaz-Barriga-Fernandez, A.D., Santibañez-Aguilar, J.E., Radwan, N., Nápoles-Rivera, F., El-Halwagi, M., and Ponce-Ortega, J.M. 2017. Strategic planning for managing municipal solid wastes with consideration of multiple stakeholders. *ACS Sustainable Chemistry and Engineering*, 2017, *5*(11), 10744–10762].

government, or the producers of the residues. Depending on the kind of residue and the intended products, the MSW may undergo treatment in various processing facilities following appropriate separation. It is necessary to optimize the processing plants' size, kind, and location. After that, the products are given to prospective customers who have particular needs. The routes for transportation must also be optimized. The profit of each stakeholder should be taken into account while designing the best MSW management system setup. The following section presents the general optimization formulation proposed to address the above-mentioned problem.

6.3 Model Formulation

The model formulation is based on the superstructure presented in Fig. 2. The index *wn* is used to describe the type of residue: plastic, metal, glass, paper, and un-recyclables. The processing plants are represented by the index *pp-wn*, the technologies for treating each subtype of residue are represented by the index *n-wn* and the products produced from the residues are given by the index *p-wn*. The index *c* is used to denote the diverse considered cities. Finally, the index *ofb* is used for the objective functions. The model formulation is presented as follows.

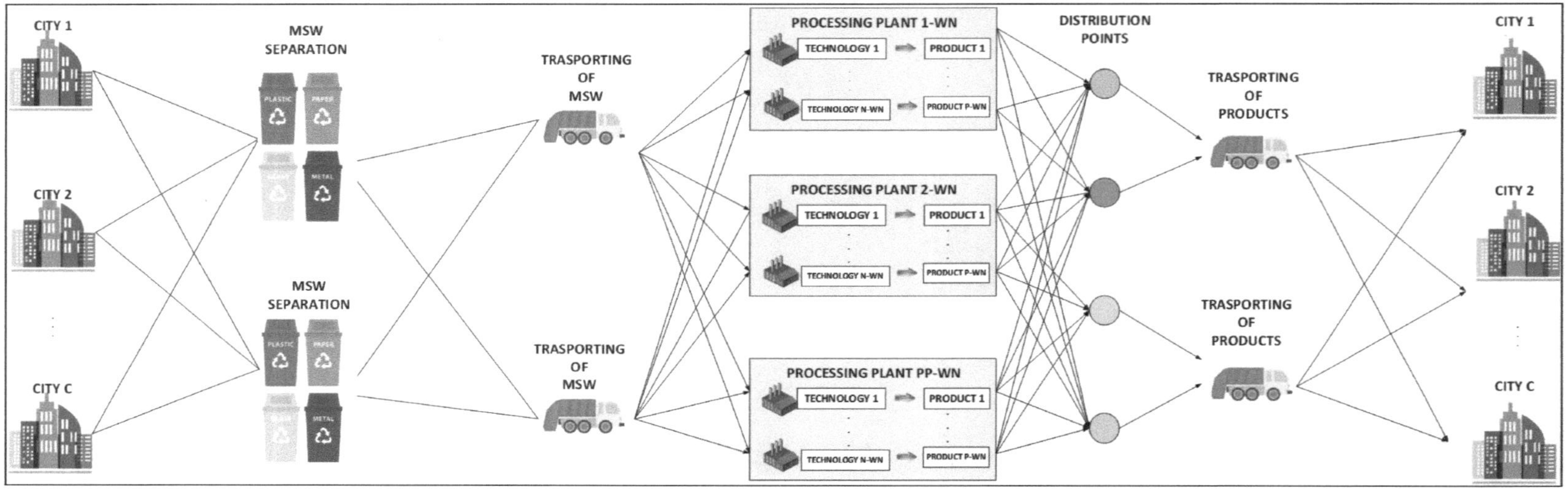

Fig. 2. Superstructure for the MSW management. [Printed (adapted) with permission from Diaz-Barriga-Fernandez, A.D., Santibañez-Aguilar, J.E., Radwan, N., Nápoles-Rivera, F., El-Halwagi, M., and Ponce-Ortega, J.M. 2017. Strategic planning for managing municipal solid wastes with consideration of multiple stakeholders. *ACS Sustainable Chemistry and Engineering*, 2017, *5*(11), 10744–10762].

6.3.1 Constraints for the Cities that Produce MSW

Each city produces residues, but only a percentage of this can be reused. Therefore, the amount of used residues *wn* to produce a certain product *p-wn* and collected from the city *c* $\left(F_{wn,c,p-wn}\right)$ should be lower than the amount of reusable residues $\left(\alpha^{sep}_{wn,c,p-wn}F^{total}_{c,p-wn}\right)$. Here, the quantity of reusable residues is equal to the total amount of produced residues $\left(F^{total}_{c,p-wn}\right)$ multiplied by a separation factor $\left(\alpha^{sep}_{wn,c,p-wn}\right)$:

$$F_{wn,c,p-wn} \leq \alpha^{sep}_{wn,c,p-wn}F^{total}_{c,p-wn}, \quad \forall wn, c, p-wn \tag{6.1}$$

6.3.2 Distribution of MSW from Cities to Processing Plants

The MSW is distributed to different processing plants located in different regions. Each plant processes one type of residue; for example, plastic, metal, glass, paper, and unrecyclables. Moreover, each processing plant has different technologies depending on the treated residue. Thus, the quantity of residues collected from the city *c* $\left(F_{wn,c,p-wn}\right)$ is equal to the sum of the residues sent from the city *c* to the processing plant *pp-wn* considering all the types of residues *wn*:

$$F_{wn,c,p-wn} = \sum_{pp-wn \in PH} F^{plant}_{wn,c,pp-wn,p-wn} \quad \forall wn, c, p-wn \tag{6.2}$$

6.3.3 Distribution of MSW into the Processing Plants to be Treated with Different Technologies

Each type of residue that arrives at the different processing plants is distributed to the different recycling technologies to obtain a specific product. Thus, the sum of the number of residues collected from all the cities $\left(F^{plant}_{wn,c,pp-wn,p-wn}\right)$ is equal to the sum of all the residues sent to each different recycling technology:

$$\sum_{c \in C} F^{plant}_{wn,c,pp-wn,p-wn} = \sum_{n-wn \in T} F^{tech}_{wn,pp-wn,n-wn,p-wn} \quad \forall wn, pp-wn, p-wn \tag{6.3}$$

6.3.4 Mass Balance in the Processing Plants

This model considers the transformation of the residues in products through several processing plants. These processes are modeled with nonlinear relationships, but in the present work they were modeled using an efficiency factor for each type of technology and treated residue $\left(\alpha^{prod}_{wn,n-wn,p-wn}\right)$, the efficiency production factor is useful to determine the amount of product that is produced $\left(P_{wn,pp-wn,n-wn,p-wn}\right)$ and this quantity is a function of the amount of residue that is treated $\left(F^{tech}_{wn,pp-wn,n-wn,p-wn,}\right)$:

$$P_{wn,pp-wn,n-wn,p-wn} = \alpha^{prod}_{wn,n-wn,p-wn}F^{tech}_{wn,pp-wn,n-wn,p-wn}, \quad \forall wn, pp-wn, n-wn, p-wn \tag{6.4}$$

The efficiency production factors are obtained through simulation or from reported literature.

6.3.5 Balance in Splitter after the Processing Plants

Once the residues are processed, the products are distributed from the processing plants to the consumers located in the different cities c. Thus, the sum of a determined product distributed in the different cities and produced in a specific processing plant $\left(P^{city}_{wn,pp-wn,c,n-wn,p-wn}\right)$ is equal to the produced product $p\text{-}wn$ in the processing plant $pp\text{-}w$:

$$P_{wn,pp-wn,n-wn,p-wn} = \sum_{c\in C} P^{city}_{wn,pp-wn,c,n-wn,p-wn} \quad \forall wn, pp-wn, n-wn, p-wn \tag{6.5}$$

6.3.6 Balance in the Product Distribution Points before the Cities

For each city c exists a product distribution point for each type of product, where the demand for the product for each city is obtained by mixing different quantities of products from different cities as can be seen in Equation (6):

$$\sum_{pp-wn\in PROCPLANT} P^{city}_{wn,pp-wn,c,n-wn,p-wn} = PT^{city}_{wn,c,n-wn,p-wn}, \quad \forall wn, n-wn, c, p-wn \tag{6.6}$$

6.3.7 Constrain of the Demand in the Cities

The total amount of each type of product in city c $\left(PT^{city}_{wn,c,n-wn,p-wn}\right)$ must be lower or equal to the demand for the products in a determined city $\left(PD^{city}_{wn,n-wn,p-wn}\right)$:

$$PT^{city}_{wn,c,n-wn,p-wn} \leq PD^{city}_{wn,n-wn,p-wn}, \quad \forall wn, c, n-wn, p-wn \tag{6.7}$$

6.3.8 MSW Separation Cost

The MSW separation process is carried out in two steps. In the first one, the residues are separated into plastic, metal, glass, paper, and un- recyclables. In the second step, the separation is more specific for each type of residue wn. The unitary cost is an estimated value for the separation cost for both separation stages.

$$\mathrm{Cos}\,tSep_{wn} = \sum_{c\in C}\sum_{p-wn\in S} C^{sep}_{wn,c,p-wn} F_{wn,c,p-wn}, \quad \forall wn \tag{6.8}$$

The total separation cost is considered, which includes the separation for each type of residue:

$$\mathrm{Cos}\,tSep^{total} = \sum_{wn\in W} \mathrm{Cos}\,tSep_{wn} \tag{6.9}$$

6.3.9 MSW Final Disposal Cost

Another cost considered is the one associated with not separating the residues, which usually is paid by the government through sanitary landfills; however, this part must be considered to obtain a more integral solution. Thus, the final disposal cost is equal to the unitary cost multiplied by the amount of residues that were not separated:

$$\mathrm{Cos}\,tNSep^{total} = \sum_{wn\in W}\sum_{c\in C}\sum_{p-wn\in S} C^{no-sep}_{wn,c,p-wn}\left(F^{total}_{wn,c,p-wn} - F_{wn,c,p-wn}\right) \tag{6.10}$$

6.3.10 Total Cost for Transporting MSW from Cities to Processing Plants

An important activity is the transportation of MSW from cities to processing plants. This activity has a cost that includes the unitary cost for transporting each type of residue wn $\left(C_{wn,p-wn}^{unit-transp}\right)$, the distance between the cities and the processing plants, and the amount of residues transported from the cities to the processing plants $\left(F_{wn,c,pp-wn,p-wn}^{plant}\right)$:

$$CostTransp_{pp}^{city} = \sum_{wn \in W} \sum_{c \in C} \sum_{pp-wn \in P} \sum_{p-wn \in S} C_{wn,p-wn}^{unit-transp} d_{c,pp-wn} F_{wn,c,pp-wn,p-wn}^{plant} \quad (6.11)$$

6.3.11 Total Transportation Cost for the Products from Processing Plants to Cities

Another activity considers the transportation of products from processing plants to the cities where they will be consumed. In this case, the unitary cost for transporting multiplied by the distance between the processing plants and the cities and the amount of generated products is equal to the total transporting cost from the processing plants to the cities:

$$CostTransp_{city}^{pp} = \sum_{wn \in W} \sum_{n-wn \in T} \sum_{c \in C} \sum_{pp-wn \in P} \sum_{p-wn \in S} C_{wn,n-wn,p-wn}^{unit-transp-product} d_{c,pp-wn} P_{wn,pp-wn,c,n-wn,p-wn}^{city} \quad (6.12)$$

6.3.12 Production Cost in the Processing Plants

Each type of technology considers the production cost, which is divided into two sections. The first one is the operating cost, which considers the unitary cost for operation $\left(C_{wn,pp-wn,n-wn,p-wn}^{processing}\right)$. Thus, the total operating cost depends on the quantity of material that is processed in each processing plant $\left(F_{wn,pp-wn,n-wn,p-wn}^{tech}\right)$:

$$CostOp^{processing} = \sum_{wn \in W} \sum_{pp-wn \in P} \sum_{n-wn \in T} \sum_{p-wn \in S} C_{wn,pp-wn,n-wn,p-wn}^{processing} F_{wn,pp-wn,n-wn,p-wn}^{tech} \quad (6.13)$$

The second one is the capital cost associated with the different processing plants. The capital cost is usually a function as $A_{wn,pp-wn,n-wn,p-wn}^{processing} + B_{wn,pp-wn,n-wn,p-wn}^{processing}\left(F_{wn,pp-wn,n-wn,p-wn}^{tech}\right)^{\alpha}$, where α is used to take the economies of scale into account. The function shown before is nonlinear and this could generate optimization problems during the solution approach, thus the equation can be linearized by assuming that the exponent is equal to one:

$$CostCap_{wn,pp-wn,n-wn,p-wn}^{processing} = A_{wn,pp-wn,n-wn,p-wn}^{processing} + B_{wn,pp-wn,n-wn,p-wn}^{processing}\left(F_{wn,pp-wn,n-wn,p-wn}^{tech}\right),$$

$$\forall wn, pp-wn, n-wn, p-wn \quad (6.14)$$

Once the individual capital cost is calculated, it is possible to obtain the total capital cost, which is the sum of the individual capital cost for each type of residue $\left(CostCap_{wn,pp-wn,n-wn,p-wn}^{processing}\right)$:

$$CostCap^{total} = \sum_{wn \in W} \sum_{pp-wn \in P} \sum_{n-wn \in T} \sum_{p-wn \in S} CostCap_{wn,pp-wn,n-wn,p-wn}^{processing} \quad (6.15)$$

The terms in the capital cost of the processing plants depend on the capacity of the plants; these are modeled through the following disjunctions, which are activated when the processed feedstock is between the lower and upper limits:

$$\vee_q \begin{bmatrix} y^{processing}_{wn,pp-wn,n-wn,p-wn,q} \\ A^{processing}_{wn,pp-wn,n-wn,p-wn} = a^{processing}_{wn,pp-wn,n-wn,p-wn,q} \\ B^{processing}_{wn,pp-wn,n-wn,p-wn} = b^{processing}_{wn,pp-wn,n-wn,p-wn,q} F^{tech}_{wn,pp-wn,n-wn,p-wn} \\ F\min^{processing}_{wn,pp-wn,n-wn,p-wn,q} \leq F^{tech}_{wn,pp-wn,n-wn,p-wn} \leq F\max^{processing}_{wn,pp-wn,n-wn,p-wn,q} \end{bmatrix}$$

$$\forall wn, pp-wn, n-wn, p-wn$$

The previous disjunction is algebraically reformulated as follows:

$$A^{processing}_{wn,pp-wn,n-wn,p-wn} = \sum_{q \in I} a^{processing}_{wn,pp-wn,n-wn,p-wn,q} y^{processing}_{wn,pp-wn,n-wn,p-wn,q}, \quad \forall wn, pp-wn, n-wn, p-wn \tag{6.16}$$

$$B^{processing}_{wn,pp-wn,n-wn,p-wn} = \sum_{q \in I} b^{processing}_{wn,pp-wn,n-wn,p-wn,q} dF^{tech}_{wn,pp-wn,n-wn,p-wn,q}, \quad \forall wn, pp-wn, n-wn, p-wn \tag{6.17}$$

Equation (6.16) is used to define the fixed capital cost $\left(A^{processing}_{wn,pp-wn,n-wn,p-wn}\right)$, which depends on the processing plant, processing technology, and type of waste. In this context, the unitary variable capital cost $\left(B^{processing}_{wn,pp-wn,n-wn,p-wn}\right)$ is defined in Equation (6.17). The value of the amount of waste in the processing plants $\left(F^{tech}_{wn,pp-wn,n-wn,p-wn,q}\right)$ is assigned through Equation (6.18), and these amounts are limited by Equations (6.19) and (6.20):

$$F^{tech}_{wn,pp-wn,n-wn,p-wn} = \sum_{q \in I} dF^{tech}_{wn,pp-wn,n-wn,p-wn,q}, \quad \forall wn, pp-wn, n-wn, p-wn \tag{6.18}$$

$$F\min^{processing}_{wn,pp-wn,n-wn,p-wn,q} y^{processing}_{wn,pp-wn,n-wn,p-wn,q} \leq dF^{tech}_{wn,pp-wn,n-wn,p-wn,q}, \quad \forall wn, pp-wn, n-wn, p-wn, q \tag{6.19}$$

$$F\max^{processing}_{wn,pp-wn,n-wn,p-wn,q} y^{processing}_{wn,pp-wn,n-wn,p-wn,q} \geq dF^{tech}_{wn,pp-wn,n-wn,p-wn,q}, \quad \forall wn, pp-wn, n-wn, p-wn, q \tag{6.20}$$

Equation (6.21) is used to define the capacity of the processing technology and the cost depending on the economy of scales by selecting the binary variable that defines the interval in which the technology operates.

$$\sum_{q \in I} y^{processing}_{wn,pp-wn,n-wn,p-wn,q} = 1, \quad \forall wn, pp-wn, n-wn, p-wn \tag{6.21}$$

6.3.13 Profit for Selling Products

The produced products can be sold in different markets, in this case, the markets are located around the cities; each product has a unitary price, which depends on the type of product and the city where it is sold. The price changes concerning the city because each city has different product demands and market conditions. The profit from the sold products (C^{sales}) is the sum of the amount of products distributed to the different markets ($P^{city}_{wn,pp-wn,c,n-wn,p-wn}$) multiplied by the corresponding unitary cost ($C^{product}_{wn,w-wn,c}$):

$$C^{sales} = \sum_{pp-wn \in P} \sum_{n-wn \in T} \sum_{c \in C} \sum_{p-wn \in S} P^{city}_{wn,pp-wn,c,n-wn,p-wn} C^{product}_{wn,w-wn,c} \tag{6.22}$$

6.3.14 Costs for Each Type of Residues

The separation cost for the type of residue *wn* must be equal to the sum of the unitary cost for separating each type of residue and multiplied by the number of residues of the type *wn* produced in a determined city *c*, which is stated through Equation (6.8). Equation (6.8) is valid for each type of residue *wn* (i.e., plastic, metal, glass, paper, and un- recyclables).

The operating cost for each type of treated residue *wn* is equal to the sum of the unitary cost for processing a certain type of residue and multiplied by the number of residues of the type *wn* treated in the processing plant *pp-wn* with the technology *n-wn*:

$$CostOp^{processing}_{wn} = \sum_{wn} \sum_{pp-wn} \sum_{n-wn} \sum_{p-wn} C^{processing}_{wn,pp-wn,n-wn,p-wn} F^{tech}_{wn,pp-wn,n-wn,p-wn} \quad , \forall wn \tag{6.23}$$

The transportation cost accounts for the sum of the unitary cost for transporting residues from the producing plant to the city *c*, where the product *p-wn* will be consumed, multiplied by the distance between the city and the processing plant, and multiplied by the amount of product produced in the plant *pp-wn*:

$$CostTransp^{pp\ city}_{wn} = \sum_{n-wn} \sum_{c} \sum_{pp-wn} \sum_{p-wn} C^{unit-transp-product}_{wn,n-wn,p-wn} d_{c,pp-wn} P^{city}_{wn,pp-wn,c,n-wn,p-wn} , \forall wn \tag{6.24}$$

6.3.15 Profit from the Sales of Each Type of Residue wn

To calculate the profit for the sale of each type of recycled residue, Equation (6.25) is used, where the sum of the unitary cost for each product is multiplied by the quantity of product sent from the processing plants *pp-wn* to the different cities where they will be consumed:

$$C^{sales}_{wn} = \sum_{pp-wn} \sum_{n-wn} \sum_{c} \sum_{p-wn} P^{city}_{wn,pp-wn,c,n-wn,p-wn} C^{product}_{wn,n-wn,c}, \quad \forall wn \tag{6.25}$$

6.3.16 Objective Functions

The proposed problem is stated as a multi-objective optimization approach and the considered objective functions are described as follows.

6.3.16.1 Overall net profit

The main objective function considers the net profit of all the residues and the costs associated with the transportation, separation, and production of the products obtained from the recycled residues. The overall net profit (ONP) is equal to the total profit from the sales of all the distributed products *p-wn* minus the total separation cost, the total operation cost, the total transportation cost for the processing plants to the cities, the total transportation cost for the collection of the residues from the cities to the processing plants and the total cost for no separation of the residues, which is stated as follows:

$$\begin{aligned}\text{NetProfit} = C^{sales} - \text{Cos}\,tSep^{total} - \text{Cos}\,tCap^{total} - \text{Cos}\,tOp^{processing} \\ - \text{Cos}\,tTransp_{city}^{pp} - \text{Cos}\,tTransp_{pp}^{city} - \text{Cos}\,tNSep^{total}\end{aligned} \quad (6.26)$$

6.3.16.2 Total recycled MSW

The proposed environmental objective function considers the maximization of the percentage of the total recycled MSW ($TotalConsume^{waste}$) based on the maximum quantity that could be reused | ($\sum_{wn\in W}\sum_{c\in C}\sum_{p-wn\in S} \alpha_{wn,c,p-wn}^{sep} F_{c,p-wn}^{total}$), and is stated as follows:

$$TotalConsume^{waste} = 100\frac{\sum_{wn\in W}\sum_{c\in C}\sum_{p-wn\in S} F_{wn,c,p-wn}}{\sum_{wn\in W}\sum_{c\in C}\sum_{p-wn\in S} \alpha_{wn,c,p-wn}^{sep} F_{c,p-wn}^{total}} \quad (6.27)$$

This is an important environmental issue because the recycled MSW positively impacts the environment and supports sustainable development.

6.3.16.3 Net profit objective function for each type of residue

The next objective functions consider the net profit for each type of recycled residue. Here, the net profit for each type of residue *wn* is equal to the benefit obtained from the sales of the residue *wn* minus the separation cost, processing cost, capital cost, and transportation cost from the processing plant to the cities for each residue *wn*:

$$\begin{aligned}Profit_{wn} = C_{wn}^{sales} - \text{Cos}\,tSep_{wn} - \text{Cos}\,tOp_{wn}^{processing} \\ - \text{Cos}\,tCap_{wn,pp-wn,n-wn,p-wn}^{processing} - \text{Cos}\,tTranp_{wn}^{pp\,city},\ \forall wn\end{aligned} \quad (6.28)$$

The objective function that takes the government expenditures into account is defined as follows:

$$Goverment\ expenditure = \text{Cos}\,tTransp_{pp}^{city} - \text{Cos}\,tNSep^{total} \quad (6.29)$$

6.3.17 Multi-stakeholder Formulation

Multi-criteria decision-making tools enable decision-makers to make informed decisions and achieve optimal results (Vucijak et al. 2016). This subject has been the reason for several publications (Soltani et al. 2016, Fuentes-Cortes et al. 2016, 2017, Dowling et al. 2016, Giri et al. 2016, González-Bravo et al. 2017). In this chapter, a multi-stakeholder approach based on a multi-objective optimization formulation was implemented and it is described as follows. First, the coordinates of the utopia point are used to represent the upper bound (UB) of the considered individual objective functions.

The lower bound (LB) represents the nadir solution for the different objective functions. For each objective function, a new variable (ρ^{obf}) is introduced. This new variable is the result of scaling the original objective function between the nadir and utopia points. Generally, ρ for each objective function is equal to the UB of the objective function minus the objective function value divided by the UB of the objective function minus the LB of the objective function for each objective:

$$\rho^{obf} = \frac{Obf^{UB} - Obf}{Obf^{UB} - Obf^{LB}} \tag{6.30}$$

6.3.17.1 Objective Functions

The objective functions for this work are the overall net profit, the percentage of MSW consumption, the government expenditures, and the net profit for plastics, metals, glass, paper, and un-recyclables.

6.3.17.1.1 Compromise Solution

Once the new variables ρ are defined, the compromise solution is stated as the sum of ρ for each objective function:

$$\begin{aligned} CS = {} & \rho^{NetPorofit} + \rho^{Consume} + \rho^{Profit_{plastics}} + \rho^{Profit_{metals}} \\ & + \rho^{Profit_{paper}} + \rho^{Profit_{glasses}} + \rho^{Profit_{no\ recyclable}} + \rho^{Govexpnd} \end{aligned} \tag{6.31}$$

Where the limits of each ρ are between one and zero because each objective function is scaled:

$$0 \le \rho^{Obf} \le 1 \tag{6.32}$$

The minimization of the CS implies that the individual ρ are also minimized, this means that the gap between the solution for each objective function and its upper bound should be reduced (i.e. each objective function will be as close as possible to its maximum profit), and there is no preference amongst the involved stakeholders.

6.3.17.1.2 Feasible Solution

In the FS, it is assumed that each objective function represents a different stakeholder and that these stakeholders have a different level of influence in the decision-making process, thus different weights are assigned for each one of them. If the FS is minimized, each of the individual objectives will try to reach its upper bound, however more importance will be given to the objectives with higher weights. In this scenario, several cases must be analyzed, *k* represents a given set of weights. Thus, for each configuration with different weights, it has an individual solution *(FS_k):*

$$\begin{aligned} FS_k = {} & w_k^{NetProfit} \rho_k^{NetProfit} + w_k^{Consume} \rho_k^{Consume} + w_k^{Profit_{plastics}} \rho_k^{Profit_{plastics}} \\ & + w_k^{Profit_{metals}} \rho_k^{Profit_{metals}} + w_k^{Profit_{paper}} \rho_k^{Profit_{paper}} + w_k^{Profit_{glasses}} \rho_k^{Profit_{glasses}} \quad \forall k \in K \\ & + w_k^{Profit_{no\ recyclables}} \rho_k^{Profit_{no\ recyclables}} + w_k^{Govexpend} \rho_k^{Govexpend} \end{aligned} \tag{6.33}$$

This way, the average feasible solution (AFS) is stated as follows:

$$AFS = \frac{\sum_{k=1}^{k} FS_k}{k} \tag{6.34}$$

The optimization formulation looks for the average of these solutions, attempting to satisfy all stakeholders on average in all potential scenarios. The AFS reflects all the various combinations of preferences or levels of influence that the stakeholders might have in the process.

In this chapter, the Latin Hypercube Sampling technique with a normal distribution was used to generate random weights for the preferences of the different stakeholders (Table A.6.1).

The resulting model is a mixed integer linear programming (MIP) model, which was coded and solved in the optimization platform GAMS® using the global solver BARON®.

Table A.6.1. Weights generated using LHS for each objective function. [Printed (adapted) with permission from Diaz-Barriga-Fernandez, A.D., Santibañez-Aguilar, J.E., Radwan, N., Nápoles-Rivera, F., El-Halwagi, M., and Ponce-Ortega, J.M. 2017. Strategic planning for managing municipal solid wastes with consideration of multiple stakeholders. *ACS Sustainable Chemistry and Engineering*, 2017, *5*(11), 10744–10762].

Case	$w^{NetProfit}$	$w^{Consume}$	$w^{Profit}_{plastics}$	w^{Profit}_{metals}	W^{Profit}_{glass}	W^{Profit}_{paper}	$w^{Profitun-}_{-recyclables}$	$w^{Profit}_{government\ expenditures}$
1	0.0855	0.134	0.1024	0.2063	0.1565	0.2261	0.0494	0.0399
2	0.0396	0.1453	0.113	0.1803	0.1534	0.1869	0.0914	0.0900
3	0.0396	0.1521	0.0955	0.1818	0.2712	0.0207	0.1167	0.1224
4	0.0828	0.1607	0.1058	0.1818	0.1944	0.0241	0.1205	0.1299
5	0.2521	0.0265	0.1701	0.0167	0.1049	0.1342	0.2136	0.0819
6	0.2184	0.1335	0.1509	0.0271	0.0835	0.1241	0.1806	0.0818
7	0.0397	0.2397	0.0786	0.1223	0.204	0.1736	0.0385	0.1034
8	0.0749	0.3244	0.056	0.1331	0.1832	0.1342	0.0232	0.0709
9	0.1665	0.0225	0.1981	0.0479	0.1279	0.3015	0.0203	0.1153
10	0.1584	0.054	0.2013	0.0454	0.1573	0.2234	0.0328	0.1275

Table A.6.2. Generated waste and sub-waste for each city and efficiency for recollection. [Printed (adapted) with permission from Diaz-Barriga-Fernandez, A.D., Santibañez-Aguilar, J.E., Radwan, N., Nápoles-Rivera, F., El-Halwagi, M., and Ponce-Ortega, J.M. 2017. Strategic planning for managing municipal solid wastes with consideration of multiple stakeholders. *ACS Sustainable Chemistry and Engineering*, 2017, *5*(11), 10744–10762].

City		**Morelia (10^{-7})[a]**	**Celaya (10^{-7})[a]**	**Apatzingan (10^{-7})[a]**	**Lazaro Cardenas(10^{-7})[a]**	**Leon(10^{-7})[a]**	**Efficiency for recollection**
MSW	**Total production**	43.50	20.0	5.49	9.720	52.30	
	Non-recyclables (70.3% MSW)	30.60	14.10	3.84	6.830	36.80	10
Plastic	**Total production**	2.520	1.160	0.316	0.564	3.030	
	HDPE(t)	0.269	0.124	0.034	6.020	0.324	6.4
	PE(t)	0.630	0.291	0.079	0.141	0.759	6.4
	PET(t)	3.680	0.017	0.004	0.001	0.004	6.4
	PP(t)	0.937	0.432	0.118	0.209	1.130	6.4
	PS(t)	0.133	0.062	0.017	29.80	0.161	6.4
Metal	**Total production**	1.350	0.062	0.169	0.301	1.620	
	Aluminum (t)	0.082	0.038	1.030	0.018	0.099	35.8
Glass	**Total production**	2.870	1.320	0.360	0.642	3.450	
	Clear glass (t)	0.521	0.240	6.540	0.117	0.627	24.7
	Green glass (t)	0.610	0.281	0.076	0.136	0.734	4.14
	Brown glass (t)	1.740	0.801	0.218	0.389	2.090	1.81
Paper	**Total production**	6.170	2.850	0.775	1.380	7.430	
	Paper useful	2.580	1.190	0.323	0.576	3.100	7.16

6.4 Case Studies

Five cities in the central-west area of Mexico were considered as sources of MSW and potential consumers of the recycled products from the treatment of the residues. Fifteen technologies were considered to treat MSW. The amount of residues produced by the type of residue from the different cities is shown in Table A.6.2 (EPA 2012, Mexican National Institute of Ecology 2013).

The transportation costs for products and waste as well as the distances between the considered places are shown in Table A.6.3 (Mexican Ministry of Transportation and Communications 2012). The separation costs for each type of residue are shown in Table A.6.4 (Mexican National Institute of Ecology 2013).

Based on this information, three different scenarios for MSW management were analyzed in this work to prove the applicability of the presented methodology, considering as stakeholders the plastic, metal, glass, and paper recycling industries, as well as the waste-to-energy industry and the government. These analyzed scenarios are described as follows.

6.4.1 Scenario 1

Scenario 1 implies the separation of the MSW as a responsibility of each industry. Here, the ONP considers the government expenditures as the costs to produce products from the residues. Similarly, the profit from the sales of those products is considered.

Table A.6.3. Distance from cities to processing facilities and unit transport costs. [Printed (adapted) with permission from Diaz-Barriga-Fernandez, A.D., Santibañez-Aguilar, J.E., Radwan, N., Nápoles-Rivera, F., El-Halwagi, M., and Ponce-Ortega, J.M. 2017. Strategic planning for managing municipal solid wastes with consideration of multiple stakeholders. *ACS Sustainable Chemistry and Engineering*, 2017, 5(11), 10744–10762].

Cities	Distances to processing facilities located in cities (km)				
	Morelia	Celaya	Apatzingan	Lazaro Cardenas	Leon
Morelia	1	147	192	144	207
Celaya	147	1	326	447	135
Apatzingan	192	326	1	186	393
Lazaro Cardenas	144	447	186	1	514
Leon	207	135	393	514	1
Unit transportation cost ($US/kg km)	1.65×10^{-5}[a]	4.0010^{-5}[b]			

[a] For plastic, MSW, paper.
[b] Metal and glass.

Table A.6.4. Unitary separation cost for type of residue. [Printed (adapted) with permission from Diaz-Barriga-Fernandez, A.D., Santibañez-Aguilar, J.E., Radwan, N., Nápoles-Rivera, F., El-Halwagi, M., and Ponce-Ortega, J.M. 2017. Strategic planning for managing municipal solid wastes with consideration of multiple stakeholders. *ACS Sustainable Chemistry and Engineering*, 2017, 5(11), 10744–10762].

Type of residue	Unitary cost for separation ($US/kg)
Plastic	0.2235
Metal	0.02
Glass	0.0008
Paper	0.04
Un- recyclables	0.026

6.4.2 Scenario 2

For Scenario 2, the government covers the cost associated with the separation of residues. This way, the profit for production facilities for each product is greater. The main idea of this assumption is that if the government takes care of the separation cost, the recycling of different materials will be more attractive. Thus, the investment of different companies that work in the recycling industry will increase and therefore less garbage will end up in the dumps.

6.4.3 Scenario 3

Scenario 3 involves mandating the separation of MSW at the household sources. Hence, the separation costs are not needed. This is an interesting scenario because there could be several laws that obligate the people to separate their residues, then the proposed approach can be helpful to identify the optimal solutions under this assumption.

6.5 Results

The objective functions are optimized individually to establish the upper and lower bounds for the optimization, then these results are used for scaling the objective functions. The results are shown in Table A.6.5.

Table A.6.5. Upper and Lower bonds for the different objective function for each scenario considered. [Printed (adapted) with permission from Diaz-Barriga-Fernandez, A.D., Santibañez-Aguilar, J.E., Radwan, N., Nápoles-Rivera, F., El-Halwagi, M., and Ponce-Ortega, J.M. 2017. Strategic planning for managing municipal solid wastes with consideration of multiple stakeholders. *ACS Sustainable Chemistry and Engineering*, 2017, *5*(11), 10744–10762].

Objective function	SCENARIO 1		SCENARIO 2		SCENARIO 3	
	UB	LB	UB	LB	UB	LB
Consumed MSW (%)	99.96	0	99.96	0	99.96	0
ONP(MM$/y)	63.18	−33.11	69.41	−19.65	74.05	−14.18
Net profit for plastics (MM$/y)	0	−6.09	0	−1.589	0	−5.196
Net profit for metals(MM$/y)	0	−17.79	1.73	0	1.730	0
Net profit for glass(MM$/y)	0	−5.70	0	−5.553	0	−5.448
Net profit for paper(MM$/y)	2.34	−17.76	9.064	0	9.065	0
Net profit for un- recyclables(MM$/y)	73.53	0	75.92	0	75.92	0
Government expenditures(MM$/y)	−12.05	−14.18	−14.09	−18.48	−12.05	−14.18

Table A.6.6. Results for different solutions under Scenario 1. [Printed (adapted) with permission from Diaz-Barriga-Fernandez, A.D., Santibañez-Aguilar, J.E., Radwan, N., Nápoles-Rivera, F., El-Halwagi, M., and Ponce-Ortega, J.M. 2017. Strategic planning for managing municipal solid wastes with consideration of multiple stakeholders. *ACS Sustainable Chemistry and Engineering*, 2017, 5(11), 10744–10762].

Objective Function/Concept	Maximizing the ONP($MM/y)	Minimizing the CS ($MM/y)	Minimizing the AFS($MM/y)
ONP	63.189	62.842	62.725
% Consumed MSW	85.17%	95.02%	95.02%
%Glass consumed	0%	81.5%	81.5%
%Paper consumed	100%	100%	100%
%Un- recyclables consumed	100%	100%	100%
Net profit for plastics	0	0	0
Net profit for metals	0	0	0
Net profit for glass	0	−0.441	−0.499
Net profit for paper	2.327	2.326	2.274
Net profit for un- recyclables	73.528	73.528	73.528
Government expenditures	−12.666	−12.570	12.578
CS	0.436	0.374	0.392
AFS	0.048	0.043	0.046

6.5.1 Scenario 1

The solution for the optimization model under the conditions of Scenario 1 is shown in Fig. 3. In this case, the separation cost is assigned to different processing plants and the government is only responsible for recollecting the MSW from the households and distributing it to the different processing plants as well as treating the part of the MSW that is not recoverable. The distribution obtained when the ONP is maximized is shown in Fig. 3. The paper residues and nonrecyclables are collected from all the cities considered in the model. After being collected, the paper is taken to a processing facility in the city of Morelia, where it is recycled and processed. The paper is delivered to the five cities that are regarded as marketplaces once it is prepared for reuse. The processing facility in Morelia receives the un-recyclables, and the energy generated there is dispersed throughout the city.

For the case when the CS is minimized, the residues collected from the five cities were glass, paper, and un-recyclables. Two types of glass were processed, green and brown glass. Both types of glass are processed in two different plants located in the city of Morelia. Then, products made from the two types of glass are distributed to Morelia. The collected paper is then sent to Morelia to be processed. Furthermore, the produced paper is distributed to Morelia, Celaya, Lazaro Cardenas, and Leon (Fig. 4).

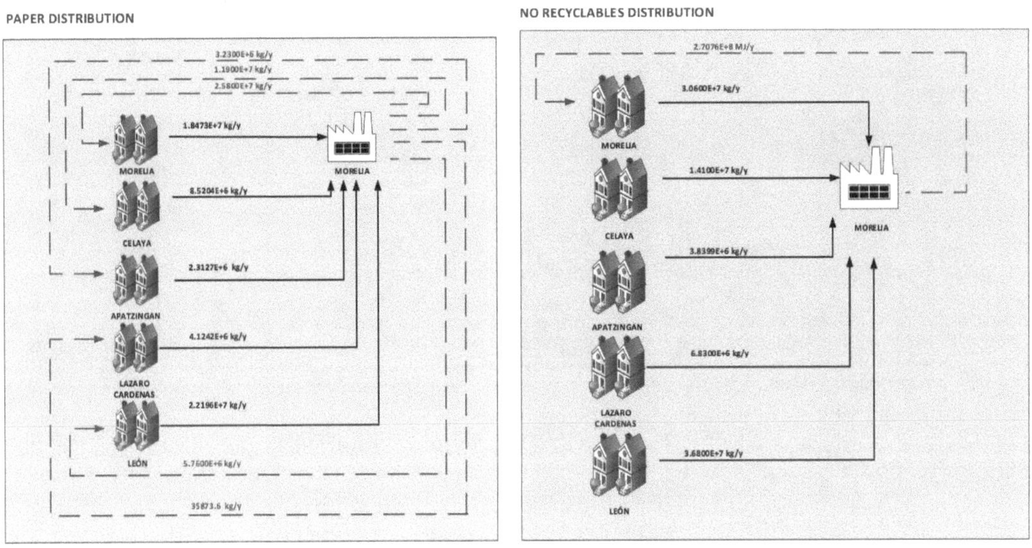

Fig. 3. Distribution for Scenario 1 obtained by maximizing the ONP. [Printed (adapted) with permission from Diaz-Barriga-Fernandez, A.D., Santibañez-Aguilar, J.E., Radwan, N., Nápoles-Rivera, F., El-Halwagi, M., and Ponce-Ortega, J.M. 2017. Strategic planning for managing municipal solid wastes with consideration of multiple stakeholders. *ACS Sustainable Chemistry and Engineering*, 2017, *5*(11), 10744–10762].

Finally, when the AFS is maximized, three types of MSW are collected from the five cities. The distribution of glass is like the case of the CS solution, the main difference is that the green glass processing plant is in Leon. The products elaborated from the green glass are distributed also in Leon. The collected paper is sent to a processing plant located in Morelia. Then, the recycled paper is distributed to Morelia and Leon. The un- recyclables are treated in a processing plant located in Morelia; the energy produced from them is consumed in the same city (Fig. 5).

The main goal of this work is to find the best solution where all the objective functions approach their maximum (or minimum in the case of government expenditures). The results obtained by maximizing the ONP and by minimizing the CS and AFS are shown in Table A.6.6. The CS has similar values that the AFS solution for all the objective functions considered. However, the results for the CS solution are better than the solutions for the ONP and AFS in terms of stakeholders' satisfaction.

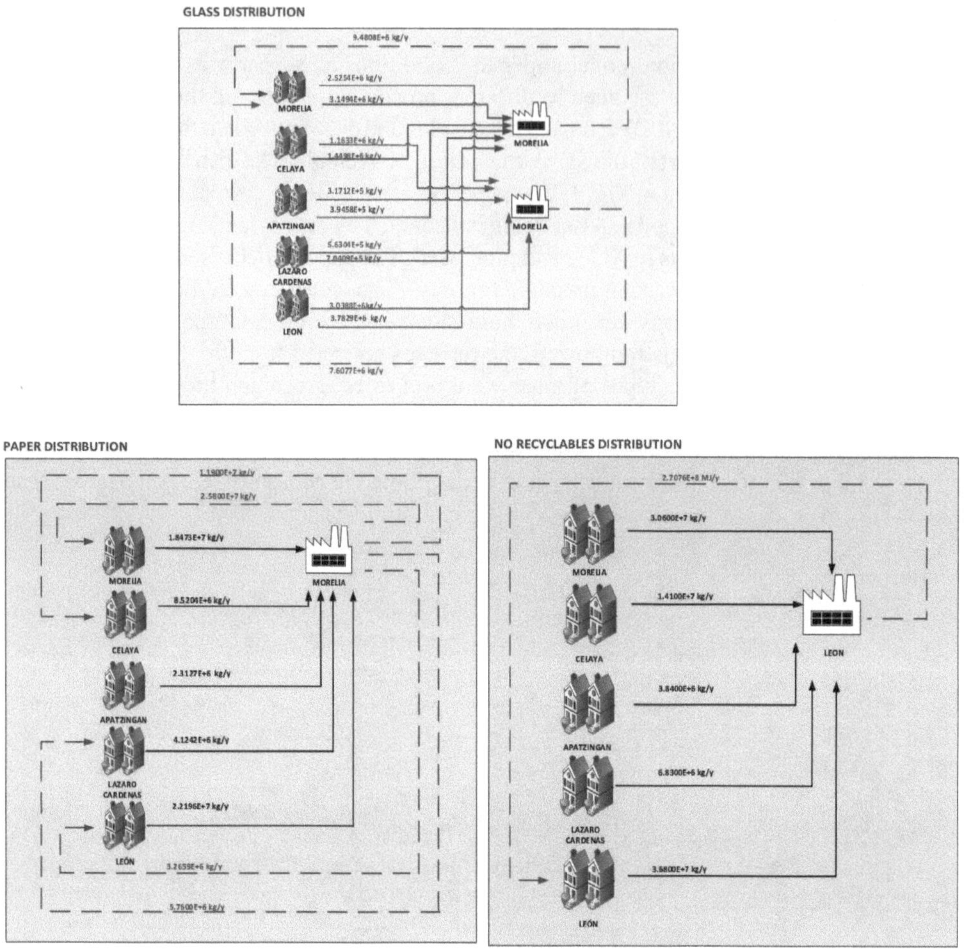

Fig. 4. Distribution for Scenario 1 obtained by minimizing the CS. [Printed (adapted) with permission from Diaz-Barriga-Fernandez, A.D., Santibañez-Aguilar, J.E., Radwan, N., Nápoles-Rivera, F., El-Halwagi, M., and Ponce-Ortega, J.M. 2017. Strategic planning for managing municipal solid wastes with consideration of multiple stakeholders. *ACS Sustainable Chemistry and Engineering*, 2017, *5*(11), 10744–10762].

This means that in the case of minimizing the CS solution, the values of the objective functions are closer to their maximum than in the ONP and AFS solutions. In Fig. 6, since the difference between the CS and the AFS for the consumption of MSW was 4.94%, the difference between the consumption of MSW and its upper bound is 14.8% when the ONP is maximized, and it is 5.10% for government expenditures (the worst value of the three cases). The percentages for the rest of the objective function were negligible except for the net profit for paper when the AFS is minimized, in which the difference with its upper bound is 2.8%, this might be the result of the weights given to the objectives. Thus, the CS yields a better compromise between all the objectives.

6.5.2 Scenario 2

Scenario 2 considered that the government is responsible for the separation cost of every residue. The results obtained for minimizing the ONP, the CS and the AFS are shown in Table A.6.7.

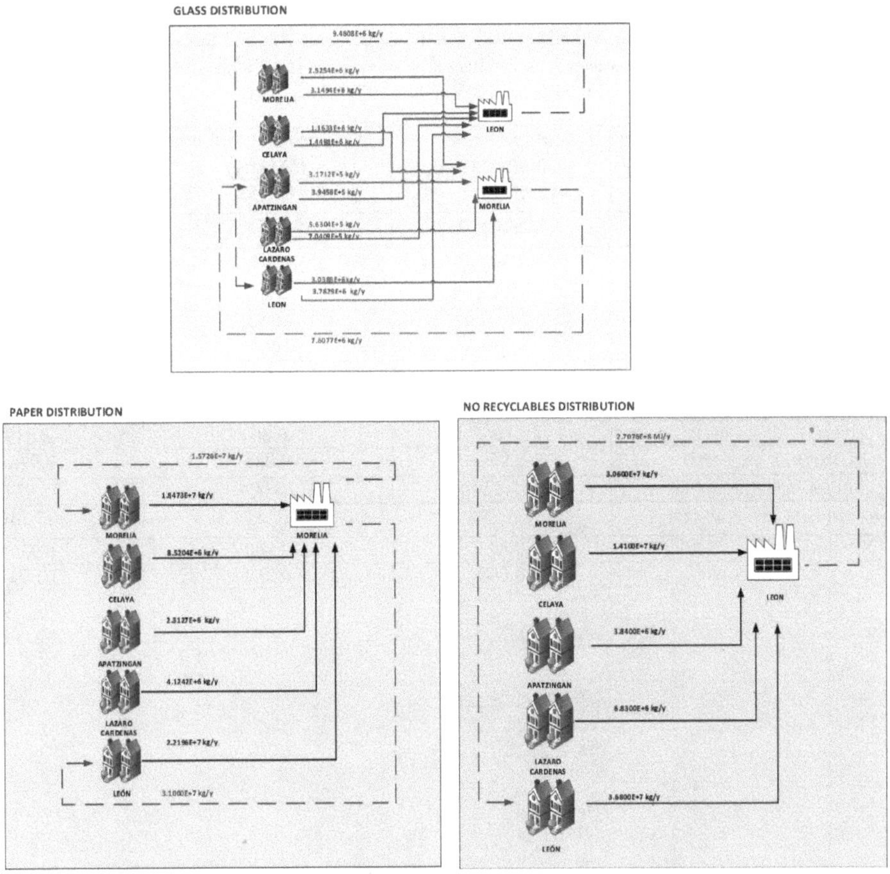

Fig. 5. Distribution for Scenario 1 by minimizing the AFS. [Printed (adapted) with permission from Diaz-Barriga-Fernandez, A.D., Santibañez-Aguilar, J.E., Radwan, N., Nápoles-Rivera, F., El-Halwagi, M., and Ponce-Ortega, J.M. 2017. Strategic planning for managing municipal solid wastes with consideration of multiple stakeholders. *ACS Sustainable Chemistry and Engineering*, 2017, *5*(11), 10744–10762].

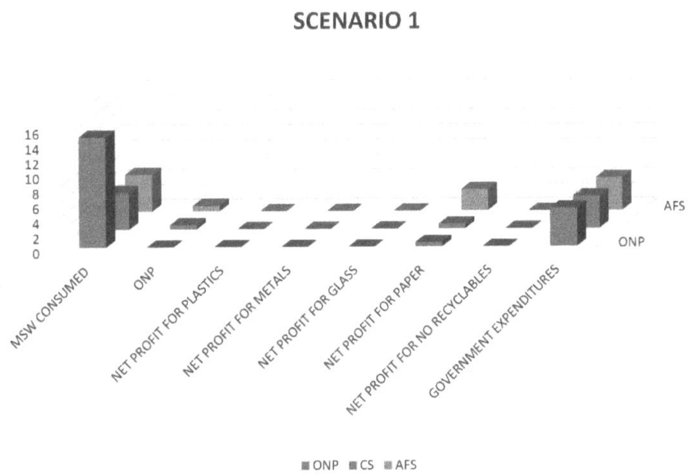

Fig. 6. Percentage difference between individual objective functions and their individual UB for Scenario 1. [Printed (adapted) with permission from Diaz-Barriga-Fernandez, A.D., Santibañez-Aguilar, J.E., Radwan, N., Nápoles-Rivera, F., El-Halwagi, M., and Ponce-Ortega, J.M. 2017. Strategic planning for managing municipal solid wastes with consideration of multiple stakeholders. *ACS Sustainable Chemistry and Engineering*, 2017, *5*(11), 10744–10762].

Table A.6.7. Results for different solutions for Scenario 2. [Printed (adapted) with permission from Diaz-Barriga-Fernandez, A.D., Santibañez-Aguilar, J.E., Radwan, N., Nápoles-Rivera, F., El-Halwagi, M., and Ponce-Ortega, J.M. 2017. Strategic planning for managing municipal solid wastes with consideration of multiple stakeholders. *ACS Sustainable Chemistry and Engineering*, 2017, *5*(11), 10744–10762].

Objective Function/Concept	Maximizing ONP($MM/y)	Compromise Solution ($MM/y)	Average Feasible Solution ($MM/y)
ONP	69.407	68.158	68.991
%Consume MSW	85.685%	95.533%	87.921%
%Consumed Metals	100%	100%	100%
%Consumed Glass	100%	100%	100%
%Consumed Paper	0%	81.5%	18.5%
%Consumed Un- Recyclables	100%	100%	100%
Net profit for the plastics	0	0	0
Net profit for the metals	1.729	1.730	1.727
Net profit for the glass	0	-0.306	–0.319
Net profit for the paper	9.052	9.046	8.998
Net profit for the un- recyclables	75.924	75.004	75.924
Government expenditures	–17.298	–17.317	–17.339
CS	0.873	0.860	0.930
AFS	0.090	0.089	0.099

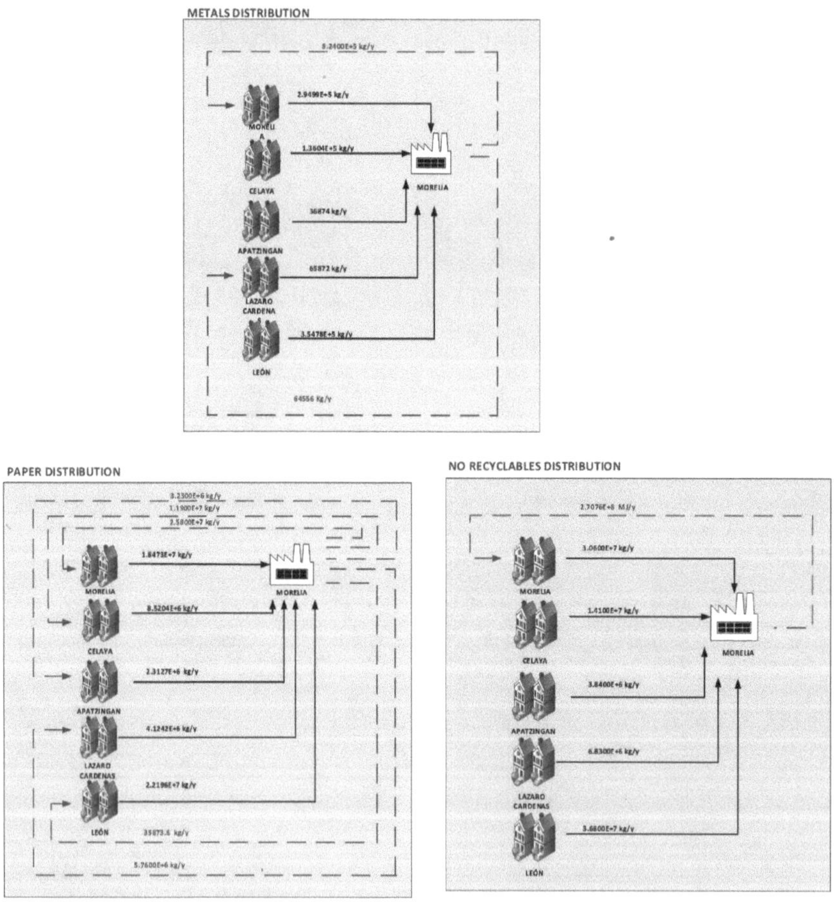

Fig. 7. Distribution for Scenario 2 obtained by maximizing the ONP. [Printed (adapted) with permission from Diaz-Barriga-Fernandez, A.D., Santibañez-Aguilar, J.E., Radwan, N., Nápoles-Rivera, F., El-Halwagi, M., and Ponce-Ortega, J.M. 2017. Strategic planning for managing municipal solid wastes with consideration of multiple stakeholders. *ACS Sustainable Chemistry and Engineering*, 2017, *5*(11), 10744–10762].

When the ONP is maximized, the total consumed MSW is 85.68% of the maximum available, and the main recycled residues are metals, paper, and un- recyclables, which correspond to 100% for both. The obtained profits from the sales of each type of produced material are 1.729$MM/y, 9052$MM/y and 75.924$MM/y, respectively. Then, the net profit is 69.407$MM/year, while the government expenditure needed is –17.298$MM/y, the value is greater than in one of Scenario 1, because in this case, the stakeholder (Government) absorbed the separation cost for each treated residue.

For the case when the CS is minimized, the results indicate a total consumption of the MSW of 95.53%, which is greater than the one in the previous case. The recycled materials include metals, glass, and un-recyclables. The individual percentages of these recycled materials are 18.5%, 100%, and 100%, respectively. Additionally, the net profit obtained for this percentage is smaller for glass, paper, and un-recyclables, while the profit for the metals is greater than the one in the previous case. These profits represent 1.730$MM/y for metals, -0.306$MM/y for glass, 9.046$MM/year for paper, and for the un- recyclables it is 75.004$MM/y.

For the AFS solution, the total consumption of MSW was 87.92%. The main recycled residues were metals, glass, paper, and non-recyclables with 100%, 18.5% 100% and 100%, respectively. The total recycled materials were lower than the ones of the previous case. However, the ONP increased to $ 68,991/y, which is the mean value of the three cases. The government expenditures were -17,339M $/y, which is the worst value for this objective function in Scenario 2. Besides, the net profit obtained from each recycled material from the residues is $ 1,727/y for metals, -0.319 $ MM/year for glass, $ 8,998 MM/year for paper, and $ 75,924 MM/y for non-recyclables. Here, the un-recyclables obtain a greater profit at the expense of the other residues.

The distribution network of the recycled residues for the case when the ONP is maximized is shown in Fig. 7. In this case, the three types of residues are collected from the five cities. The metals are processed in a processing plant located in Morelia, then the recycled metals are distributed to Morelia and Lazaro Cardenas. The collected paper is sent to a processing plant located in Morelia. This plant distributes the recycled paper to the five cities considered as markets. The un-recyclables are processed in Morelia and the energy produced is distributed in the same city.

Evaluating the case when the CS is maximized, the four types of residues were collected from the five cities. The metals are sent to a processing plant located in Leon. The products made from metals are distributed in the same city. Two types of glass are recycled, green and brown glass. In the processing plant located in Morelia, the green glass is processed, while in Leon the brown glass is processed. The products obtained from the two types of glass are distributed in the same cities where they were produced, respectively. The collected paper is treated in the processing plant located in Lazaro Cardenas. The recycled paper is distributed in four cities (Morelia, Celaya, Apatzingan, and Leon). The un- recyclables are treated in a processing plant located in Lazaro Cardenas and the generated energy is distributed in Leon (Fig. 8).

Analyzing the distribution of the residues when the AFS was minimized in Scenario 2 (Fig. 9), it was noted that four types of residues were collected from the five cities. The metals are processed in Morelia and the products obtained from them are distributed in Morelia, Celaya, and Apatzingan. Only clear glass is recycled in a processing plant located in Morelia. The products obtained from clear glass are distributed in Leon. The collected paper is sent to a processing plant located in Morelia. The recycled paper is distributed in Morelia and Leon. In the processing plant located in Apatzingan, un-recyclables are treated. The energy produced from the un-recyclables is consumed in Celaya.

The percentages for each objective function difference concerning its individual UB for scenario 2 are shown in Fig. 10. The percentage of deviation for the consumed MSW when the ONP is maximized is 14.2% and for the government expenditures are 22.7%. In the case when CS was minimized, the consumption of MSW, the net profit, the profit of un -recyclables, and the government expenditures present a deviation of 4.4%, 1.8%, 1.2%, and 22.8% respectively. For the last case when the AFS is minimized, the values are like the case when the ONP is maximized. Although there are four objectives that deviate from their UB when minimizing the CS as opposed to two when reducing the ONP, the

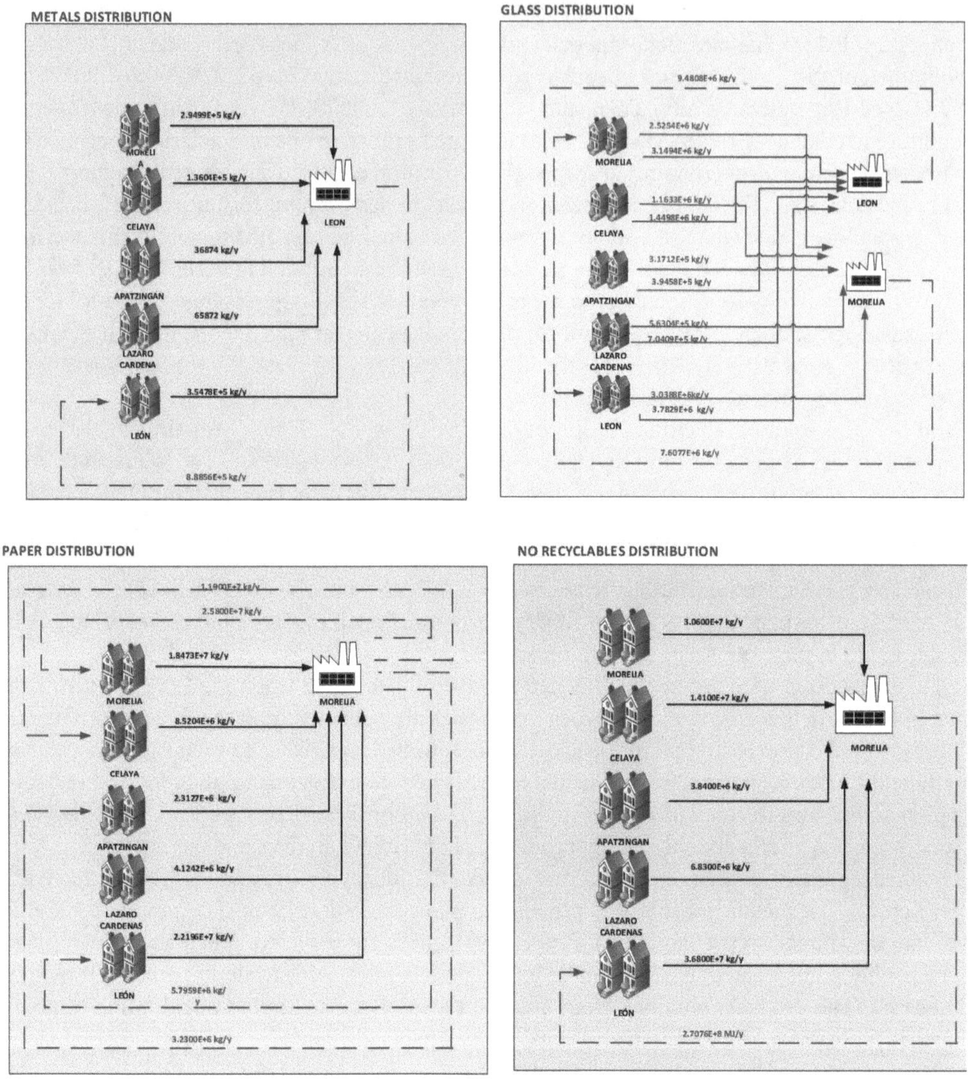

Fig. 8. Distribution of glass for Scenario 2 obtained by maximizing the CS. [Printed (adapted) with permission from Diaz-Barriga-Fernandez, A.D., Santibañez-Aguilar, J.E., Radwan, N., Nápoles-Rivera, F., El-Halwagi, M., and Ponce-Ortega, J.M. 2017. Strategic planning for managing municipal solid wastes with consideration of multiple stakeholders. *ACS Sustainable Chemistry and Engineering*, 2017, *5*(11), 10744–10762].

aggregate of the variations is smaller, meaning that the CS also produces a better compromise for every stakeholder involved.

6.5.3 Scenario 3

In this scenario, the separation costs are not considered, the main assumption is that the separation of the different residues is made at the houses and buildings where they are generated. The results for the ONP, CS and AFS solutions are shown in Table A.6.8. For the first case, when the ONP is maximized, the total consumed MSW was 85.68%. The main residues reused in this case were metals, paper, and un-recyclables. The percentage of each residue reused was 100%. The solution for the ONP was greater than the ones of the previous scenarios, with a value of 74.04$MM/y and the government expenditures decreased to -12.659$MM/y.

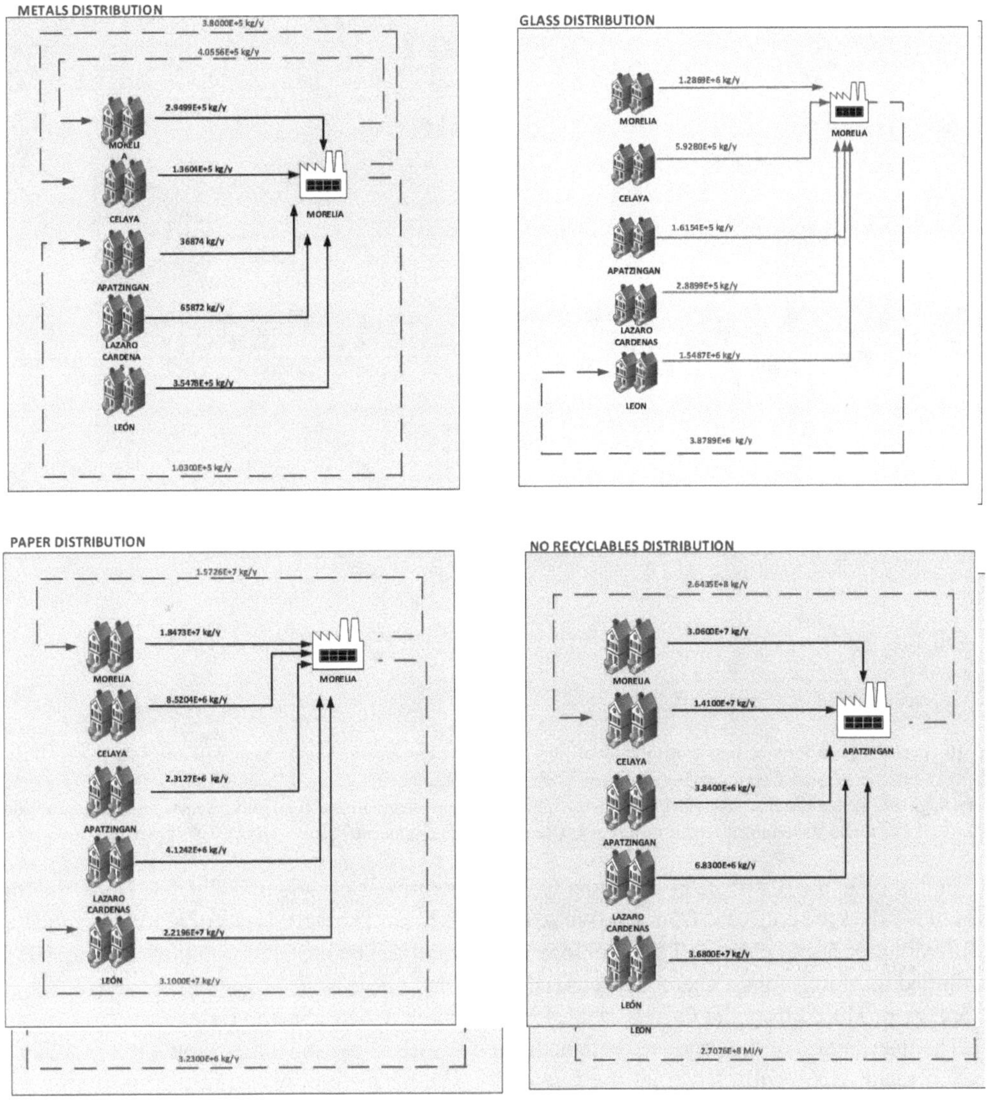

Fig. 9. Distribution for Scenario 2 obtained by maximizing the AFS. [Printed (adapted) with permission from Diaz-Barriga-Fernandez, A.D., Santibañez-Aguilar, J.E., Radwan, N., Nápoles-Rivera, F., El-Halwagi, M., and Ponce-Ortega, J.M. 2017. Strategic planning for managing municipal solid wastes with consideration of multiple stakeholders. ACS Sustainable Chemistry and Engineering, 2017, *5*(11), 10744–10762].

Analyzing the case when the CS was minimized, the total consumed MSW was 95.53%. The main reused residues were glass, paper, and un- recyclables, each one with a percentage of 100%, 81.5%, and 100%, respectively. In this case, the ONP was 73.861$MM/y, which is lower than the one in the previous case. However, in this case, glass was consumed, and the government expenditures decreased to -12.533$MM/y.

For the case when the AFS was minimized, the total consumed MSW was 95.53%. The reused residues were glass, paper, and un recyclables. The used percentage of these residues was the same as in the previous case. The ONP was the lowest of the three cases with 73.819$MM/y. Nevertheless, in this case, various residues were reused, although this increases the government expenditures to -12.533$MM/y, which is greater than the one of the previous cases.

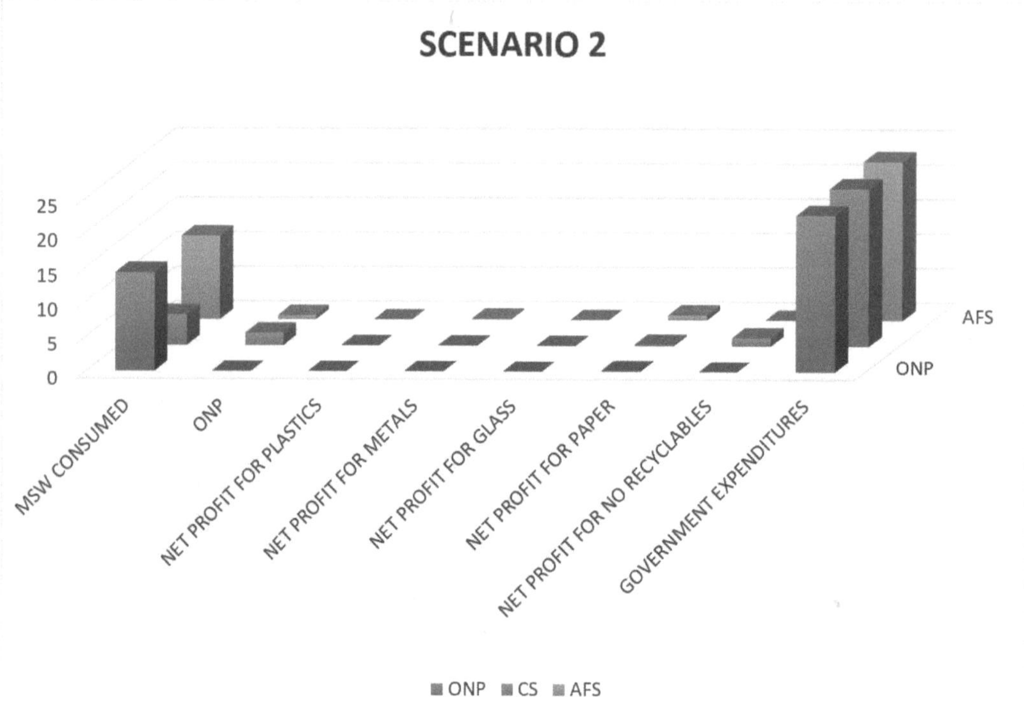

Fig. 10. Percentage difference between individual objective functions and their individual UB for Scenario 2. [Printed (adapted) with permission from Diaz-Barriga-Fernandez, A.D., Santibañez-Aguilar, J.E., Radwan, N., Nápoles-Rivera, F., El-Halwagi, M., Ponce-Ortega, J.M. (2017). Strategic planning for managing municipal solid wastes with consideration of multiple stakeholders. *ACS Sustainable Chemistry and Engineering*, 2017, *5*(11), 10744–10762].

In this case, the distribution when the ONP is maximized is shown in Fig. 11. Here, the three types of residues are collected from the five considered cities. The three types of MSW are processed in different processing plants, all of them located in Morelia. The products obtained from metals are distributed to Morelia and Lazaro Cardenas, the recycled paper is distributed to the five cities, and the energy produced from the un-recyclables is distributed to Morelia and Celaya.

The distribution of the products obtained when the CS is minimized is shown in Fig. 12. Four types of residues are collected from the five cities. The metals are treated in the processing plant located in Leon and the products obtained from them are distributed in the same city. Only green and brown glass are recycled, both are processed in Morelia in different plants. The two types of products produced from the two types of glass are distributed in Morelia. The collected paper is processed in a plant in Morelia. The recycled paper is then distributed in Morelia, Celaya, and Leon. The un- recyclables are treated in Lazaro Cardenas and the energy produced is distributed in Morelia.

For the case when the AFS is minimized, the four types of reused residues are collected from the five cities. In a processing plant located in Morelia, metals are processed, and the main products obtained from them are distributed to four cities (Celaya, Apatzingan, Lazaro Cardenas, and Leon). Two types of glass (green and brown glass) are treated in Morelia in two different plants, the green recycled glass is distributed to Lazaro Cardenas, while the brown glass is distributed to Morelia. The collected paper is sent to a processing plant located in Celaya. The recycled paper is distributed in the five cities involved. In a processing plant located in Lazaro Cardenas, there are processed un-recyclables, and the generated energy is distributed in Morelia (Fig. 13).

Fig. 14 shows the percentage difference between the obtained objective and the upper bound. Under these conditions, the minimization of CS and AFS have similar results, with a marginal improvement for using the CS.

Table A.6.8. Results for different solutions for Scenario 3. [Printed (adapted) with permission from Diaz-Barriga-Fernandez, A.D., Santibañez-Aguilar, J.E., Radwan, N., Nápoles-Rivera, F., El-Halwagi, M., and Ponce-Ortega, J.M. 2017. Strategic planning for managing municipal solid wastes with consideration of multiple stakeholders. *ACS Sustainable Chemistry and Engineering*, 2017, 5(11), 10744–10762].

Objective Function/Concept	Maximizing ONP($MM/y)	Minimizing CS($MM/y)	Minimizing AFS ($MM/y)
ONP	74.047	73.861	73.819
%Consumed MSW	85.685%	95.533%	95.533%
%Consumed Metals	100%	100%	100%
%Consumed Glass	0%	81.5%	81.5%
%Consumed Paper	100%	100%	100%
%Consumed Un- Recyclables	100%	100%	100%
Net profit for the plastics	0	0	0
Net profit for the metals	1.729	1.729	1.724
Net profit for the glass	0	–0.304	–0.348
Net profit for the paper	9.052	9.045	9.052
Net profit for the un- recyclables	75.924	75.924	75.924
Government expenditures	–12.659	–12.533	–12.533
CS	0.429	0.337	0.341
AFS	0.048	0.038	0.039

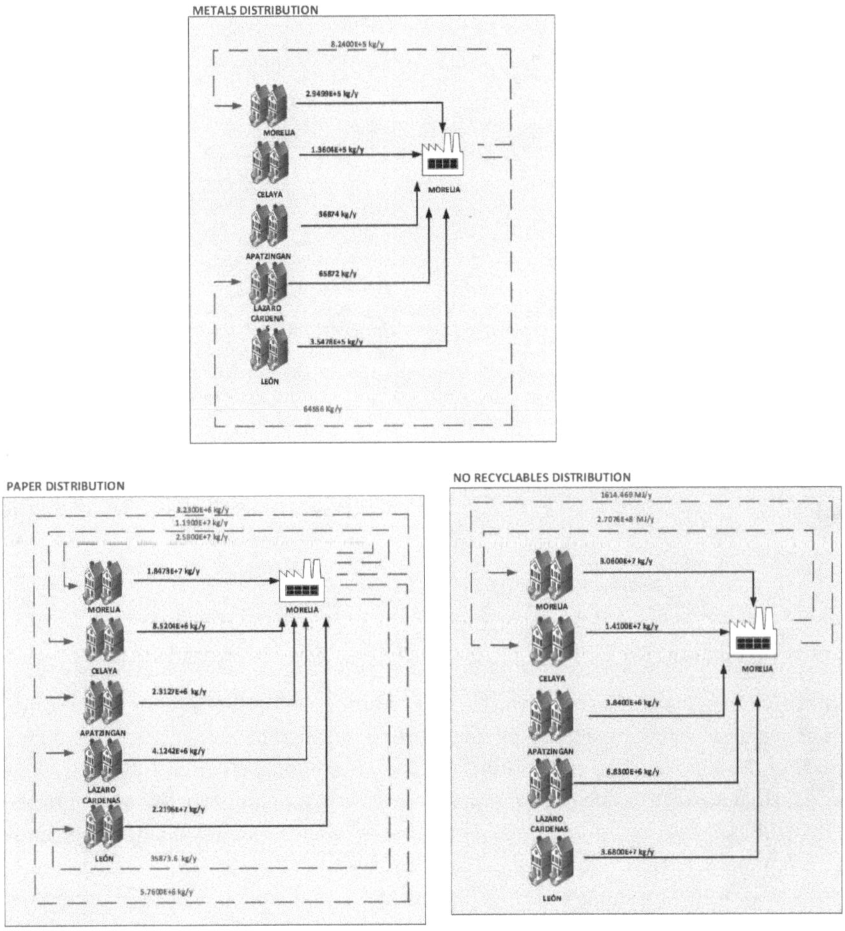

Fig. 11. Distribution for Scenario 3 obtained by maximizing the ONP. [Printed (adapted) with permission from Diaz-Barriga-Fernandez, A.D., Santibañez-Aguilar, J.E., Radwan, N., Nápoles-Rivera, F., El-Halwagi, M., and Ponce-Ortega, J.M. 2017. Strategic planning for managing municipal solid wastes with consideration of multiple stakeholders. *ACS Sustainable Chemistry and Engineering*, 2017, 5(11), 10744–10762].

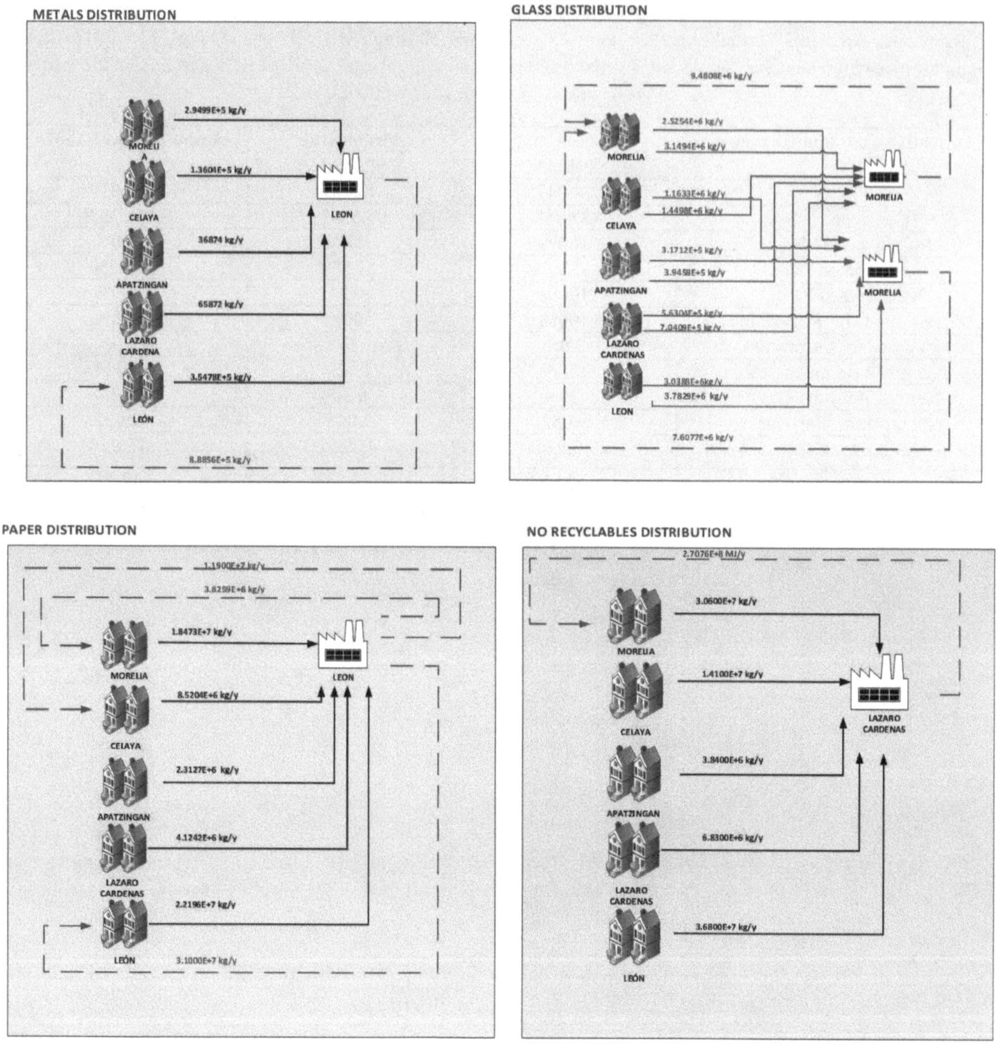

Fig. 12. Distribution for Scenario 3 obtained by maximizing the CS. [Printed (adapted) with permission from Diaz-Barriga-Fernandez, A.D., Santibañez-Aguilar, J.E., Radwan, N., Nápoles-Rivera, F., El-Halwagi, M., and Ponce-Ortega, J.M. 2017. Strategic planning for managing municipal solid wastes with consideration of multiple stakeholders. *ACS Sustainable Chemistry and Engineering*, 2017, *5*(11), 10744–10762].

6.6 Concluding Remarks

This chapter has presented a novel optimization approach for the strategic planning of MSW management system. The proposed model accounts for all the activities associated with the supply chain including the recollection, transportation separation, and processing of residues to produce useful products that are distributed to different markets. A mathematical optimization model has been proposed to model all these activities and to determine the optimal supply chain design and operation, including the optimal selection of processing facilities as well as their location. To account for the profit of the various stakeholders, a multi-objective optimization approach has been developed to reconcile the benefits of all involved parts (i.e., government, suppliers, consumers, and treatment providers).

A case study was employed for a region in Mexico involving five neighboring cities. The problem was analyzed under different scenarios and the solutions for the proposed multi-stakeholder approach showed for all the scenarios considered that the best solution was obtained when the CS was minimized,

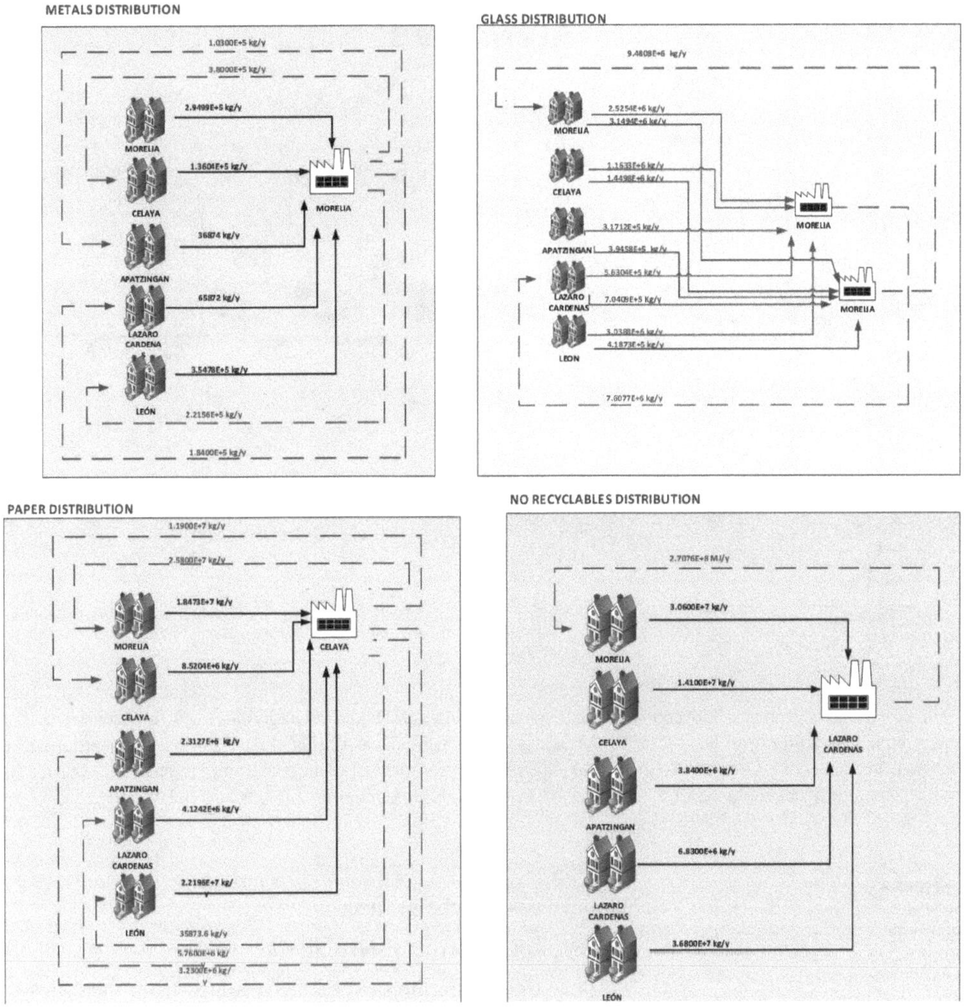

Fig. 13. Distribution for Scenario 3 obtained by maximizing the AFS. [Printed (adapted) with permission from Diaz-Barriga-Fernandez, A.D., Santibañez-Aguilar, J.E., Radwan, N., Nápoles-Rivera, F., El-Halwagi, M., and Ponce-Ortega, J.M. 2017. Strategic planning for managing municipal solid wastes with consideration of multiple stakeholders. *ACS Sustainable Chemistry and Engineering*, 2017, 5(11), 10744–10762].

which yields a better compromise between the considered objectives. Although a particular case study was employed and solved, the proposed approach is general and it can be used for planning different MSW management systems in different and broader regions considering multiple stakeholders, such as recycling industry owners, the government, or any other actor involved in the decisions of MSW management. In future works, the presented model can also be extended to consider the associated uncertainty (feedstock availability, and feedstock and product costs) and the generation of dynamic models to consider the dependence of the variables over time.

6.7 Nomenclature

6.7.1 Parameters

$C_{wn,pp-wn,n-wn,p-wn}^{processing}$ Unitary cost for processing the residue *wn* with the technology *n-wn* to produce a product *p-wn*.

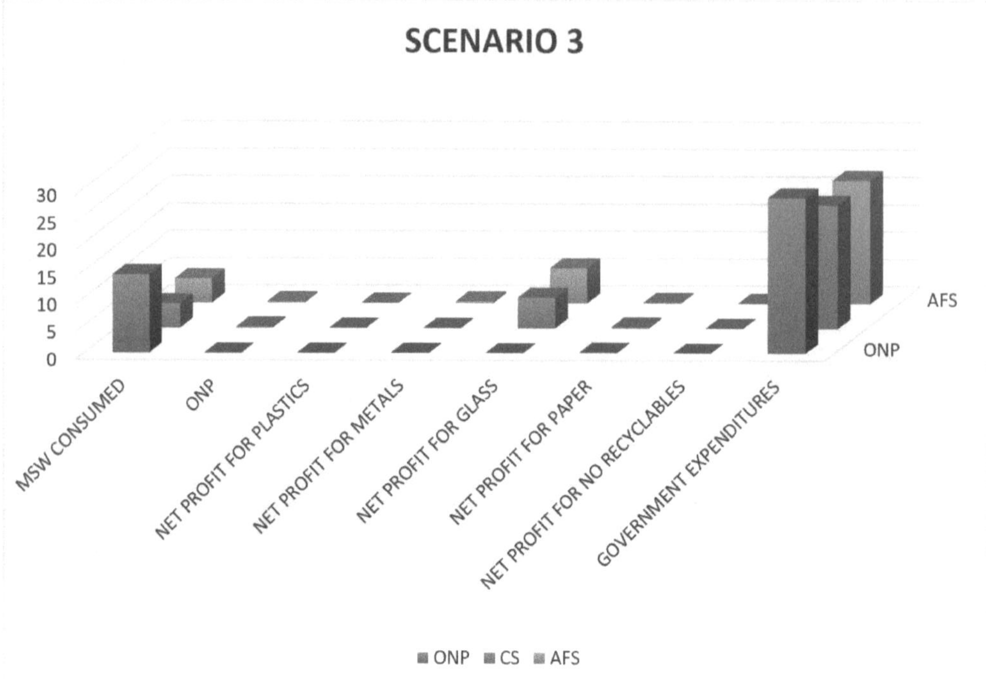

Fig. 14. Percentage difference between individual objective functions and their individual UB for Scenario 3. [Printed (adapted) with permission from Diaz-Barriga-Fernandez, A.D., Santibañez-Aguilar, J.E., Radwan, N., Nápoles-Rivera, F., El-Halwagi, M., and Ponce-Ortega, J.M. 2017. Strategic planning for managing municipal solid wastes with consideration of multiple stakeholders. *ACS Sustainable Chemistry and Engineering*, 2017, *5*(11), 10744–10762].

$C^{no-sep}_{wn,c,p-wn}$	Unitary cost for no separation.
$C^{product}_{wn,w-wn,c}$	Unitary cost for the product *p-wn*.
$C^{sep}_{wn,c,p-wn}$	Unitary cost for type of residue *wn*.
$C^{unit-transp}_{wn,p-wn}$	Unitary transportation cost of the residues from the cities to the processing plants.
$C^{unit-transp-product}_{wn,n-wn,p-wn}$	Unitary transportation cost of the products from the processing plants to the cities.
$d_{c,pp-wn}$	Distance between the city c and the processing plant *pp-wn*.
$F\max^{processing}_{wn,pp-wn,n-wn,p-wn,q}$	Lower limit for the amount of residues processed in the processing plant *pp-wn* with the technology *n-wn*.
$F\min^{processing}_{wn,pp-wn,n-wn,p-wn,q}$	Upper limit for the amount of residues processed in the processing plant *pp-wn* with the technology *n-wn*.
$PD^{city}_{wn,n-wn,p-wn}$	Demand of product *p-wn* in the city *c*.
Obf^{LB}	Lower bound of the objective function *obf*.
Obf^{UB}	Upper bound of the objective function *obf*.
$\alpha^{sep}_{wn,c,p-wn}$	Separation factor.
$\alpha^{prod}_{wn,n-wn,p-wn}$	Efficiency production factor.

6.7.2 Variables

AFS	Average feasible solution.
$A^{processing}_{wn,pp-wn,n-wn,p-wn}$	Fixed part of the capital cost for processing facilities.
$B^{cao}_{wn,pp-wn,n-wn,p-wn}$	Part of the capital cost that depends on the capacity of the processing facilities.
$CostCap^{processing}_{wn,pp-wn,n-wn,p-wn}$	Capital cost for processing the residues.
$CostNSep^{total}$	Total capital cost.
$CostNSep^{total}$	Total no separation cost for all the residues that is not reused.
$CostOp^{processing}$	Operation cost for processing all the residues.
$CostOp^{processing}_{wn}$	Operation cost for processing the residues of the type *wn*.
$CostSep_{wn}$	Separation cost for the residues *wn*.
$CostSep^{total}$	Total separation cost for all the residues.
$CostTransp^{city}_{pp}$	Transportation cost of the residues from the cities to the processing plants.
$CostTransp^{pp}_{city,wn}$	Transportation cost of the residues of the type *wn* from the cities to the processing plants.
$CostTransp^{pp}_{city}$	Transportation cost for the products from the processing plants to the cities.
CS	Compromise solution.
C^{sales}	Profit obtained from sales of the produced products.
C^{sales}_{wn}	Profit from the sales of the products obtained from residues of the type *wn*.
$F^{plant}_{wn,c,pp-wn,p-wn}$	Amount of residues sent from the city c to the processing plant *pp-wn*.
FS_k	Feasible solution for the stakeholder *k*.
$F^{tech}_{wn,pp-wn,n-wn,p-wn}$	Amount of residues sent from the processing plant *pp-wn* to the technology *n-wn*.
$F^{total}_{c,p-wn}$	Total amount of residues produced from all the cities.
$F_{wn,c,p-wn}$	Amount of residues used *wn* from city *c* to produce a certain product *p-wn*.
$Goverment\ expenditure$	Government expenditures.
NetProfit	Overall net profit.
$P_{wn,pp-wn,n-wn,p-wn}$	Amount of products *p-wn* produced in the processing plant *pp-wn* and with the technology *n-wn*.
$P^{city}_{wn,pp-wn,c,n-wn,p-wn}$	Amount of product *p-wn* distributed in the city *c*.
$Profit_{wn}$	Net profit obtained from the residues of the type *wn*.
$PT^{city}_{wn,c,n-wn,p-wn}$	Total amount of product *p-wn* in a determine city *c*.
$TotalConsume^{waste}$	Percentage of the total consumed waste.
$W^{NetProfit}_k$	Weight assigned to the decision-making variable for the objective function *obf*.
ρ^{obf}	Decision-making variable for the objective function *obf*.

6.7.3 Binary Variables

$y^{processing}_{wn,pp-wn,n-wn,p-wn}$ Binary variables used to define the existence of the processing plants.

6.8 Appendix

Table A.6.1 shows the weights used in the feasible solution objective function. Table A.6.2 shows the information for the MSW used in the case study, Table A.6.3 presents the data for the transportation costs, and Table A.6.4 shows the data for the separation costs. Table A.6.5 shows the bounds for the considered scenarios. Tables A.6.6, A.6.7, and A.68 show the results for Scenarios 1, 2, and 3, respectively.

6.9 References

Achillas, C., Moussiopoulos, N., Karagiannidis, A., Banias, G., and Perkoulidis, G. 2013. The use of multi-criteria decision analysis to tackle waste management problems: a literature review. Waste Management, *31*(2), 115–129.

Carvalho, P., Marques, R.C., and Dollery, B. 2015. Is bigger better? An empirical analysis of waste management in New South Wales. Waste Management, *39*, 277–286.

Chen, Y.T., and Chang, D.S. 2010. Diffusion effect and learning effect: an examination on MSW recycling. *Journal of* Cleaner Production, *18*(5), 496–503.

Chifari, R., Lo Piano, S., Bukkens, S.G.F., and Giampietro, M. 2016. A holistic framework for the integrated assessment of urban waste management systems. Ecological Indicators, *94*, 24–36. doi.org/10.1016/j.ecolind.2016.03.006.

Dewi, O., Koerner, I., and Harjoko, T. 2017. A review on decision support models for regional sustainable waste management. In: The International Solid Waste Association World Conference Retrieved from http://www.iswa.org/uploads/tx_iswaknowledgebase/Candra_Dewi.pdf (Accessed January 2017).

Diaz-Barriga-Fernandez, A.D., Santibañez-Aguilar, J.E., Radwan, N., Nápoles-Rivera, F., El-Halwagi, M., and Ponce-Ortega, J.M. 2017. Strategic planning for managing municipal solid wastes with consideration of multiple stakeholders. ACS Sustainable Chemistry and Engineering, 2017, *5*(11), 10744–10762.

Dowling, A.W., Ruiz-Mercado, G., and Zavala, V.M. 2016. A framework for multi-stakeholder decision-making and conflict resolution. Computers and Chemical Engineering, *90*, 136–150.

EPA. 2012. Environmental Protection Agency. Office of Resource Conservation and Recovery. Municipal Solid Waste Generation, Recycling and Disposal in the United States Tables and Figures for 2012, USA. https://www.epa.gov/sites/production/files/2015-09/documents/2012_msw_dat_tbls.pdf (Accessed January 2017)

Fuentes-Cortés, L.F., Dowling, A.W., Rubio-Maya, C., Zavala, V.M., and Ponce-Ortega, J. M. 2016. Integrated design and control of multigeneration systems for building complexes. Energy, *116*, 1403–1416.

Fuentes-Cortes, L.F., Serna-Gonzalez, M., and Ponce-Ortega, J.M. 2017. Analysis of carbon policies in the optimal design of domestic cogeneration systems involving biogas consumption. ACS Sustainable Chemistry and Engineering, *5*(5), 4429–4442.

Gilardino, A., Rojas, J., Mattos, H., Larrea-Gallegos, G., and Vazquez-Rowe, I. 2017. Combining operational research and Life Cycle Assessment to optimize municipal solid waste collection in a district in Lima (Peru). Journal of Cleaner Production, *156*, 589–603.

Giri, S., Nejadhashemi, A.P., and Woznicki, S.A. 2016. Regulators' and stakeholders' perspectives in a framework for bioenergy development. Land Use and Policy, *59*, 143–153.

Gonzalez-Bravo, R., Fuentes-Cortes, L.F., and Ponce-Ortega, J.M. 2017. Defining priorities in the design of power and water distribution networks. Energy, *137*, 1026–1040. DOI:10.1016/j.energy.2017.03.079.

Gracida-Alvarez, U.R., Keenan, L.M., Sacramento-Rivero, J.C., and Shonnard, D.R. 2016. Resource and greenhouse gas assessment of the thermochemical conversion of municipal solid waste in Mexico. ACS Sustainable Chemistry and Engineering, *4*(11), 5972–5978.

Harijani, A.M., Mansour, S., Karimi, B., and Lee, C.G. 2017. Multi-period sustainable and integrated recycling network for municipal solid waste –A case study in Tehran. Journal of Cleaner Production, *151*, 96–108.

Ho, W.S., Hashim, H., Lim, J.S., Lee, C.T., Sam, K.C., and Tan, S.T. 2017. Waste management pinch analysis (WAMPA): Application of pinch analysis for greenhouse gas (GHG) emission reduction in municipal solid waste management. Applied Energy, *185*, 1481–1489.

Hoornweg, D., and Bhada-Tata, P. 2012. What a waste a global review of solid waste management, *Urban* development series knowledge papers. The World Bank. New York, USA.

Korai, M.S., Mahar, R.B., and Uqaili, M.A. 2017. The feasibility of municipal solid waste for energy generation and its existing management practices in Pakistan. Renewable and Sustainable Energy Reviews, *72*, 338–353.

Lee, C.K.M., Yeung, C.L., Xiong, Z.R., and Chung, S.H. 2016. A mathematical model for municipal solid waste management –A case study in Hong Kong. Waste Management, *58*, 430–441.

Lemme, M.M.V., Rocha, M.H., Lora, E.E.S., Venturini, O.J., Lopes, B.M., and Ferreira, C.H. 2014. Technoeconomic analysis and environmental impact assessment of energy recovery from Municipal Solid Waste (MSW) in Brazil. Resources, Conservation and. Recycling, *87*, 8–20.

Mexican National Institute of Ecology. 2012. Generation and Composition of the Municipal Solid Wastes, Mexico City, Mexico. http://www2.ine.gob.mx/publicaciones/libros/128/cap3.html (Accessed January 2017).

Mexican Ministry of Transportation and Communications. 2012. http://elreciclaje.org/content/reciclaje-de-vidrio (Accessed January 2017).

Mirmajlesi, S.R., and Shafaei R. 2016. An integrated approach to solve a robust forward/reverse supply chain for short lifetime products. Computers and Industrial Engineering, *97*, 222–239.

Morrissey, A.J., and Browne, J. 2004. Waste management models and their application to sustainable waste management. Waste Management, *24*(3), 297–308.

Ng, W.P.Q., Lam, H.L., Varbanov, P.S., and Klemeš, J.J. 2014. Waste-to-energy (WTE) network synthesis for municipal solid waste (MSW). Energy Conversion and Management, *85*, 866–874.

Rada, E.C., Zatelli, C., and Mattolin, P. 2014. Municipal solid waste collection and tourism. *WIT Press*. UK.

Santibañez-Aguilar, J.E., Ponce-Ortega, J.M., González-Campos, J.B., Serna-González, M., and El-Halwagi, M.M. 2013. Optimal planning for the sustainable utilization of municipal solid waste. Waste Management, *33*(12), 2607–2622.

Santibañez-Aguilar, J.E., Martínez-Gómez, J., Ponce-Ortega, J.M., Nápoles-Rivera, F., González-Campos, J.B., Serna-González, M., and El-Halwagi, M.M. 2015. An optimal planning for the reuse of municipal solid waste considering economic, environmental and safety objectives. AIChE Journal, *61*(6), 1881–1899.

Santibañez-Aguiilar, J.E., Flores-Tlacuahuac, A., Rivera-Toledo, M., and Ponce-Ortega, J.M. 2017. Dynamic optimization for the planning of a waste management system involving multiple cities. Journal of Cleaner Production, *165*, 190–203. DOI: 10.1016/j.jclepro.2017.07.063.

Soltani, A., Hewage, K., Reza, B., and Sadiq, R. 2015. Multiple stakeholders in multi-criteria decision-making in the context of Municipal Solid Waste Management: A review. Waste Management, *35*, 318–328.

Struk, M. 2017. Distance and incentives matter: The separation of recyclable municipal waste. Resources, Conservation and Recycling, *122*, 155–162.

Sukholthaman, P., and Shirahada, K. 2015. Technological challenges for effective development towards sustainable waste management in developing countries: Case study of Bangkok, Thailand. Technology in Society, 43, 231–239.

Sukholthaman, P., and Sharp, A. 2016. A system dynamics model to evaluate effects of source separation of municipal solid waste management: A case of Bangkok, Thailand. Waste Management, *52*, 50–61.

Tan S.T., Lee, C.T., Hashim, H., Ho, W.S., and Lim, J.S. 2014. Optimal process network for municipal solid waste management in Iskandar Malaysia. Journal of Cleaner Production, *71*, 48–58.

Vadenbo, C., Hellweg, S., and Guillén-Gosálbez, G. 2014. Multi-objective optimization of waste and resource management in industrial networks –Part I: Model description. Resources, Conservation and Recycling, *89*, 52–63.

Vucijak, B., Kurtagic, S.M., and Silajdzic, I. 2016. Multicriteria decision making in selecting best solid waste management scenario: a municipal case study from Bosnia and Herzegovina. Journal of Cleaner Production, *130*, 166–174.

Warren, A., El-Halwagi, M.M. 1996. An economic study on the co-generation of liquid fuel and hydrogen from coal and municipal solid waste. Fuel Processing and Technology, *49*, 157–166.

Yadav, V., Bhurjee, A.K., Karmakar, S., and Dikshit, A.K. 2017. A facility location model for municipal solid waste management system under uncertain environment. Science of the Total Environment, 603–604, 760–771. DOI: 10.1016/j.scitotenv.2017.02.207.

7

Financial Risk Analysis in Waste Management

This chapter presents a multi-objective optimization approach for the strategic planning of a municipal solid waste (MSW) management system under uncertainty. The approach considers the involved tasks such as recycling, reuse, transportation, separation, and distribution. It also accounts for the multiple involved stakeholders to maximize the benefit to all the participants. The main variables that are considered under uncertainty were the MSW availability and the prices of the products made from the recovered MSW. A case study for Mexico is analyzed, where a random generation approach based on historical data is used to introduce three different financial risk levels: the optimistic, mean, and pessimistic cases. This way, the results obtained provide additional information about the system. Then, the stakeholders will be more certain of making any decision.

7.1 Introduction

Since residues are continuously produced in all cities, managing municipal solid waste (MSW) is a major challenge worldwide. If residues are not properly managed, they may have negative effects on the environment and those exposed to them. Some of these affections are global warming, soil contamination, underwater contamination, and health hazards such as the spread of diseases, fires and explosions, vegetation damage, and unpleasant odors. The most common way to dispose of municipal solid waste in developing countries is to bury or burn them without any previous separation. The main consequence of this is the generation of leachate by the excess rainwater that percolates through the waste layers of the landfill. In this context, El-Fadel et al. (1997) reported an overview of the leachate formation mechanisms in landfills and the adverse environmental impacts, and they described some methods to eliminate or minimize these impacts. Kjeldsen et al. (2010) presented a review of the composition of the leachate in the long term. However, this scenario could be averted if the MSW is separated, primarily the organic matter from other residues. Additionally, a variety of harmful pollutants are produced, which can seriously harm those who are exposed.

Ding et al. (2018) determined the optimal location for landfills is another useful strategy to reduce problems, such as the scarcity of land, high operation costs, and serious environmental impacts caused by the inappropriate management of MSW. Thus, Javaheri et al. (2006) utilized a geographic information system as a useful tool to assess the ideal landfill location utilizing a multi-criteria evaluation method known as weighted linear combination. Sumathi et al. (2008) addressed the siting of a new landfill using multi-criteria decision analysis and overlay analysis using a geographic information system, considering several factors such as geology, water supply resources, land use, sensitive sites, air quality, and groundwater quality. Moeinaddini et al. (2010) evaluated the suitability of a landfill site for MSW in Karaj using a weighted linear combination method and spatial cluster analysis. Gilardino et al. (2017) developed an optimization model to find the best sites for MSW collection. Jamshidi-Zanjani and Rezaei (2017) applied 12 factors to select the landfill site in Markazi province in Iran, employing techniques such as the weighted linear combination approach, ordered weighted average, and the analytic network process in conjunction with a fuzzy language quantifier in a geographic information system to locate a waste site that has been appropriated.

Other authors have studied how to transform MSW into value-added products. Warren and El-Halwagi (1996) presented a techno-economic analysis of a new process for co-liquefying coal and plastic wastes to produce hydrocarbon gases, naphtha, jet fuel, and diesel fuel. Ng et al. (2014) presented a model for the waste-to-energy supply chain. Korai et al. (2017) evaluated the feasibility of using MSW for the generation of energy in Pakistan. Taifouris and Martín (2017) proposed a multiscale scheme because of its availability and high fertilizer yield, MSW is the waste type with the most potential for use in Spain's biogas production process. An MSW system must be properly planned in order to maximize all available resources and prevent environmental harm. Santibañez-Aguilar et al. (2013) presented a mathematical programming model for the optimal planning of the supply chain associated with the MSW system to maximize the economic benefit while accounting for technical and environmental issues and Santibañez-Aguilar et al. (2015) included safety issues in this model. Tan et al. (2013) implemented a study for synthesizing an MSW processing network to produce energy and other value-added products. Zhang et al. (2014) reported a multi-echelon multi-period solid waste management system, where waste managers, suppliers, and distributors can be engaged in a joint strategic planning and operational execution. Pan et al. (2015) evaluated several waste-to-energy technologies, which include combustion, gasification, and anaerobic digestion. Niziolek et al. (2017) reported a superstructure for the sustainable production of liquid transportation fuels, olefins, and aromatics from MSW. Subramanyam et al. (2017) presented an optimization approach to solve the multi-period vehicle routing problem; these approaches could be applied for the optimal planning of MSW recollection. Habibi et al. (2017) developed a multi-objective optimization model for the supply chain of MSW in Tehran. Entezaminia et al. (2016) presented a model for production planning in a green supply chain considering the traditional production system with recycling centers. Sun and Chungpaibulpatana (2017) focused on predicting the generation of MSW through linear and nonlinear models that include certain variables that affect the amount of MSW generation in Bangkok. Pin et al. (2017) analyzed the economic viability of a power generation project from MSW in Brazil. Additionally, the dependence on time in the supply chain of MSW is an aspect that makes the system more realistic (Mirmajlesi and Shafaei 2016, Sukholthaman and Sharp 2016, Santibañez-Aguilar et al. 2017, Nguyen-Tron et al. 2017, Harijani et al. 2017, Zhang and Huang 2013, Dai et al. 2012).

Furthermore, as the production and the composition of MSW depend on many factors like population growth, consumption patterns, and climatic conditions, the uncertainties associated with MSW management must be considered. McLean and Li (2013) presented a novel strategy for supply chain optimization under uncertainty. Laokhongthavorn and Chalida (2017) reported a model to optimize two objective functions, including the expected operational cost (maximizing revenues) and the expected net carbon dioxide equivalent emissions in the solid waste disposal system. Xu et al. (2017) investigated the design of a global reverse supply chain that involves environmental protection, resource conservation, and the development of logistic operations and manufacturing technology accounting for the uncertainty involved during the transportation process. Ma et al. (2017) presented a model for MSW management that integrates two types of fuzzy sets involving chance-constrained programming. Yadav et al. (2017) presented an algorithm to tackle uncertainties in the MSW management system and select the economically best location in a hypothetical urban center. Mishra et al. (2017) quantified the uncertainty about the human health risk from municipal solid waste landfill leachate contamination. Private investors' involvement in waste-to-energy projects sometimes entails uncertainty surrounding the creation of MSW, posing a significant financial risk to both the government and the private investors. As a result, decision assistance is essential and required for the waste-to-energy facilities' location planning. Thus, Hu et al. (2017) proposed a bi-objective two-stage robust model to help governments identify cost-effective and environmental waste-to-energy facility locations under uncertainty.

Correct MSW separation and treatment are not done since doing so would cost the government money for which it would receive no compensation. Most MSW management initiatives rely on a fantastic method for the stakeholders willing to fund the project to reap the benefits. Furthermore, the primary route to sustainable MSW management is the involvement of all relevant stakeholders in the decision-making process. (Joseph 2006, 2012). Soltani et al. (2015) reported a review of the application of multi-criteria decision analysis for solving MSW management problems in studies that have considered multiple stakeholders. Xu et al. (2016) applied the methodology of stakeholder analysis and social network model for food waste recycling and composting. López-Toro et al. (2016) identified the effects of waste management activities and assessed their impact on stakeholders. Dos Muchangos et al. (2017) combined the complementing stakeholder analysis and the social network analysis to add to the decision-making process to improve the engagement and interaction of stakeholders in the MSW management. Goulart Coelho et al. (2017) presented a literature review of multi-criteria decision-making applications used in solid waste management, offering a critical assessment of the current practices, and providing suggestions for future works. Diaz-Barriga-Fernandez et al. (2017) presented a mathematical model for the strategic planning of MSW management considering the different stakeholders involved. Wang et al. (2018) introduced a multi-attribute decision analysis method for prioritizing the MSW treatment alternatives based on the interval-valued fuzzy set theory.

To provide tools for decision-makers, Vandenbo et al. (2014) presented a general multi-objective mixed-integer linear programming optimization problem. Sukholthaman and Shirahada (2015) proposed approaches for increasing the effectiveness of waste management services and success factors to overcome technological problems. Chifari et al. (2016) developed an approach to improve socio-economic indicators for MSW management systems. Lee et al. (2016) reported a mathematical model for MSW management, which was applied to the case of Hong Kong. Achillas et al. (2013) presented a literature review of multi-criteria decision-making approaches for MSW management. Furthermore, financial analysis has been considered, Carvalho et al. (2015) presented an approach that considers the economies of scale in MSW management, and Gaeta-Bernardi and Parente (2016) introduced a financial feasibility study of using organic solid waste as a feedstock to produce biodiesel through the conversion of volatile fatty acids into lipids.

To look for incentives for the government and recycling industries to invest in MSW management, Zheng et al. (2014) analyzed the application of landfill gas-fired and MSW incineration to produce power. He et al. (2017) examined the policy-making coordination of MSW policies. Jafari et al. (2017) identified factors affecting residents and incentive dependency to participate in electronic waste recycling programs. Ho et al. (2017) used the waste management pinch analysis to examine the implication of two objectives, landfill, and GHG emissions as targets in sustainable waste management.

Nevertheless, previously reported approaches for the strategic planning of MSW management have not considered the uncertainty in the availability of MSW and the demand for the products obtained from them, as well as the multi-stakeholders involved in three different cases of risk (optimistic, mean, and pessimistic). This chapter presents an optimization model for the strategic planning of MSW management that takes into consideration the unpredictability involved as well as the competing interests of the many supply chain stakeholders. Furthermore, this chapter offers a novel method for resolving this kind of stochastic multi-objective issue. The application of the created solution approach and suggested model to a case study from Mexico is presented in this chapter. (Diaz-Barriga-Fernandez et al. 2018).

7.2 Problem Statement

The main objective of the addressed problem is determining the optimal solution for the MSW management system considering the financial risk through three levels of perspectives (optimistic, pessimistic, and average), while endeavoring to attain the maximum benefit for the different stakeholders and considering the uncertainty in the availability of the residues and the product prices. The proposed model is based on the superstructure presented in Fig. 1. The problem includes several

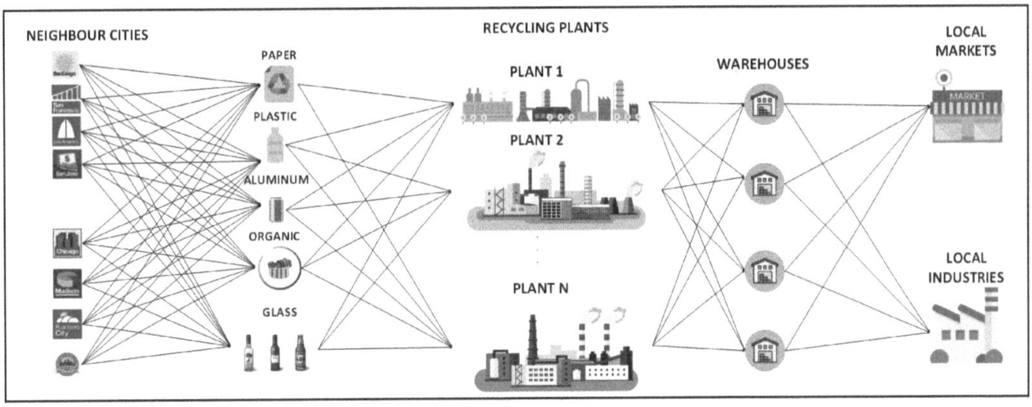

Fig. 1. Superstructure for the municipal solid waste management system. [Printed (adapted) with permission from Diaz-Barriga-Fernandez, A.D., Santibañez-Aguilar, J.E., Nápoles-Rivera, F., and Ponce-Ortega, J.M. 2018. Analysis of the financial risk under uncertainty in the municipal solid waste management involving multiple stakeholders. *Computers and Chemical Engineering*, 117, 433–450].

neighboring cities that generate different quantities and qualities of MSW. The solution must determine the optimal tasks for recollection, transportation, separation, and treatment for the MSWs and the sitting for the processing facilities. The considered stakeholders include the government, suppliers, consumers, and service providers for treating/recycling MSW.

A government-run system is responsible for collecting MSW. The recoverable part of the MSW is separated into five main materials: plastics, metals, glass, paper, and organics. Residues that are not suitable for recycling or for use as waste-to-energy are sent to be landfilled. The separation of the MSW is carried out by each processing industry. MSW, after being properly separated, may be treated in different industries depending on the type of residue and the desired products. The location, type, and size of the processing plants are variables to be optimized. Products are then distributed to potential consumers with specific demands. Transportation routes are variables to be optimized. The following section presents the proposed general optimization formulation to address the above-mentioned problem.

7.3 Model Formulation

The proposed mathematical model for the MSW management system is based on the superstructure shown in Fig. 1, which includes a set of neighboring cities, where the MSW is collected and then sent to different processing industries. Each industry has a set of technologies to treat a certain type of residue. It is assumed that only one type of product is produced from each type of technology to simplify the modeling. Then, the products are sent to different warehouses located in a specific city to be sent to a market or an industry as feedstock. The mathematical model is presented as follows.

7.3.1 Balance for the MSW sent to Different Industries

The balance for the MSW sent to different industries is shown in Equation (7.1). The municipal solid waste of type r from city c for each scenario e, which is suitable for recycling, must be equal to the sum of each flux entering every industry considered.

$$F_{r,c,e} = \sum_i F_{r,c,i,e}^{industry} \qquad \forall r,c,e \tag{7.1}$$

7.3.2 Restriction on the Quantity of Recycled Materials

Not all the residues are available to be recycled, so the term is included to represent the municipal solid waste to be recycled. Equation (7.2) states that the quantity of residues of type *r* from city *c* that is suitable to be recycled is limited by the fraction of solid waste that can be recycled multiplied by the total amount of residues of type *r* from city *c*.

$$F_{r,c,e} \le \theta^{msr}_{r,c} F^{Total}_{r,c,e} \qquad \forall r,c,e \tag{7.2}$$

7.3.3 Balance in the Technologies used in Each Industry

Equation (7.3) describes the balance for the technologies used in each industry, which is the sum of the number of residues that come from each city *c*, and it is equal to the sum of the amount of residues that feeds each type of technology in all the industries in the model.

$$\sum_{c} F^{industry}_{r,c,i,e} = \sum_{t} T_{r,i,t,e} \qquad \forall r,i,e \tag{7.3}$$

7.3.4 Obtained Products

The obtained products for each industry *i* are stated in Equation (7.4), which are equal to the amount of residue of the type *r* treated in a certain technology multiplied by the term $x^{conv}_{r,t}$ that represents the conversion fraction of the product obtained from the type of residue *r*.

$$P_{r,i,t,e} = x^{conv}_{r,t} T_{r,i,t,e} \qquad \forall r,i,t,e \tag{7.4}$$

7.3.5 Balance for the products in the warehouses

The product obtained from the residue type *r* and treated in plant *i* by technology *t* is equal to the sum of the product sent to the different warehouses located in the considered cities as Equation (7.5) describes.

$$P_{r,i,t,e} = \sum_{c} W_{r,i,t,c,e} \qquad \forall r,i,t,e \tag{7.5}$$

7.3.6 Balance in the Distribution of the Product in the Warehouses to the Markets or Industries

Equation (7.6) describes the distribution of the different products in the warehouses and sent to various markets or industries as a feedstock. The sum of the obtained products from technology *t*, at the warehouse located in the city *c* are equal to the amount of the product sent to a certain market or industry.

$$\sum_{i} W_{r,c,i,t,e} = F^{market/industry}_{r,c,t,e} \qquad \forall r,c,t,e \tag{7.6}$$

7.3.7 Production Constrained by the Demand in Each City

The amount of product sent to each market is limited by the demand for the same product in the city where the market or industry is located, (Equation (7.7)), where the amount of product received in each market/industry must be lower than the demand in each city *c*.

$$F_{r,c,t,e}^{market/industry} \leq Demand_{r,c,t} \qquad \forall r,c,t,e \tag{7.7}$$

7.3.8 Total Separation Cost

The total separation cost is given by Equation (7.8), where the sum of the unitary cost for the separation of the residue of type *r* multiplied by the amount of residue of type *r* susceptible to being recycled is equal to the total separation cost.

$$TotalSepCost_e = \sum_r \sum_c C_{r,c}^{separation} F_{r,c,e} \qquad \forall e \tag{7.8}$$

7.3.9 Individual Separation Costs

Equation (7.9) defines the separation cost by type of residue *r* that is equal to the sum of the unitary cost for the separation cost multiplied by the amount of residue that can be recycled in each city.

$$SepCost_{r,e} = \sum_c C_{r,c}^{separation} F_{r,c,e} \qquad \forall r,e \tag{7.9}$$

7.3.10 Cost for Landfilling MSW

The amount of residues that is not used must be treated and sent to controlled landfills considering the cost stated in Equation (7.10), which is equal to the sum of the unitary cost for landfilling multiplied by the amount of residues that is not recycled.

$$Landfillcost_e = \sum_r \sum_c C_{r,c}^{landfill} \left(F_{r,c,e}^{total} - F_{r,c,e} \right) \qquad \forall e \tag{7.10}$$

7.3.11 Total MSW Recollection Cost

It is necessary to consider the cost of the recollection of the residues from the cities to the industries. Equation (7.11) states that the recollection cost is equal to the sum of the unitary recollection cost multiplied by the distance between the cities where the residues are collected to the industries where they will be treated and multiplied by the amount of residues collected from that city.

$$TotalRecollectionCost_e = \sum_r \sum_c C_r^{recollection} d_{c,i} F_{r,c,i,e}^{industry} \qquad \forall e \tag{7.11}$$

7.3.12 Total Transportation Cost for the Distribution of the Products from the Warehouses to the Markets and Industries

Another cost considered is the total transportation cost, which is the cost of sending the product from the warehouses to the cities. Equation (7.12) states that the total transportation cost must be equal to the sum of the unitary cost for transporting the products multiplied by the distance between the warehouse and the city and by the number of residues in the warehouse.

$$TotalTransportCost_e = \sum_r \sum_c \sum_i \sum_t C_{r,t}^{Transport} d_{c,i} W_{r,c,i,t,e} \qquad \forall e \tag{7.12}$$

7.3.13 Transportation Cost for the Type of Residue Treated in the Industries

The equation for the transportation cost by type of residue r is like Equation (7.12), the difference lies in the sum, where the terms on the right-hand side consider the city, industries, and technologies (Equation (7.13)).

$$TransportCost_{r,e} = \sum_{c}\sum_{i}\sum_{t} C^{Transport}_{r,c,i,t}\, d_{c,i} W_{r,c,i,t,e} \qquad \forall r,e \tag{7.13}$$

7.3.14 Total Operating Cost

The total processing cost is represented in Equation (7.14), which is equal to the sum of the product between the unitary cost for processing the residues and the number of residues treated in each technology.

$$TotalProcessingCost_{e} = \sum_{r}\sum_{i}\sum_{t} C^{procesiing}_{r,i,t,} T_{r,i,t,e} \quad \forall e \tag{7.14}$$

7.3.15 Operating Cost for the Type of Residue

Equation (7.15) is formulated for the processing cost by type of residue, where the unitary cost is multiplied by the amount of residue in a technology and summed for all the industries and technologies.

$$ProcessingCost_{r,e} = \sum_{i}\sum_{t} C^{procesing}_{r,i,t} T_{r,i,t,e} \quad \forall r,e \tag{7.15}$$

7.3.16 Capital cost for the type of residue

The capital cost by type of residue is equal to the sum of the fixed cost plus the variable cost per industry and technology.

$$CapitalCost_{r,e} = \sum_{i}\sum_{t}\left(FC_{r,i,t,e} + VC_{r,i,t,e}\right) \qquad \forall r,e \tag{7.16}$$

7.3.17 Total Capital Cost

The total capital cost is stated by Equation (7.17), which is equal to the sum of the fixed cost plus the variable cost. The cost in Equation (7.17) considers all types of residues.

$$TotalCapitalCost_{e} = \sum_{r}\sum_{i}\sum_{t}\left(FC_{r,i,t,e} + VC_{r,i,t,e}\right) \qquad \forall e \tag{7.17}$$

7.3.18 Disjunctions for the Industries

The model was formulated to accomplish the optimum configuration for the management of the residues, so the disjunction shown below is used to account for the economies of scale. The binary variable $y^{industry}_{r,i,t,s,e}$ is stated for the existence of the industry, if the industry exists the equation for the economies of scale is applied.

$$\vee_{s} \begin{bmatrix} y^{industry}_{r,i,t,s,e} \\ FC_{r,i,t,e} = fc_{r,i,t,s} \\ VC_{r,i,t,e} = vc_{r,i,t,s} T_{r,i,t,s,e} \\ T^{min}_{r,i,t,q} \le T_{r,i,t,e} \le T^{max}_{r,i,t,q} \end{bmatrix} \forall r,i,t,s$$

Equations (7.18)–(7.23) show the reformulation of the disjunction presented above (Vecchietti et al., 2003). Equation (7.18) states that the fixed cost is equal to the binary variable for the existence of the industry multiplied by the fixed cost.

$$FC_{r,i,t,e} = \sum_{s} y^{industry}_{r,i,t,s,e} fc_{r,i,t,s} \qquad \forall r,i,t,e \tag{7.18}$$

The variable cost must be equal to the variable cost multiplied by the disaggregated variables for the amount of residues treated for a certain technology.

$$VC_{r,i,t,e} = \sum_{s} vc_{r,i,t,s} dT_{r,i,t,s,e} \qquad \forall r,i,t,e \tag{7.19}$$

The amount of waste treated by a determined technology is equal to the sum of all the disaggregated variables for waste treated.

$$T_{r,i,t,e} = \sum_{s} dT_{r,i,t,s,e} \qquad \forall r,i,t,e \tag{7.20}$$

Equations (7.21) and (7.22) are formulated to limit the number of residues treated by a technology t between the upper $T^{max}_{r,i,t,q}$ and lower $T^{min}_{r,i,t,q}$ limits.

$$T^{min}_{r,i,t,s} y^{industry}_{r,i,t,s,e} \leq dT_{r,i,t,s,e} \qquad \forall r,i,t,s,e \tag{7.21}$$

$$T^{max}_{r,i,t,s} y^{industry}_{r,i,t,s,e} \geq dT_{r,i,t,s,e} \qquad \forall r,i,t,s,e \tag{7.22}$$

Equation (7.23) defines the capacity of the industries with a certain technology and the cost depending on the economies of scale by selecting the binary variable that defines the interval in which the technology operates.

$$\sum_{s} y^{industry}_{r,i,t,s,e} = 1 \qquad \forall r,i,t,e \tag{7.23}$$

7.3.19 Total Profit from Sales of the Products

The total profit obtained from the sales of the obtained products is equal to the sum of the unitary cost for a type of product multiplied by the amount of this product in a warehouse.

$$TotalSalesProfit_{e} = \sum_{r}\sum_{c}\sum_{i}\sum_{t} C^{product}_{r,c,t,e} W_{r,c,i,t,e} \qquad \forall e \tag{7.23}$$

7.3.20 Profit from Sales for Each Type of Residue

The profit from sales by type of residue is presented in Equation (7.24), which is similar to Equation (7.23), but in this case, the sum is by city, industry, and technology.

$$SalesProfit_{r,e} = \sum_{c}\sum_{i}\sum_{t} C^{product}_{r,c,t,e} W_{r,c,i,t,e} \qquad \forall r,e \tag{7.24}$$

7.3.21 Objective Functions

The overall net profit (ONP) considers the total profit and all the costs of the whole system. ONP is equal to the total profit obtained by sales of all the products, minus all the costs: total separation cost, landfilling cost, total recollection cost, total transportation cost, total processing cost, and total capital cost.

$$OverallNetProfit_e = TotalSalesProfit_e - TotalSep\,Cost_e - Lanfillcost_e - TotalRecolectionCost_e - TotalTransportCost_e - TotalProcessingCost_e - TotalCapitalCost_e \quad \forall e \quad (7.25)$$

7.3.22 Total Percentage of Consumed MSW

The total percentage of MSW consumed is calculated by Equation (7.26), which is the ratio between the number of recycled residues and the amount of residues that can be recycled.

$$\%Consumed_e = 100\frac{\sum_r \sum_c F_{r,c,e}}{\sum_r \sum_c \theta^{msr}_{r,c} F^{Total}_{r,c,e}} \quad (7.26)$$

7.3.23 Profit for each type of residue

Each type of residue has a different economic performance; thus, the profit for each type of waste can be different. This way, the profit for each type of residue is equal to the profit obtained from the sales of the obtained product from the residue of type r, minus the transportation cost, the processing cost, and capital cost for the type of residue (Equation (7.27)).

$$Profit_{r,e} = SalesProfit_{r,e} - SepCost_{r,e} - TransportCost_{r,e} - ProcessingCost_{r,e} - CapitalCost_{r,e}, \quad \forall r,e \quad (7.27)$$

7.3.24 Government Expenditures

In several countries, the Government is the institution that must provide a waste management system. For the presented model, the Government expenditures are calculated by the sum of the cost for landfilling, plus the total recollection cost as is stated in Equation (7.28).

$$Govexpnd_e = Lanfillcost_e + TotalRecolectionCost_e \quad \forall e \quad (7.28)$$

7.3.25 Optimistic, Mean, and Worst Scenarios under Uncertainty

The terms optimistic, mean, and worst scenarios are introduced to refer to the three financial risks considered. All the scenarios are solved simultaneously, in the optimistic case the objective is to obtain the configuration that maximizes a single objective from all scenarios. The mean value maximizes the mean value of the objectives from all scenarios and the worst case maximizes the worst possible value of all scenarios. To analyze the involved financial risk and help the decision-makers select the best options, the next equations were stated. The optimistic function for each objective function o must be the greatest of all the objective functions o in the different scenarios e randomly generated.

$$OptFunction_o \geq Function_{o,e} \quad \forall o \quad (7.29)$$

7.3.26 Disjunction for the Optimistic Scenarios

It is necessary to limit the value of the optimistic function through the following disjunction.

$$\begin{bmatrix} y^{op}_{o,e} \\ OptFunction_{o,e} \leq Function_{o,e} \end{bmatrix} \quad \forall o,e$$

The previous disjunction was implemented in the model through Equations (7.30) and (7.31) (Vecchietti et al., 2003). Equation (7.30) states that there must be only one binary variable for each objective function.

$$\sum_{e} y_{o,e}^{op} = 1 \qquad \forall o \tag{7.30}$$

Equation (7.31) states that if the value of the binary variable is 0, then the optimistic function will always be lower than the right-hand side of the equation. Here, the parameter $OptF^{max}$ is a large enough value. Otherwise, if the binary variable takes a value of 1, the disjunction formulated above will be activated.

$$OptFunction_{o,e} \leq Function_{o,e} + OptF^{max}\left(1 - y_{o,e}^{op}\right) \qquad \forall o,e \tag{7.31}$$

7.3.27 Mean scenario

The mean case is the average of each objective function, it is equal to the sum of the objective functions for each scenario divided by the cardinal of e, which is the number of scenarios (Equation (7.32)).

$$MeanFunction_{o} = \frac{\sum_{e} Function_{o,e}}{card(e)} \qquad \forall o \tag{7.32}$$

7.3.28 Pessimistic Scenario

The pessimistic case is the lowest value for the objective function *o* between all the objective functions of a certain type *o* for all the scenarios.

$$PessimisticFunction_{o} \leq Function_{o,e} \qquad \forall o \tag{7.33}$$

7.3.29 Normalized Objectives

Once the optimistic, mean, and pessimistic cases are analyzed to formulate an equation that involves the multi-stakeholders, it is important to normalize the three cases. Equation (7.34) describes the normalized optimistic function, which is obtained by the difference between the greatest value obtained for the objective function when it was maximized individually and the optimistic function, divided by the difference between the upper bound for the optimistic objective function minus the lower bound for the optimistic function.

$$NOF_{obj} = \frac{OF_{obj}^{UB} - OptFunction_{obj}}{OF_{obj}^{UB} - OF_{obj}^{LB}} \qquad \forall obj \tag{7.34}$$

The normalized mean function is equal to the ratio between the upper bound for the mean function minus the mean function and the upper bound for the mean function minus the lower bound for the mean function.

$$NMF_{obj} = \frac{MF_{obj}^{UB} - MeanFunction_{obj}}{MF_{obj}^{UB} - MF_{obj}^{LB}} \qquad \forall obj \tag{7.35}$$

Equation (7.36) states the normalized pessimistic function that is equal to the difference between the upper bound and the nominal value for the pessimistic function, divided by the difference between the upper and lower bounds for the pessimistic function.

$$NPF_{obj} = \frac{PF_{obj}^{UB} - PessimisticFunction_{obj}}{PF_{obj}^{UB} - PF_{obj}^{LB}} \quad \forall obj \tag{7.36}$$

7.3.30 Compromise One-objective Solution

The compromise solution for each objective function is equal to the sum of the optimistic normalized function, plus the mean normalized function and the pessimistic normalized function.

$$CUF_o = NOF_o + NMF_o + NPF_o \quad \forall o \tag{7.37}$$

7.3.31 Overall Compromise Solution

The total compromise function is the sum of all the normalized objective functions for the three main cases (optimistic, mean, and pessimistic).

$$TCS = \sum_{o} CUF_o \tag{7.38}$$

The resulting formulation is a mixed integer linear programming (MIP) model, which was coded and solved in the optimization platform GAMS® using the solver CPLEX®. The number of equations is 54,958, the single variables are 64,607, and the discrete variables 8,080.

7.4 Case Study

The presented mathematical model was implemented for five neighboring cities located in the central-western part of Mexico to prove the applicability of this approach. The number of residues produced in each city and product prices are generated randomly based on historical data using a deterministic algorithm (called a generator) inside the software GAMS®. This is used to introduce the uncertainty in the model. Fifteen technologies were considered to treat MSW. The production factors are shown in Table 1. Additional details for the addressed problem can be found in Diaz-Barriga-Fernandez et al. (2017).

7.5 Results and Discussion

The three cases considered in this work (optimistic, mean, and pessimistic) for the objective function ONP with three different standard deviations (σ= 0.05, 0.1, and 0.2) were considered to obtain the results shown in Fig. 2. Different standard deviations were used to consider different values for the variability of the three generated scenarios since the standard deviation is a measure of data distribution. The lines O1, M1, and W1 were obtained by maximizing the optimistic, mean, and pessimistic ONP objectives, respectively, with a standard deviation of 0.05. Lines O2, M2 and W2, as lines O3, M3 and W3, were obtained with a standard deviation of 0.1 and 0.2, respectively.

Results obtained for the optimistic, mean, and pessimistic cases with σ= 0.05 correspond to the lines O1, M1 and W1. Point A with a value of 82.72 $MM/y is the biggest value for the optimistic line O1, which has less probability to occur among those that could happen. Point B with 75.89$MM/y is one of the values that takes line M1, in this case the results for the mean value are always above the pessimistic and optimistic values. Accordingly, point C, with a value of 64.35 $MM/y, is obtained when the pessimistic case is maximized, this means that this scenario is more likely to occur even if it is worse than the previous ones (point A and B). Moreover, this point is better than most of the optimistic scenarios with the exception of point A.

Basically, in the cases when σ is 0.1 and 0.2, the optimistic, mean, and pessimistic lines follow the same trend as when σ is 0.05. One of the main differences in the cases mentioned before and the

Table 1. Efficiency factor for the processing technologies to treat MSW. [Printed (adapted) with permission from Diaz-Barriga-Fernandez, A.D., Santibañez-Aguilar, J.E., Nápoles-Rivera, F., and Ponce-Ortega, J.M. 2018. Analysis of the financial risk under uncertainty in the municipal solid waste management involving multiple stakeholders. *Computers and Chemical Engineering*, 117, 433–450].

Type of waste	Subwaste	Technology	Product	Production factor (kg product/kg waste)
PLASTIC	PP	Technology 1 for nanotubes	Nanotubes of 140 μm diameter	0.11
		Technology 2 for nanotubes	Nanotubes of 252 μm diameter	0.51
		Material recycling 1	Pellet	0.672
		Pyrolysis	Branched paraffins	0.304
		Pyrolysis	Branched paraffins	333
	PE	Technology 1 for nanotubes	Nanotubes of 140 μm diameter	0.16
		Technology 2 for nanotubes	Nanotubes of 252 μm diameter	0.32
		Material recycling 1	Pellet	0.672
		Thermal recycling	Gasoline	0.299
		Thermal recycling	Diesel	0.089
		Thermal recycling	Heavy Oil	0.068
		Chemical recycling	Heat energy	19.44
	PET	Technology 1 for nanotubes	Nanotubes of 140 μm diameter	0.19
	PS	Material recycling 1	Pellet	0.672
	HDPE	Pyrolysis	n-Paraffins	0.432
		Pyrolysis	n-Olefines	0.446
PAPER	PAPER	Material recycling 2	New paper	1
METAL	ALUMINUM	Material recycling 3	Useful aluminum	1
GLASS	CLEAR GLASS	Material recycling 4	New Glass	1
	GREEN GLASS	Material recycling 4	New Glass	1
	BROWN GLASS	Material recycling 4	New Glass	1
ORGANIC	MSW	Mass burn (incineration)	Electricity	1.9584
		Pyrolysis	Electricity	2.0556
		Pyrolysis/gasification	Electricity	2.466
		Plasma arc gasification	Electricity	2.9376
		Conventional gasification	Electricity	2.466

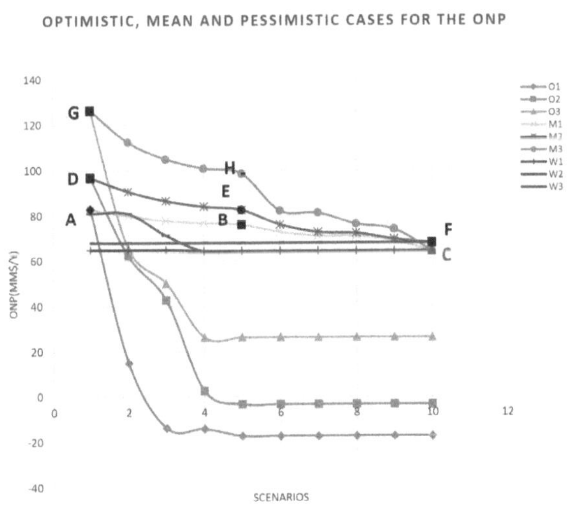

Fig. 2. Optimistic, mean, and pessimistic for the objective function ONP with three different standard deviations (σ= 0.05, 0.1, and 0.2). [Printed (adapted) with permission from Diaz-Barriga-Fernandez, A.D., Santibañez-Aguilar, J.E., Nápoles-Rivera, F., and Ponce-Ortega, J.M. 2018. Analysis of the financial risk under uncertainty in the municipal solid waste management involving multiple stakeholders. *Computers and Chemical Engineering*, 117, 433–450].

case when sigma is 0.05, is that as the value of σ increases the values of the ONP increase for each scenario. The other difference lies in line W3, which does not follow the same trend as lines W1 and W2. Besides, from the fourth scenario, line W3 gives similar values as line W1. Point D is 96.67 and point G 126.31$MM/y, they represent the optimistic values for lines O2 and O3, respectively. Points E and H are examples of the mean scenarios of lines M2 and M3, their values were 82.47 and 98.37$MM/y, respectively. Finally, points F and C with values of 67.85 and 64.25 $MM/y represent the pessimistic scenarios in lines W2 and W3, respectively.

The risk is associated with the standard deviation, if the standard deviation is lower the risk in the decision will be lower in terms of the behavior of data as for a normal distribution 95% of the values will be within 2 standard deviations, however, the data cannot be controlled thus, in terms of decision making if the stakeholders decide to implement a low-risk solution, the pessimistic case should be used, as all the cases report positive profits despite of the standard deviation. This means that from all scenarios the minimum profit is at least $64 MM, and in some cases, it could reach $80.1 MM. For the optimistic case, 7 out of 10 scenarios report negative profit, for and 6 out of 10 for . If the standard deviation is the highest all scenarios have positive profits, with the potential to earn as much as $126 MM. If the mean case is selected for the design, all the scenarios report profit and only 1 gives less profit than the pessimistic case, thus this will also be an attractive option for the stakeholders, however, the single objectives must also be analyzed to define the best alternative for the investors.

The percentage of consumption was examined for the optimistic, mean, and pessimistic cases as well. Figure 3 shows the results obtained for the three σ considered. Point A reaches the biggest consumption of 99.9% for all the standard deviations in the optimistic case. Now, analyzing point B, which is obtained in a scenario for the mean (σ = 0.05), the optimistic and mean lines join so there is a larger chance to have a consumption between 60–63%. Point C, with a value of 34% of the consumption of waste, is the lowest for σ = 0.05. Furthermore, line O1 is always above line W1. This

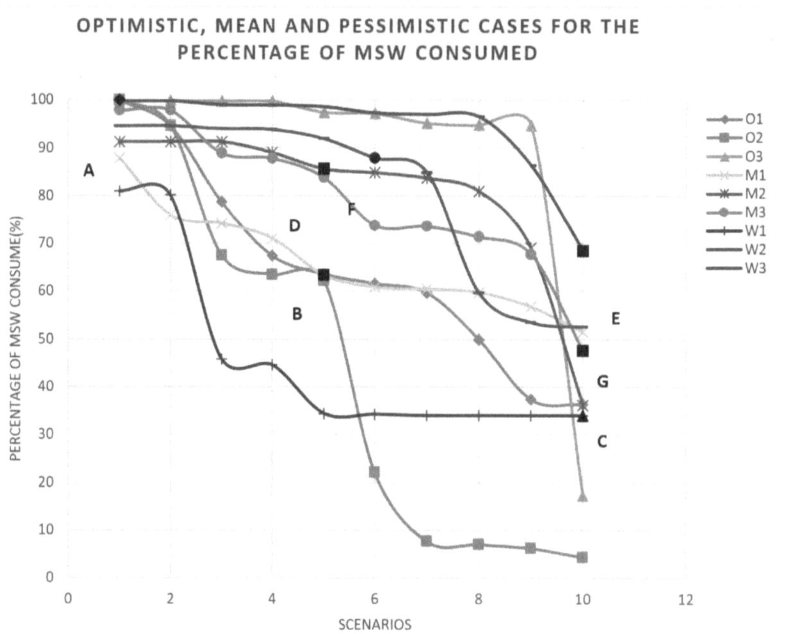

Fig. 3. Optimistic, Mean, and pessimistic cases for three different standard deviations (σ=0.05, 0.1, 0.2) when the percentage of consumption was maximized. [Printed (adapted) with permission from Diaz-Barriga-Fernandez, A.D., Santibañez-Aguilar, J.E., Nápoles-Rivera, F., and Ponce-Ortega, J.M. 2018. Analysis of the financial risk under uncertainty in the municipal solid waste management involving multiple stakeholders. *Computers and Chemical Engineering*, 117, 433–450].

means that in all the scenarios for the optimistic cases, we will always obtain greater consumptions than if we take the pessimistic case.

Studying the cases with σ=0.1, the lines show a drastic change, especially line W2 (pessimistic case), which is above lines O2 and M2 (optimistic and mean cases, respectively). So, there are more possibilities to obtain bigger consumptions for the pessimistic case than in the other cases. Point E, with 68% of consumption of waste, is the lowest value for the pessimistic case and the only point worse than one scenario for the mean case. Consequently, the optimistic line (O2) tends to be below the mean (M2). Point D, with 73.8% of consumption, is one of the scenarios for the mean which is like the pessimistic case.

If we analyze the different cases when σ=0.2, the optimistic line O3 is above the mean and pessimistic lines (M3 and W3). Point A shows the greatest value for the optimistic case, just one of the scenarios tends to decrease drastically. Line W3 (pessimistic case) tends to be above line M3 (mean) in most of the scenarios, like point F with 87.8%, and the lowest value is for point G with 47.7% for the mean.

Analyzing the objective function for the net profit from metal (NPM) in the considered cases (optimistic, mean, and pessimistic). Figure 4 shows how this objective function behaves at different standard deviations (σ=0.05, 0.1, and 0.2). Point A is the net profit for metals reached for the optimistic case with σ=0.05, its value is 1.93$MM/y of profit. The lines M1 and W1 are overlapped having in

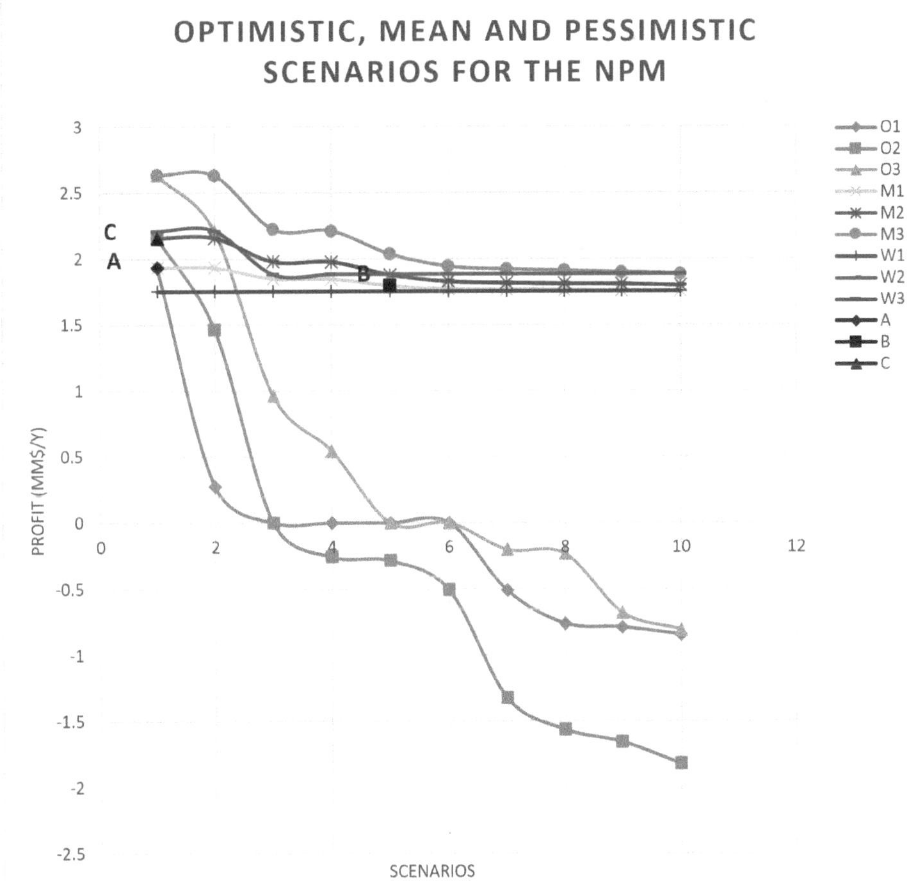

Fig. 4. Optimistic, Mean, and pessimistic cases for three different standard deviations (σ=0.05, 0.1, 0.2) when the NPM was maximized. [Printed (adapted) with permission from Diaz-Barriga-Fernandez, A.D., Santibañez-Aguilar, J.E., Nápoles-Rivera, F., and Ponce-Ortega, J.M. 2018. Analysis of the financial risk under uncertainty in the municipal solid waste management involving multiple stakeholders. *Computers and Chemical Engineering*, 117, 433–450].

common point B with 1.79\$MM/y. The line for the pessimistic case always has the same value 1.75 \$MM/y. For the investors it would be better to choose avoiding the risk since the optimistic case is not much better than the pessimistic case. In Fig. 4, lines O2, M2, and W2 follow the same trend that lines O1, M1, and W1, respectively. The differences are the values, for example, the highest value for line O2 is point C with 2.16 \$MM/y of profit, and point D with 1.88 \$MM/y and point E with 1.79 \$MM/y for the mean and pessimistic cases, respectively.

For lines M3 and W3, the trend is like the mean and pessimistic lines described before. But in this case, the optimistic line (O3) has more scenarios with positive profits than lines O1 and O2. Point F with 2.63 \$MM/y, point G with 2.03 \$MM/y, and point H with 1.88 \$MM/y, are the profits obtained for the optimistic, mean, and pessimistic cases respectively (σ=0.2).

The net profit for paper waste is analyzed for the three different risks considered (Fig. 5). The mean lines (M1, M2 and M3) are above the optimistic and pessimistic lines for the three standard deviations. The optimistic line (O1) tends to have the biggest value (7.39 \$MM/y) and then decreases drastically. While the pessimistic line (W1) is just a little below the mean line (M1), with a constant value of 6.66 \$MM/y.

Essentially, the rest of the lines with different standard deviations keep the same pattern that the one mentioned before. One more time, the main difference is the value that takes the objective function. Point A has 7.39 \$MM/y in line O1, then point B increases to 9.06 \$MM/y, and at last point C to 10.15 \$MM/y. The increase in lines M1, M2, M3 W1, W2, and W3 is similar as can be seen in Fig. 5.

Evaluating the optimistic, mean, and pessimistic cases for the net profit from the organics (NPO) waste with different standard deviations (σ=0.05, 0.1, 0.2) like in the previous cases (Fig. 6). First,

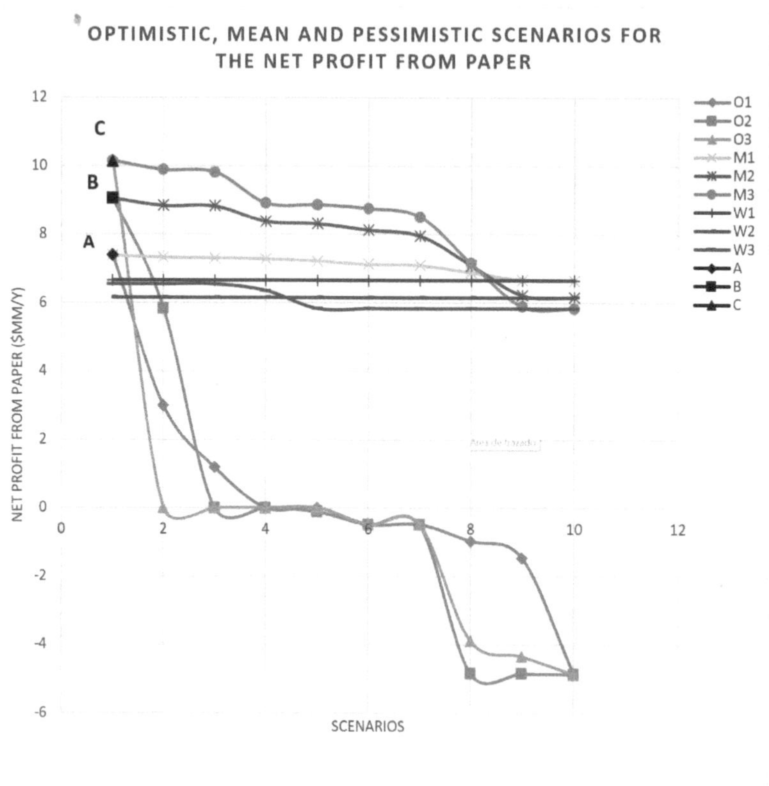

Fig. 5. Optimistic, Mean, and pessimistic cases for three different standard deviations (σ=0.05, 0.1, 0.2) when the NPRP was maximized. [Printed (adapted) with permission from Diaz-Barriga-Fernandez, A.D., Santibañez-Aguilar, J.E., Nápoles-Rivera, F., and Ponce-Ortega, J.M. 2018. Analysis of the financial risk under uncertainty in the municipal solid waste management involving multiple stakeholders. *Computers and Chemical Engineering*, 117, 433–450].

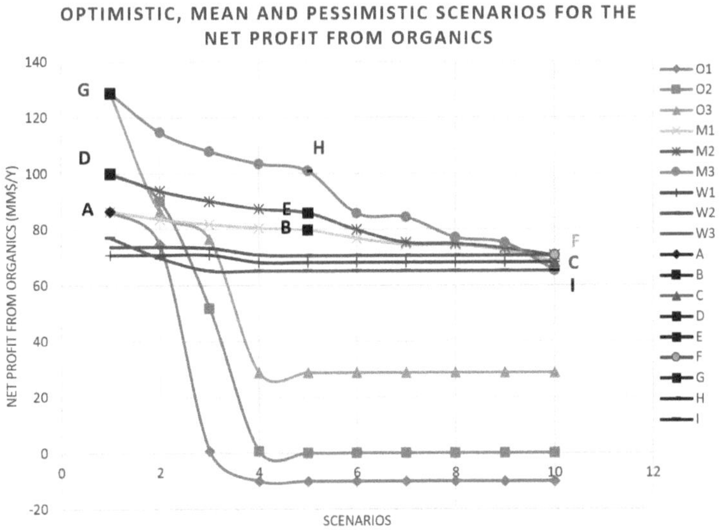

Fig. 6. Optimistic, Mean, and pessimistic cases for three different standard deviations (σ=0.05, 0.1, 0.2) when the NPO was maximized. [Printed (adapted) with permission from Diaz-Barriga-Fernandez, A.D., Santibañez-Aguilar, J.E., Nápoles-Rivera, F., and Ponce-Ortega, J.M. 2018. Analysis of the financial risk under uncertainty in the municipal solid waste management involving multiple stakeholders. *Computers and Chemical Engineering*, 117, 433–450].

it is important to note that for the three standard deviation values σ, the mean lines (M1, M2, and M3) are always above the optimistic and pessimistic lines. Some examples are points B, E, and H, whose values are 79.63, 85.83, and 100.89$MM/y. Moreover, the optimistic lines with different σ's have an upper value and then tend to abruptly decrease. The highest values for this case are points A with 86.45, D with 99.95, and G with 128.67$MM/y. For the pessimistic cases, the lines keep as a constant line, this can be seen in points F, C and I with profits of 70.82, 68.18, and 65.18 $MM/y respectively, all very similar.

The results obtained when the optimistic, mean, and pessimistic cases were evaluated for the different standard deviations considered for the government expenditures are shown in Fig. 7. We can see that for the three σ, all the lines are in the same range of values. For instance, when σ=0.05 point A shows the greatest government expenditures for the optimistic case with -11.33 $MM/y. On the other hand, point B, with -11.89 $MM/y, shows one of the mean values for this objective function. Points A and B are not so much different than point C, with -12.33$MM/y, which is the value obtained for the pessimistic case.

In the case when σ=0.1, the lowest value for line O2 was -10.61 $MM/y, which corresponds to point D. Point E is one of the scenarios for line M2 with a value of -11.09$MM/y. Whereas point F, with -12.60$MM/y, is the greatest value for the government expenditures.

The best value for the optimistic case was obtained when σ=0.2, like we can see in point G with a value of -9.18$MM/y. The mean and pessimistic lines have the worst values, like point H with -11.41 for line M3 and point I with -13.15 $MM/y for line W3. For the three standard deviations, the mean lines are over the optimistic and pessimistic lines. Moreover, when σ=0.2 the optimistic and pessimistic lines come together in most of the scenarios.

Once the different risk cases are analyzed, the upper and lower values for each objective function are calculated (Table 2). Then, these values are used to obtain the multi-stakeholder model.

The results for the total compromise solution (TCS) for a value of σ=0.1 are shown in Fig. 8. The higher objective function values are associated with the percentage of MSW consumed (%consumed), the ONP, and the net profit from waste to energy (NPW) with 97.66%, 73.69$MM/y and 81.22$MM/y,

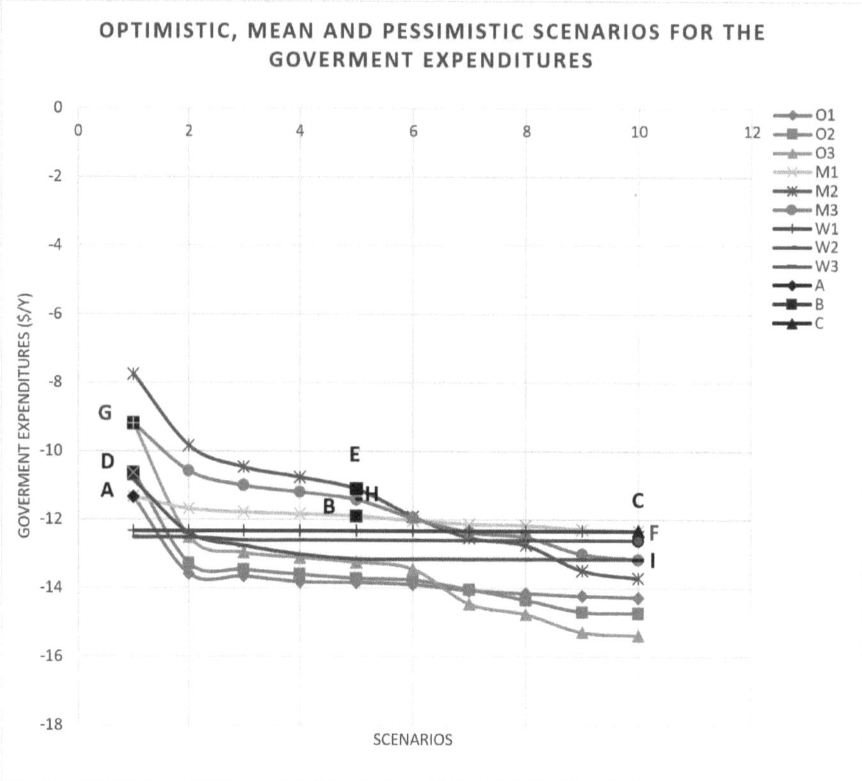

Fig. 7. Optimistic, Mean, and pessimistic cases for three different standard deviations (σ=0.05, 0.1, 0.2) when the government expenditures were minimized. [Printed (adapted) with permission from Diaz-Barriga-Fernandez, A.D., Santibañez-Aguilar, J.E., Nápoles-Rivera, F., and Ponce-Ortega, J.M. 2018. Analysis of the financial risk under uncertainty in the municipal solid waste management involving multiple stakeholders. *Computers and Chemical Engineering*, 117, 433–450]

Table 2. Upper and Lower bounds for the objective functions according to the risk case. [Printed (adapted) with permission from Diaz-Barriga-Fernandez, A.D., Santibañez-Aguilar, J.E., Nápoles-Rivera, F., and Ponce-Ortega, J.M. 2018. Analysis of the financial risk under uncertainty in the municipal solid waste management involving multiple stakeholders. *Computers and Chemical Engineering*, 117, 433–450].

OBJECTIVE FUNCTION	OPTIMISTIC		MEAN		PESSIMISTIC	
	UPPER BOUND	LOWER BOUND	UPPER BOUND	LOWER BOUND	UPPER BOUND	LOWER BOUND
%CONSUMED	99.9645	0.4699	99.9645	4.1053	99.9580	4.0309
GOVEXP (MM$/y)	–10.615	–14.7547	–11.2445	–14.7022	–11.3286	–14.7214
ONP (MM$/y)	96.6714	–525.2891	96.6714	–517.557	67.8545	–524.354
NPG (MM$/y)	0.0000	–11.9438	0.0000	–11.9485	0.0000	–11.9546
NPM (MM$/y)	2.1594	–2.5072	2.1594	–2.5074	2.1594	–2.5059
NPP (MM$/y)	7.9448	–4.8760	7.9448	–4.8746	7.2454	–4.8672
NPRP (MM$/y)	0.0000	–14.2616	0.0000	–14.2620	0.0000	–16.2039
NPO (MM$/y)	99.9519	–483.7516	99.9519	–483.807	80.0193	–483.752

respectively. This indicates that these variables (%consumed, ONP and NPW) promote a better value of the overall objective function.

The line for the percentages of consumed waste (%consumed) is constant at first, and just two of ten scenarios show a different value with consumptions of 65% and 45%. While lines ONP and NPW are similar, they tend to decrease almost constantly, except for one scenario that shows the lowest value for this function with zero for the NPW and a negative value for the ONP. Besides, the lines with the lowest values for objective function, but still with profits are the net profit for paper (NPP) and the NPM. These lines keep almost a constant linear trend, the highest values from them are 7.3 and 1.7$MM/y. For most of the scenarios, these lines take values very similar at the highest values. The line for the net profit for recycled plastics (NPRP) takes a value of zero since recycling of plastic is not feasible for this problem, this also can be seen in Table 2, where the upper bound for NPRP is zero. Now, studying the results for the government expenditures (GOVEXP), this line shows a linear behavior with a smooth slope. The range of values taken for this line is between –10.9 and –13.48$MM/y.

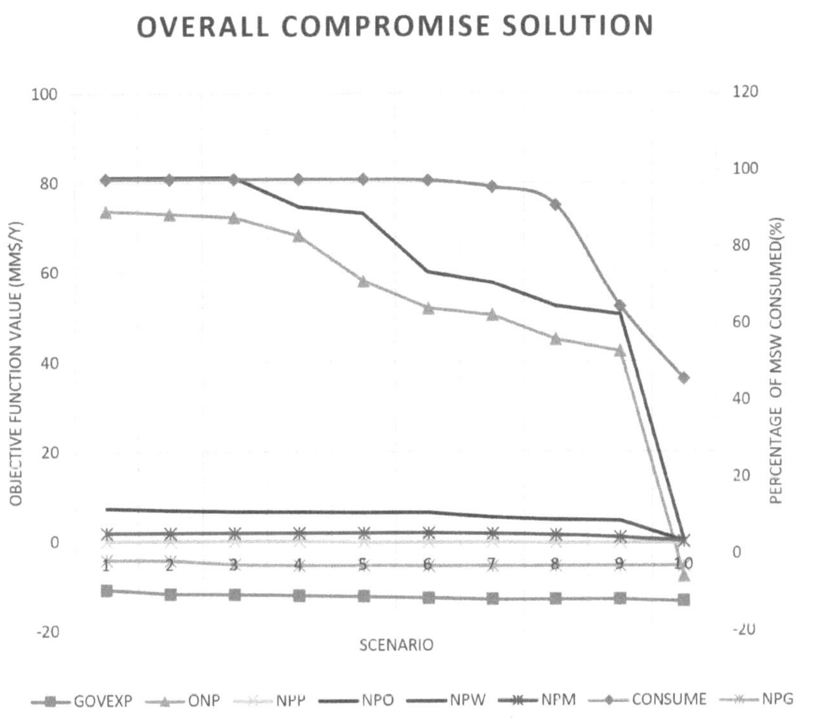

Fig. 8. Results obtained for each objective function when the TCS was minimized. [Printed (adapted) with permission from Diaz-Barriga-Fernandez, A.D., Santibañez-Aguilar, J.E., Nápoles-Rivera, F., and Ponce-Ortega, J.M. 2018. Analysis of the financial risk under uncertainty in the municipal solid waste management involving multiple stakeholders. *Computers and Chemical Engineering*, 117, 433–450].

From the results, it is possible to calculate the dissatisfaction percentage for each objective function. The dissatisfaction ratio is equal to the difference between the value of a specific objective and the desired value; divided by the desired value. Dissatisfaction value is a measure to determine the value of an objective variable concerning a target value. Figure 9 shows the dissatisfaction percentage for each objective function when the TCS was minimized. The functions with the lowest percentage of dissatisfaction in most of the scenarios were the NPRP, the NPM and the government expenditures. The best values for them were 0%, 3.2% and 3.7%, respectively, and the worst ones were 2.7%, 17

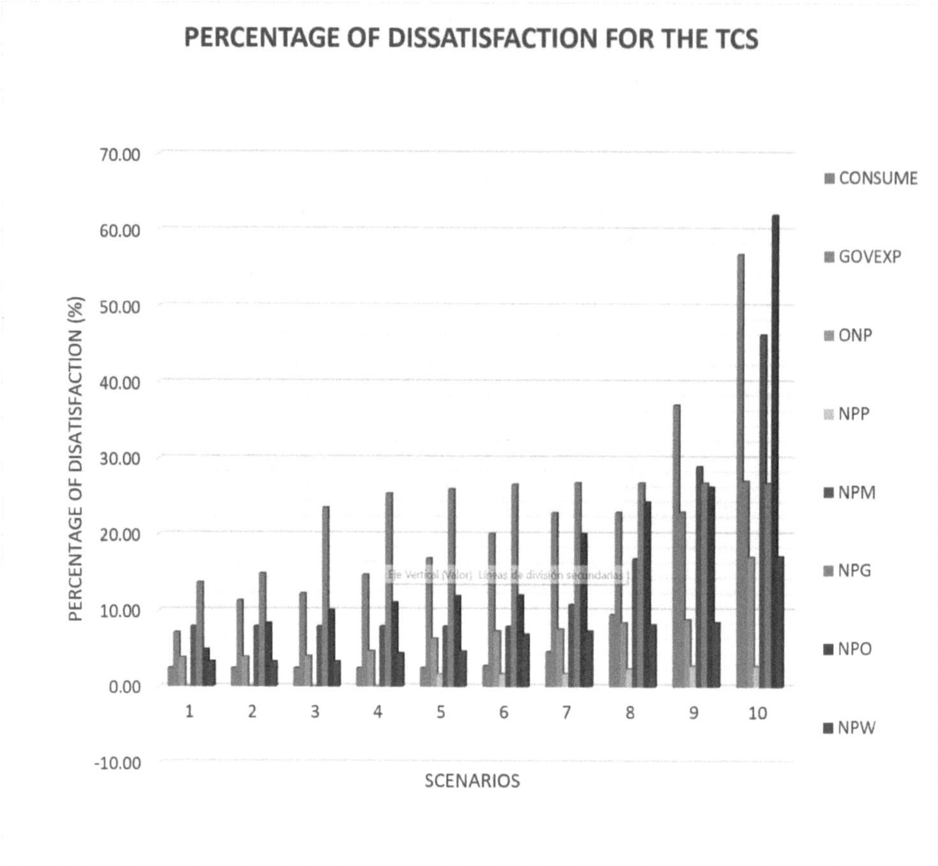

Fig. 9. Percentage of dissatisfaction obtained for each objective function when the TCS was minimized. [Printed (adapted) with permission from Diaz-Barriga-Fernandez, A.D., Santibañez-Aguilar, J.E., Nápoles-Rivera, F., and Ponce-Ortega, J.M. 2018. Analysis of the financial risk under uncertainty in the municipal solid waste management involving multiple stakeholders. *Computers and Chemical Engineering*, 117, 433–450].

and 17.1%. For the %CONSUMED, the percentage is low in most of the scenarios, just in two of ten the percentage increases drastically from 2.4% to 56.8%. Similarly, the NPM and NPP have the lowest values of 7.8% and 4.81%, respectively; then, they tend to increase to 46.3% and 61.9%.

Previously, the TCS was obtained, which produces the best solution for all the objective functions considered. However, to get a value for each objective function closer to its maximum certain limits were included for each objective function, enforcing to satisfy this limit for each objective and then all of them simultaneously Fig. 10 shows the value of the TCS vs. the limits given to the different objective functions. It is interesting to notice that if the limits increase, the TCS decreases, but this only happens if we increase the limits for all the objective functions at the same time. The limits taken for each normalized objective function are shown in Table 3.

Another important issue is about how the limits were established, in this work the limits were obtained from the results obtained for the TCS for the three financial risks and for the different normalized objective functions. For each objective function, the greatest value obtained among the three cases. The maximum values were the first limits taken, which correspond to point 1 in Fig. 10. The next limits are generated from these values.

Figure 11 shows these results for the optimistic case, where most of the objective function values are closer to their maximum. The percentage of MSW consumed is the main objective function that does not decrease with the greatest restricted limit. This makes sense because the consumption

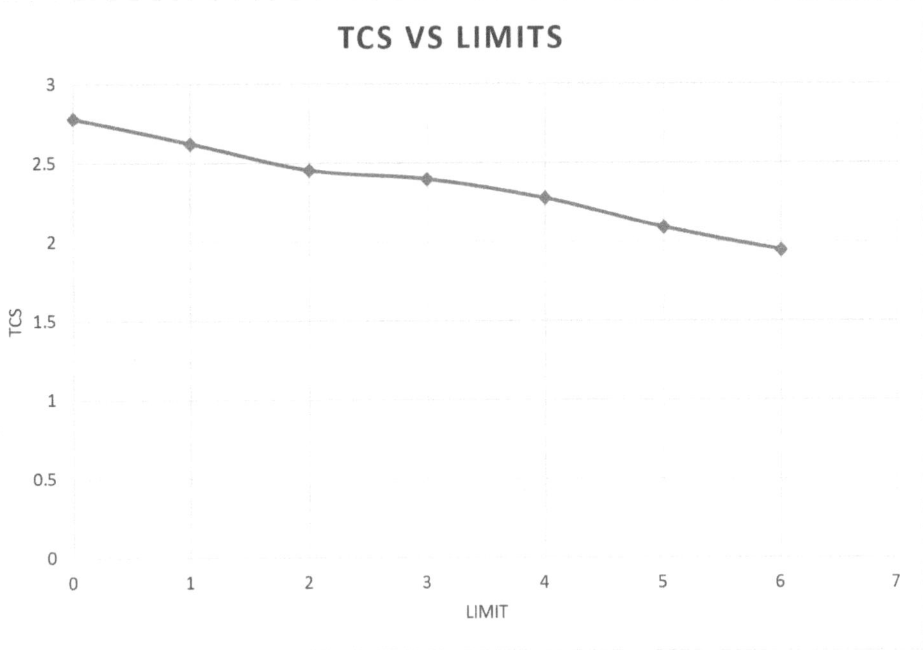

Fig. 10. Total compromise solution obtained for each limit proposed. [Printed (adapted) with permission from Diaz-Barriga-Fernandez, A.D., Santibañez-Aguilar, J.E., Nápoles-Rivera, F., and Ponce-Ortega, J.M. 2018. Analysis of the financial risk under uncertainty in the municipal solid waste management involving multiple stakeholders. *Computers and Chemical Engineering*, 117, 433–450].

Table 3. Limits taken for the normalized objective function. [Printed (adapted) with permission from Diaz-Barriga-Fernandez, A.D., Santibañez-Aguilar, J.E., Nápoles-Rivera, F., and Ponce-Ortega, J.M. 2018. Analysis of the financial risk under uncertainty in the municipal solid waste management involving multiple stakeholders. *Computers and Chemical Engineering*, 117, 433–450].

Normalized Objective Function	**Limit 1**	**Limit 2**	**Limit 3**	**Limit 4**	**Limit 5**	**Limit 6**	**Limit 7**
%CONSUMED	0.08510731	0.07659657	0.06808584	0.05957511	0.05106438	0.04255365	0.034042922
GOVEXP	0.42898944	0.42469954	0.42469954	0.42469954	0.42469954	0.42469954	0.424699543
ONP	0.11548797	0.10393917	0.09239037	0.08084158	0.06929278	0.05774398	0.046195187
NPG	0.16455633	0.14810069	0.13164506	0.11518943	0.0987338	0.08227816	0.06582253
NPM	0.10733398	0.09660059	0.08586719	0.08586719	0.08586719	0.08586719	0.085867188
NPP	0.19283851	0.17355466	0.15427081	0.13498696	0.11570311	0.09641925	0.077135403
NPRP	0.05038418	0.04534576	0.04030734	0.03526892	0.03023051	0.02519209	0.020153671
NPO	0.11633867	0.1047048	0.09307093	0.08143707	0.0698032	0.05816933	0.046535467

decreases as the other functions get closer to their maximum. In other words, the optimum profit is given by the demand for the products, so it does not depend on the availability of the residues. In the case of government expenditures, it is different, when the number of residues is to be recollected and sent to the different industries, the transportation costs are reduced. Thus, the government expenditures get closer to their minimum, while the consumed MSW gets away from its maximum. In Fig. 11 not all the objective functions decrease with the limits used.

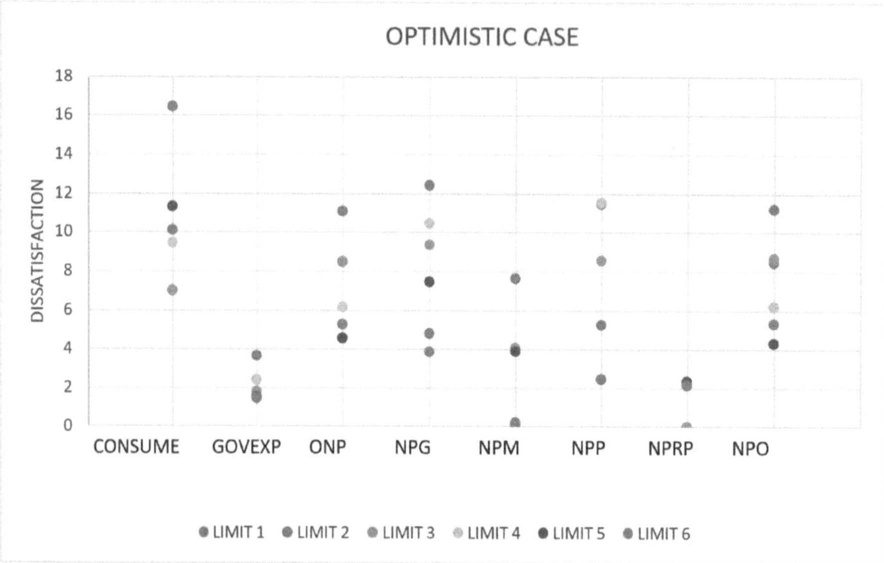

Fig. 11. Dissatisfaction for each objective function with different limits, for the optimistic case. [Printed (adapted) with permission from Diaz-Barriga-Fernandez, A.D., Santibañez-Aguilar, J.E., Nápoles-Rivera, F., and Ponce-Ortega, J.M. 2018. Analysis of the financial risk under uncertainty in the municipal solid waste management involving multiple stakeholders. *Computers and Chemical Engineering*, 117, 433–450].

When the mean financial risk is analyzed, the results are very similar to those found in the first case (Optimistic). The only difference lies in the results for the percentage of consumed MSW and the government expenditure. In this case, the government expenditures get further away from their minimum as the function is more limited, and vice versa for the percentage of MSW consumed (Fig. 12).

For the pessimistic case, the results obtained are shown in Fig. 13, which shows the same trend for all the objective functions, the limit decreases as the function gets closer to its maximum (minimum

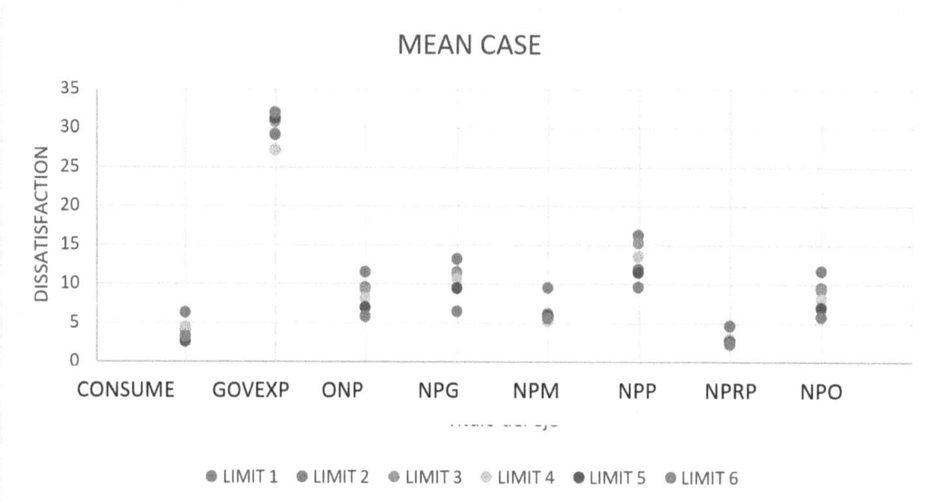

Fig. 12. Dissatisfaction for each objective function with different limits, for the mean case. [Printed (adapted) with permission from Diaz-Barriga-Fernandez, A.D., Santibañez-Aguilar, J.E., Nápoles-Rivera, F., and Ponce-Ortega, J.M. 2018. Analysis of the financial risk under uncertainty in the municipal solid waste management involving multiple stakeholders. *Computers and Chemical Engineering*, 117, 433–450].

in the case of the government expenditure). This trend is to be expected due to the limits proposed from the pessimistic case for the TCS.

Figure 14 shows the dissatisfaction for all the objective functions for the different analyzed scenarios when the smallest limits (the last limits used) for the functions is applied. These results compared with those from Fig. 9 show a considerable improvement. All the objective functions, with exception of the government expenditures, show a dissatisfaction lower than 0.11%. Although the dissatisfaction for the government expenditures is under 50%, which is worse than the results for this function when the TCS was optimized, the difference is just 20% approximately, which is not too much. Therefore, it obtained the best result for most of the objective functions at the expense of a small increase in government expenditures.

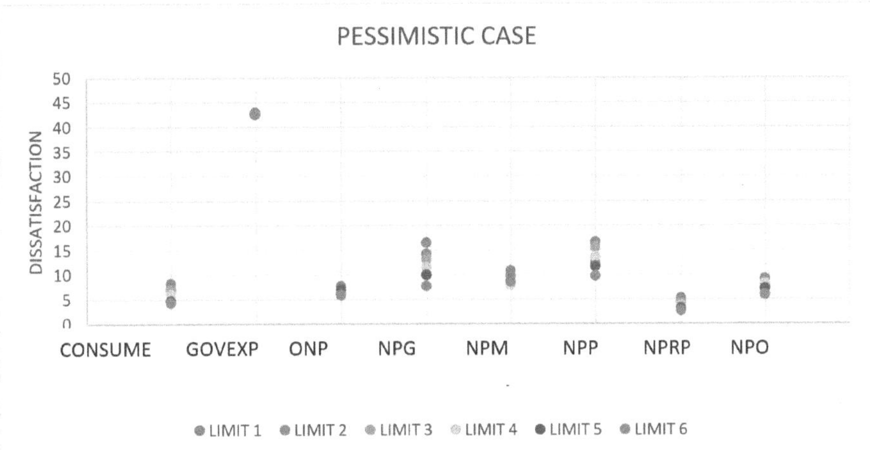

Fig. 13. Dissatisfaction for each objective function with different limits, for the pessimistic case. [Printed (adapted) with permission from Diaz-Barriga-Fernandez, A.D., Santibañez-Aguilar, J.E., Nápoles-Rivera, F., and Ponce-Ortega, J.M. 2018. Analysis of the financial risk under uncertainty in the municipal solid waste management involving multiple stakeholders. *Computers and Chemical Engineering*, 117, 433–450].

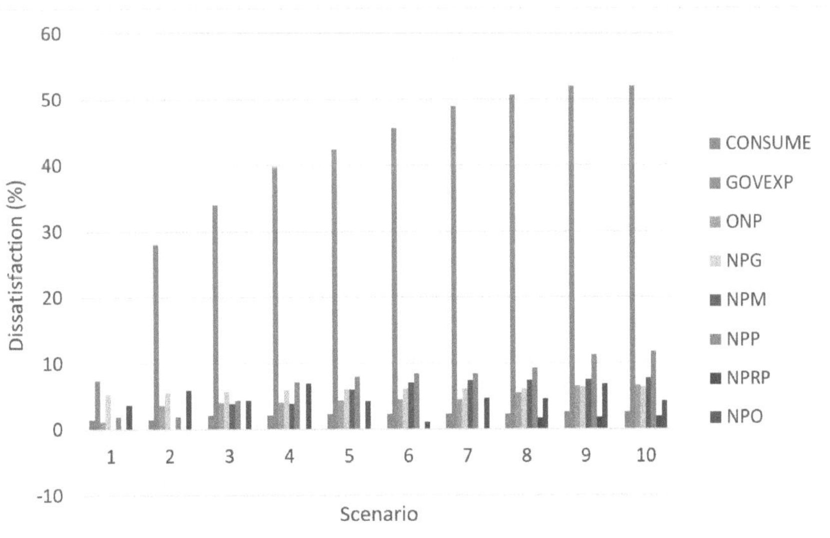

Fig. 14. Percentage of dissatisfaction obtained for each objective function when the last limit was implemented. [Printed (adapted) with permission from Diaz-Barriga-Fernandez, A.D., Santibañez-Aguilar, J.E., Nápoles-Rivera, F., Ponce-Ortega, J.M. (2018). Analysis of the financial risk under uncertainty in the municipal solid waste management involving multiple stakeholders. *Computers and Chemical Engineering*, 117, 433–450].

For all the scenarios the cost of separation, landfilling, and other activities related to the supply chain were considered constant.

7.6 Concluding Remarks

This chapter has presented an optimization formulation for strategic planning in the municipal solid waste management system, the model accounts for all the activities involved in the entire supply chain. The presented model is formulated as multi-objective optimization under uncertainty, which accounts for different levels of financial risks, and has been applied to a case study from Mexico.

As has been shown that the analysis of MSW management in the face of uncertainty in three distinct risk scenarios is a helpful tool for stakeholders' decision-making. This study shows when it makes sense to take a chance and when it is not necessary. Additionally, it provides us with a broader perspective of the most likely outcomes.

A solution that seeks to satisfy all parties involved can be obtained by applying multi-stakeholder optimization to our model. As a result, this encourages MSW management through financial and ecological incentives.

The software compels the goal function to take a value close to its maximum when the objective functions are restricted to a specific value. According to the results, this might be an intriguing tactic to ensure that all parties involved are satisfied. The decision-makers, however, have the last say, and the suggested method aids in decision-making by giving stakeholders relevant information.

7.7 Nomenclature

7.7.1 Indices

c	Index to denote the city where the residues come from.
e	Index to denote a specific scenario.
i	Index to denote the industry where the residues are treated.
obj	Index to denote the objective function.
r	Index to denote the type of residue.
s	Index to denote the economy of scale.
t	Index to denote the technology used to treat each type of residue.

7.7.2 Abbreviations

%CONSUMED	Total percentage of MSW consume (%).
GOVEXP	Government expenditures ($US/y).
NPP	Net profit for paper ($US/y).
NPM	Net profit for metal ($US/y).
NPG	Net profit for glass ($US/y).
NPRP	Net profit for recycled plastic ($US/y).
NPO	Net profit for organics ($US/y).

7.7.3 Parameters

$C_{r,c}^{separation}$	Unitary cost for separation of the residue r (*$US/kg*).
$C_{r,c}^{landfill}$	Unitary cost for landfill residues r from city c (*$US/kg*).

$C_{r}^{recollection}$	Unitary recollection cost of the residues from the cities to the processing plants (*$US/kg*).
$C_{r,t}^{Transport}$	Unitary transportation cost of the products from the industries to the cities (*$US/kg*).
$C_{r,i,t}^{processing}$	Unitary cost for processing the residue *r* through the technology *t*(*$US/kg*).
$C_{r,c,t,e}^{product}$	Unitary cost for the product obtained from the residue *r*(*$US/kg*).
$d_{c,i}$	Distance between the city *c* and the industry *i* (*km*).
$Demand_{r,c,t}$	Demand of product obtained from the type of residue *r* through technology *t* in the city *c* (*kg*).
OF_{obj}^{UB}	Upper bound of the optimistic objective function *obj*.
OF_{obj}^{LB}	Lower bound of the optimistic objective function *obj*.
MF_{obj}^{UB}	Upper bound of the mean objective function *obj*.
MF_{obj}^{LB}	Lower bound of the mean objective function *obj*.
PF_{obj}^{UB}	Upper bound of the pessimistic objective function *obj*.
PF_{obj}^{LB}	Lower bound of the pessimistic objective function *obj*.
$T_{r,i,t,s}^{min}$	Upper limit for the amount of residues processed in the industry *i* with the technology *t* (*kg*).
$T_{r,i,t,s}^{max}$	Lower limit for the amount of residues processed in the industry *i* with the technology *t* (*kg*).
$x_{r,t}^{conv}$	Efficiency production factor.
$\theta_{r,c}^{msr}$	Separation factor.

7.7.4 Variables

$CapitalCost_{r,e}$	Capital cost for processing residues *r* (*$US/y*).
$RecycledMSW_{e}$	Percentage of the total recycled wastes.
CUF_{obj}	Compromise solution for each objective function *obj*.
$F_{r,c,e}$	Amount of used residues *r* from city *c* to produce a certain product (kg).
$F_{r,c,e}^{Total}$	Total amount of residues produced from all the cities (kg).
$F_{r,c,i,e}^{industry}$	Amount of residues sent from city *c* to industry *i* (kg).
$F_{r,c,t,e}^{market/industry}$	Total amount of product sent to a market or industry in a determine city *c* (kg).
$Function_{obj,e}$	Objective function *obj*.
$Landfillcost_{e}$	Total landfilling cost for all the residues that was not treated ($US/kg).
$FC_{r,i,t,e}$	Fixed part of the capital cost for processing facilities ($US).

$Govexpnd_e$	Government expenditures ($US/y).
$MeanFunction_{obj}$	Mean objective function.
NOF_{obj}	Normalized optimistic objective function *obj*.
NMF_{obj}	Normalized mean objective function *obj*.
NPF_{obj}	Normalized pessimistic objective function *obj*.
TCS	Total compromise solution.
$OptFunction_{obj}$	Optimistic objective function.
OverallNetProfit	Overall net profit (kg/y).
$P_{r,i,t,e}$	Amount of products obtained in industry *i* with the technology *t* (kg).
$PessimisticFunction_{obj}$	Pessimistic objective function.
$Profit_{r,e}$	Net profit obtained from the residues of the type *wn* (kg/y).
$ProcessingCost_{r,e}$	Operation cost for processing residue *r* (*$US/kg*).
$SalesProfit$	Profit from the sales of the products obtained from residues of type *r* ($US/y).
$SepCost_{r,e}$	Separation cost for residues *r* ($US/kg).
$TotalSepCost_e$	Total separation cost for all the residues ($US/y).
$TotalCapitalCost_e$	Total capital cost ($US/y).
$TotalProcesingCost_e$	Total operation cost for processing all the residues ($US/y).
$TotalRecollectionCost_e$	Total Recollection cost for all the residues ($US/y).
$TotalTransportCost_e$	Total transportation cost for the products from the industries to the markets ($US/y).
$TotalSalesProfit_e$	Profit obtained from sales of the produced products ($US/y).
$TransportCost_{r,e}$	Transportation cost for the products from the industries to the markets ($US/y).
$T_{r,i,t,e}$	Amount of residues sent from industry *i* to the technology *t* (kg).
$VC_{r,i,t,e}$	Variable part of the capital cost that depends on the capacity of the processing facilities ($US/kg y).
$W_{r,c,i,t,e}$	Amount of product in a warehouse distributed in the city *c* (kg).

7.7.5 Binary Variables

$y^{industry}_{r,i,t,s,e}$	Binary variable used to define the existence of the processing plants.
$y^{op}_{obj,e}$	Binary variable used to limit the value of the optimistic objective function.

7.8 References

Achillas, C., Moussiopoulos, N., Karagiannidis, A., Banias, G., and Perkoulidis, G. 2013. The use of multi-criteria decision analysis to tackle waste management problems: a literature review. *Waste Management & Research*, 31(2), 115–129.

Carvalho, P., Marques, R. C., and Dollery, B. 2015. Is bigger better? An empirical analysis of waste management in New South Wales. Waste Management, *39*, 277–286.

Chifari, R., Piano, S. L., Bukkens, S. G., and Giampietro, M. 2016. A holistic framework for the integrated assessment of urban waste management systems. Ecological Indicators, *94*, 24–36. DOI:10.1016/j.ecolind.2016.03.006.

Christensen, T. H., and Kjeldsen, P. 1989. Basic biochemical processes in landfills. *Sanitary Landfilling. Process, Technology, and Environmental Impact.* Academic Press, New York, USA.

Dai, C., Li, Y.P., and Huang, G. H. 2012. An interval parameter chance-constrained dynamic programming approach for capacity planning under uncertainty. Resources, Conservation and Recycling, *62*, 37–50.

Diaz-Barriga-Fernandez, A.D., Santibañez-Aguilar, J.E., Radwan, N., Nápoles-Rivera, F., El-Halwagi, M.M., and Ponce-Ortega, J.M. 2017. Strategic planning for managing municipal solid wastes with consideration of multiple stakeholders. ACS Sustainable Chemistry and Engineering, *5*(11), 10744–10762.

Diaz-Barriga-Fernandez, A.D., Santibañez-Aguilar, J.E., Nápoles-Rivera, F., and Ponce-Ortega, J.M. 2018. Analysis of the financial risk under uncertainty in the municipal solid waste management involving multiple stakeholders. Computers and Chemical Engineering, *117*, 433–450.

Ding, Z., Zhu, M., Wang, Y., and Zhu, J. 2018. An AHP-GIS based model of C&D waste landfill site selection: A triangulation of critical factors. Proceedings of the 21st International Symposium on Advancement of Construction Management and Real Estate. Springer, Singapore.

Dos Muchangos, L.S., Tokai, A., and Hanashima, A. 2017. Stakeholder analysis and social network analysis to evaluate the stakeholders of a MSWM system–A pilot study of Maputo City. Environmental Development, *24*, 124–135

El-Fadel, M., Findikakis, A.N., and Leckie, J.O. 1997. Environmental impacts of solid waste landfilling. Journal of Environmental Management. *50*(1), 1–25.

Gaeta-Bernardi, A., and Parente, V. 2016. Organic municipal solid waste (MSW) as feedstock for biodiesel production: a financial feasibility analysis. Renewable Energy, *86*, 1422–1432.

Gilardino, A., Rojas, J., Mattos, H., Larrea-Gallegos, G., and Vázquez-Rowe, I. 2017. Combining operational research and Life Cycle Assessment to optimize municipal solid waste collection in a district in Lima (Peru). Journal of Cleaner Production, *156*, 589–603.

Goulart Coelho, L.M., Lange, L.C., and Coelho, H.M. 2017. Multi-criteria decision making to support waste management: A critical review of current practices and methods. Waste Management and Research, *35*(1), 3–28.

Habibi, F., Asadi, E., Sadjadi, S.J., and Banrzinpour, F. 2017. A multi-objective robust optimization model for site-selection and capacity allocation of municipal solid waste facilities: A case study in Tehran. Journal of Cleaner Production, 166, 816–834.

Harijani, A.M., Mansour, S., Karimi, B., and Lee, C.G. 2017. Multi-period sustainable and integrated recycling network for municipal solid waste–A case study in Tehran. Journal of Cleaner Production, *151*, 96–108.

He, Z., Chu, Z., Zhao, M., Zhuang, J., and Liu, F. 2017. Policy-making coordination of municipal solid waste policies in China: a content analysis. Journal of Material Cycles and Waste Management, *20*, 1073–1084.

Ho, W.S., Hashim, H., Lim, J.S., Lee, C.T., Sam, K.C., and Tan, S.T. 2017. Waste management pinch analysis (WAMPA): Application of pinch analysis for greenhouse gas (GHG) emission reduction in municipal solid waste management. Applied Energy, *185*, 1481–1489.

Hu, C., Liu, X., and Lu, J. 2017. A bi-objective two-stage robust location model for waste-to-energy facilities under uncertainty. *Decision Support Systems*, 99, 37–50.

Jafari, A., Heydari, J., and Keramati, A. 2017. Factors affecting incentive dependency of residents to participate in e-waste recycling: a case study on adoption of e-waste reverse supply chain in Iran. Environment, Development and Sustainability, *19*(1), 325–338.

Jamshidi-Zanjani, A., and Rezaei, M. 2017. Landfill site selection using combination of fuzzy logic and multi-attribute decision-making approach. Environmental Earth Sciences, *76*(13), 448. DOI: 10.1007/s12665-017-6774-7.

Javaheri, H., Nasrabadi, T., Jafarian, M.H., Rowshan, G.R., and Khoshnam, H. 2006. Site selection of municipal solid waste landfills using analytical hierarchy process method in a geographical information technology environment in Giroft. Journal of Environmental Health Science and Engineering, *3*(3), 177–184.

Kjeldsen, P., Barlaz, M.A., Rooker, A.P., Baun, A., Ledin, A., and Christensen, T.H. 2002. Present and long-term composition of MSW landfill leachate: a review. Critical Reviews in Environmental Science and Technology, *32*(4), 297–336.

Korai, M.S., Mahar, R.B., and Uqaili, M.A. 2017. The feasibility of municipal solid waste for energy generation and its existing management practices in Pakistan. Renewable and Sustainable Energy Reviews, *72*, 338–353.

Laokhongthavorn, L., and Chalida, U. 2017. A multi-objective optimization model for solid waste disposal under uncertainty: a case study of Bangkok, Thailand. International Journal of Civil Engineering, *15*(2), 205–212.

Lee, C.K.M., Yeung, C.L., Xiong, Z.R., and Chung, S.H. 2016. A mathematical model for municipal solid waste management–A case study in Hong Kong. Waste Management, *58*, 430–441.

López-Toro, A.A., Rubio-Romero, J.C., Suárez-Cebador, M., and Arjona-Jiménez, R. 2016. Consideration of stakeholder interests in the planning of sustainable waste management programmes. Waste Management and Research, *34*(10), 1036–1046.

Ma, X., Ma, C., Wan, Z., and Wang, K. 2017. A fuzzy chance-constrained programming model with type 1 and type 2 fuzzy sets for solid waste management under uncertainty. Engineering Optimization, *49*(6), 1040–1056.

McLean, K., and Li, X. 2013. Robust scenario formulations for strategic supply chain optimization under uncertainty. Industrial and Engineering Chemistry Research, *52*(16), 5721–5734.

Mirmajlesi, S.R., and Shafaei, R. 2016. An integrated approach to solve a robust forward/reverse supply chain for short lifetime products. Computers and Industrial Engineering, *97*, 222–239.

Mishra, H., Karmakar, S., Kumar, R., and Singh, J. 2017. A framework for assessing uncertainty associated with human health risks from MSW landfill leachate contamination. Risk Analysis, *37*(7), 1237–1255.

Moeinaddini, M., Khorasani, N., Danehkar, A., and Darvishsefat, A.A. 2010. Siting MSW landfill using weighted linear combination and analytical hierarchy process (AHP) methodology in GIS environment (case study: Karaj). Waste Management, *30*(5), 912–920.

Morrissey, A.J., and Browne, J. 2004. Waste management models and their application to sustainable waste management. Waste Management, *24*(3), 297–308.

Ng, W.P Q., Lam, H.L., Varbanov, P.S., and Klemeš, J.J. 2014. Waste-to-energy (WTE) network synthesis for municipal solid waste (MSW). Energy Conversion and Management, *85*, 866–874.

Nguyen-Trong, K., Nguyen-Thi-Ngoc, A., Nguyen-Ngoc, D., and Dinh-Thi-Hai, V. 2017. Optimization of municipal solid waste transportation by integrating GIS analysis, equation-based, and agent-based model. Waste Management, *59*, 14–22.

Niziolek, A.M., Onel, O., and Floudas, C.A. 2017. Municipal solid waste to liquid transportation fuels, olefins, and aromatics: Process synthesis and deterministic global optimization. Computers and Chemical Engineering, *102*, 169–187.

Pan, S.Y., Du, M.A., Huang, I.T., Liu, I.H., Chang, E.E., and Chiang, P.C. (2015). Strategies on implementation of waste-to-energy (WTE) supply chain for circular economy system: a review. Journal of Cleaner Production, *108*, 409–421.

Pavão, L.V., Pozo, C., Costa, C.B., Ravagnani, M.A., and Jiménez, L. 2017. Financial risks management of heat exchanger networks under uncertain utility costs via multi-objective optimization. *Energy*, 139, 98–117.

Pin, B.V.R., Barros, R.M., Lora, E.E.S., and Dos Santos, I.F.S. 2017. Waste management studies in a Brazilian microregion: GHG emissions balance and LFG energy project economic feasibility analysis. Energy Strategy Reviews, DOI:10.1016/j.esr.2017.11.002.

Santibañez-Aguilar, J.E., Flores-Tlacuahuac, A., Rivera-Toledo, M., and Ponce-Ortega, J.M. 2017. Dynamic optimization for the planning of a waste management system involving multiple cities. Journal of Cleaner Production, 165, 190–203.

Santibañez-Aguilar, J.E., Martínez-Gómez, J., Ponce-Ortega, J.M., Nápoles-Rivera, F., Serna-González, M., and El-Halwagi, M.M. 2014. An optimal planning for the reuse of municipal solid waste considering economic, environmental and safety objectives. Computer Aided Chemical Engineering, 33, 1027–1032.

Santibañez-Aguilar, J.E., Ponce-Ortega, J.M., González-Campos, J.B., Serna-González, M., and El-Halwagi, M.M. 2013. Optimal planning for the sustainable utilization of municipal solid waste. Waste Management, *33*(12), 2607–2622.

Soltani, A., Hewage, K., Reza, B., and Sadiq, R. 2015. Multiple stakeholders in multi-criteria decision-making in the context of municipal solid waste management: a review. Waste Management, *35*, 318–328.

Subramanyam, A., Mufalli,F, Pinto J.M., and Gounaris, C. E. 2017. Robust Multi-Period Vehicle Routing under Customer Order Uncertainty. http://www.optimization-online.org/DB_HTML/2017/04/5947.html

Sukholthaman, P., and Sharp, A. 2016. A system dynamics model to evaluate effects of source separation of municipal solid waste management: A case of Bangkok, Thailand. Waste Management, *52*, 50–61.

Sukholthaman, P., and Shirahada, K. 2015. Technological challenges for effective development towards sustainable waste management in developing countries: Case study of Bangkok, Thailand. Technology in Society, *43*, 231–239.

Sumathi, V. R., Natesan, U., and Sarkar, C. 2008. GIS-based approach for optimized siting of municipal solid waste landfill. Waste Management, *28*(11), 2146–2160.

Sun, N., and Chungpaibulpatana, S. 2017. Development of an appropriate model for forecasting municipal solid waste generation in Bangkok. Energy Procedia, *138*, 907–912.

Tan, S.T., Lee, C.T., Hashim, H., Ho, W.S., and Lim, J.S. 2014. Optimal process network for municipal solid waste management in Iskandar Malaysia. Journal of Cleaner Production, *71*, 48–58.

Taifouris, M. R, and Martín, M. 2017. Multiscale scheme for the optimal use of residues for the production of biogas across Castile and Leon. Journal of Cleaner Production, *185*, 239–251.

Vadenbo, C., Hellweg, S., and Guillén-Gosálbez, G. 2014. Multi-objective optimization of waste and resource management in industrial networks–Part I: Model description. Resources, Conservation and Recycling, *89*, 52–63.

Veccietti, A., Lee, S., and Grossmann, I. 2003. Modeling of discrete/continuous optimization problems: characterization and formulation of disjunctions and their relaxations. Computers and Chemical Engineering, *27*(3), 433–448

Wang, Z., Ren, J., Goodsite, M.E., and Xu, G. 2017. Waste-to-energy, municipal solid waste treatment, and best available technology: Comprehensive evaluation by an interval-valued fuzzy multi-criteria decision making method. Journal of Cleaner Production, *172*, 887–899.

Warren, A., and El-Halwagi, M. 1996. An economic study for the co-generation of liquid fuel and hydrogen from coal and municipal solid waste. Fuel Processing Technology. *49*(1–3), 157–166.

Xu, Z., Elomri, A., Pokharel, S., Zhang, Q., Ming, X.G., and Liu, W. 2017. Global reverse supply chain design for solid waste recycling under uncertainties and carbon emission constraint. Waste Management, *64*, 358–370.

Yadav, V., Bhurjee, A.K., Karmakar, S., and Dikshit, A.K. 2017. A facility location model for municipal solid waste management system under uncertain environment. Science of the Total Environment, *603*, 760–771.

Zhang, X., and Huang, G. 2013. Optimization of environmental management strategies through a dynamic stochastic probabilistic multiobjective program. Journal of Hazardous Materials, 246–247, 257–266.
Zhang, Y., Huang, G.H., and He, L. 2014. A multi-echelon supply chain model for municipal solid waste management system. Waste Management, 34(2), 553–561.
Zheng, L., Song, J., Li, C., Gao, Y., Geng, P., and Qu, B. 2014. Preferential policies promote municipal solid waste (MSW) to energy in China: Current status and prospect. Renewable and Sustainable Energy Reviews, 36, 135–148.

8

Optimization Framework for Solid Waste Management

Municipal solid waste management (MSW) is an important but complex logistical problem. The deployment of MSW management systems is hindered by the ever-growing generation of waste and the insufficient infrastructure to manage, process, and dispose of waste. This chapter presents a coordinated framework for complex MSW management systems. The framework accommodates multiple key stakeholders in MSW systems, such as suppliers of waste, consumers of waste and derived products, and providers of transportation and processing services. Here, the stakeholders submit bids to a coordinator that solves an optimization problem to determine allocations and clearing prices that maximize the collective profit for all stakeholders and that balance supply and demand for waste and products. Furthermore, the clearing process guarantees that the individual profits are non-negative (no stakeholder loses money). Notably, the framework operates as a competitive market that accelerates transactions between stakeholders and that handles complex logistical constraints that would be difficult to handle in peer-to-peer transactions. The framework also facilitates the integration of policy incentives and the monetization of environmental impacts. In this regard, we evaluate a tax applied to open dump disposal. To illustrate the applicability, an MSW system in Mexico was analyzed as a case study. Results reveal that taxation can be used to incentivize the provision of services for all stakeholders. Specifically, we found that an appropriate tax can completely avoid disposal in open dumps. A tax of 5.1 USD/tonne was identified as the minimum penalization that avoids diverting waste to open dumps.

8.1 Introduction

Around the world, waste management is becoming an increasingly pressing issue. This is especially true in developing nations, where trash volume is rapidly rising and there is insufficient infrastructure for collection and processing. (Yousefloo and Babazadeh 2020). The lack of these types of systems has led to significant social and environmental issues that increase year by year with the generation of waste. According to the World Bank, the annual solid waste generation globally was 1.3 billion tons in 2012 and is expected to grow to 2.2 billion tons by 2025 (Hoornweg and Bhada-Tata 2012). In the USA, the waste generation per day is approximately 0.64 MT, followed by Germany with 0.14 MT, Mexico with 0.13 MT, and Japan with 0.10 MT (Das et al. 2019).Large urban centers are experiencing a growing shortage of landfill space and collection and processing facilities, which may necessitate the use of uncontrolled (open dump) sites for disposal (Ojeda-Benitez and Beraud-Lozano 2003). The use of open dump systems is a common practice in developing countries such as Mexico. The Mexican environmental protection agency (SEMARNAT) reported in 2012 that, of all waste generated in the country, 72% was disposed of at sanitary landfills and regulated sites, 23% was disposed at open dumps, and only 5% was recycled (Semarnat 2012).

Unfortunately, the environmental, social, and safety impacts of open dump systems have not received as much attention from policymakers and academics (Medina 2010). These systems do not provide technologies for controlled landfills, such as leachate treatment, geological protection, and gas treatment (Ojeda-Benitez and Beraud-Lozano 2003). Methane, which is created when organic materials break down, can therefore seep into the environment and start fires. Strong leachates can also contaminate groundwater and surface water. Food scraps can draw wildlife, which can infect people with diseases. Furthermore, it might take years for organic waste to biodegrade, which restricts how much area the open system can use in the future (Medina 2010).

Recently, the collection, storage, and recycling of solid wastes in Mexico have started to be incentivized. Recycling has increased from 2.3% in 2000 to 5% in 2012 (Botello-Álvarez et al. 2018). However, this percentage remains very small compared to waste recycled worldwide (approximately 19%) (Gutberlet 2015). Several policies and financing strategies are currently being investigated to accelerate the deployment of more advanced waste management systems around the world. A hypothetical scenario reported in Korea considers a tax collection framework that incentivizes the deployment of recycling facilities. This analysis indicates that people are willing to bear the cost of discarding recyclable waste (Ko et al. 2020). Similarly, studies in Serbia have revealed that residents are in general willing to pay for a pharmaceutical disposal program (Paut Kusturica et al. 2020). These studies also contend that financial resources for waste collection and policy regulations around waste dumping are necessary. However, a fictitious landfill tax scenario in Israel demonstrates that adding externality costs to recycling does not significantly improve it (Lavee 2007). Furthermore, a management cycle for waste, including the participation of infrastructure managers, workers, and households, was recently proposed by Jiang et al. (2020). Here, the waste-dumping behavior and empirical decisions of each part are considered to guide policy regulations.

Financial sustainability is a major issue in the design of solid waste management (SWM) systems. In this regard, a cost-revenue analysis for a SWM system in Bahir Dar revealed that the cost-structure of waste services must be improved to enable sustainability (Lohri et al. 2014). Improving fee collection, expanding the value chain through sales of recyclables and derived products, incorporating finance mechanisms (such polluter payments and cross-subsidies), and enhancing overall efficiencies are some of the solutions this study suggested to accomplish this. Additionally, there have been reports of models for industrial symbiosis-based trash recycling between businesses. Here, different supply-demand relationships were considered to find optimal waste pricing decisions. The results show that industries tend to cooperate when the marginal cost of recycling is lower than the cost of the raw materials (He et al. 2020). An important observation is that this symbiosis is effectively a coordinated market framework. Recently, a general coordinated market framework for organic waste that facilitates transactions between multiple stakeholders (suppliers of waste, consumers of waste and derived products, transportation providers, and processing facilities) was proposed by Sampat et al. (2019). This strategy looks for defined prices for waste and derived products as well as allocations for all parties involved in order to optimize their combined profit. Notably, it is demonstrated that the management framework is comparable to a market with competition and inherits important economic characteristics of such markets, including revenue adequacy (the costs of garbage supply, collection, and processing are covered by consumer payments). Moreover, this framework can help monetize environmental impacts to foster investment and development efforts. The authors argue that the framework can be used to predict the impact of distinct regulations or incentives and is scalable since it provides open access that promotes transactions between large numbers of stakeholders.

Other optimization models have been previously proposed for MSW management systems. A supply chain optimization model for an MSW system that considers economic, social, and environmental factors was proposed by Mohammadi et al. (2019). Furthermore, a mathematical optimization approach that includes waste reduction processes and landfilling has been reported by Garibay-Rodriguez et al. (2018). The results show that the deployment of a landfill gas-to-energy system and a material recycling facility can improve the overall economics of the MSW system (a step towards achieving financial sustainability). Regarding the economic efficiency of recycling, an approach that reveals its viability has been reported by Lavee (2007). The presented results show that 51% of the municipalities of Israel can benefit from adopting recycling. The beneficiaries were mostly large municipalities. An important issue here is the uncertainty related to the price of recycled materials, which may discourage municipalities and make them prefer landfill disposal (Lavee et al. 2004). To overcome this problem, options to stabilize prices and ensure long-term contracts with recycling plants have been proposed. This approach avoids unnecessary investments caused by the repeated change of disposal methods (Lavee et al. 2009). Moreover, a multi-objective model including the minimization of costs and risk objectives for an MSW supply chain network has been developed by Yousefloo and Babazadeh (2020). The risk function captures the population affected by waste treatment centers and emissions that result from waste processing. This work involves a real case study in Iran. Another multi-objective model to allocate waste to treatment technologies was proposed by Minoglou and Komilis (2013). Here, operational and transportation costs are considered, as well as different processing options (incineration, compost, anaerobic digestion, and landfilling). The minimization of costs and emissions is used as an objective function. Distinct methods to capture the collection and transportation of waste were reported by Paul et al. (2018). In their linear programming model, the optimal allocation of waste to different processing technologies was evaluated. Mixed-integer programming models have also been proposed to find the optimal number of collection trucks in an MSW system in Hong Kong (Lee et al. 2016). A mathematical model to perform optimal planning that maximizes the profit has been reported by Santibañez-Aguilar et al. (2013). The maximizing of recycled garbage is taken into consideration here. The model selects the best technologies and their geographic locations, as well as how waste and goods are distributed from and to various places. The findings also point to trade-offs between environmental and economic goals. These studies frequently have the flaw of adopting a "central" perspective on the issue, where one stakeholder controls the entire MSW system. Consequently, these models don't offer any information about how various system components ought to be compensated. Additionally, prior methods have not taken into account the revenues of each stakeholder involved in the MSW system. In particular, there hasn't been any reporting of a system that takes into account the overall well-being or profit of all stakeholders in addition to their individual gains and allocations. These factors are crucial since every member matters in an actual MSW management program. Furthermore, because consumer revenue supports supplier and provider payments, balancing supply and demand through a coordinated structure prevents financial losses.

This chapter presents an optimization formulation for MSW management systems within a coordinated market framework. The proposed model includes the optimal planning for MSW considering different alternatives for the disposal of waste: sanitary landfills, open dumps, and reuse. Specifically, the mathematical model involves several stakeholders: urban centers that produce waste, sanitary landfills and open dump systems, processing technologies for different types of waste, and financial consumers of derived products. From a coordinated market perspective, suppliers, and consumers (demand) of waste and products, as well as providers (both transportation and technology), can be identified in the MSW system. Additional details about coordinated management are explained in section 8.2. Section 8.4 shows how to use the coordinated framework to identify suitable tax structures for the disposal of waste in open dumps. The results reveal that taxation can effectively incentivize the provision of services by all stakeholders. Specifically, we found that an appropriate tax can avoid open dump disposal completely. Moreover, the optimal allocation that provides positive profits for each stakeholder was identified through the coordinated framework. To illustrate the applicability of this approach, an MSW system in Mexico was analyzed as a case study. However, the formulation is general and can be applied to any MSW system. In this regard, our objective is to provide a coordinated management perspective for MSW systems by including taxation to avoid negative environmental impacts. Particularly, we focused on the impact of open dump systems, but the approach can be extended to include other environmental issues.

The presented approach is structured as follows: section 8.2 describes the elements of the MSW system and additional considerations for coordinated management. Section 8.3 describes the mathematical model formulation of the proposed system. Section 8.4 refers to the applicability of the formulation through a case study in the central-west region of Mexico.

8.2 Coordinated MSW Management System

The proposed system comprises different suppliers, consumers, and providers of transportation and processing (transformation) services involved in MSW management. As mentioned in section 8.1, these stakeholders can be identified from a coordinated market perspective. Furthermore, each of these stakeholders manages distinct waste or products at a particular geographical location. These considerations can be visualized in the superstructure shown in Fig. 1. Here, we can see that several urban centers (geographical locations) are considered, and each of them involves a specific generation rate of different types of waste (plastic, metal, organic, glass, and non-recyclables). Moreover, these types can be classified into subtypes (such as clear, green, and brown glass). Note that the urban centers act as suppliers of waste for the sanitary landfills (which act as consumers) and the processing facilities. Urban centers also act as consumers who purchase products from the processing facilities. It is considered that the waste that is not allocated to consumers is sent to open dumps. This is a common practice in developing countries, despite the environmental, social, and safety impacts (Medina 2010). In a typical situation, it is assumed that this can be done at no cost (or at a small cost). This cost is economic and does not involve the environmental cost, which can be greatly expensive. Therefore, to prevent open dump disposal, we include a tax for the waste disposed of at such systems. This tax can be interpreted as a service that the environment provides to society (the environment absorbs the impacts of the waste but at a cost). The MSW system also includes sanitary landfills, open dumps, and processing facilities in each urban center. The processing providers offer different types of treatment and technologies for each waste type. The transportation providers can move waste to sanitary landfills, to processing facilities, and final consumers. The stakeholders are categorized by the type of waste they handle (plastic, metal, organic, glass, and non-recyclables).

Some additional considerations of the coordination system are explained in the following. First, suppliers, consumers, and service providers submit bids into the coordinated system. Then,

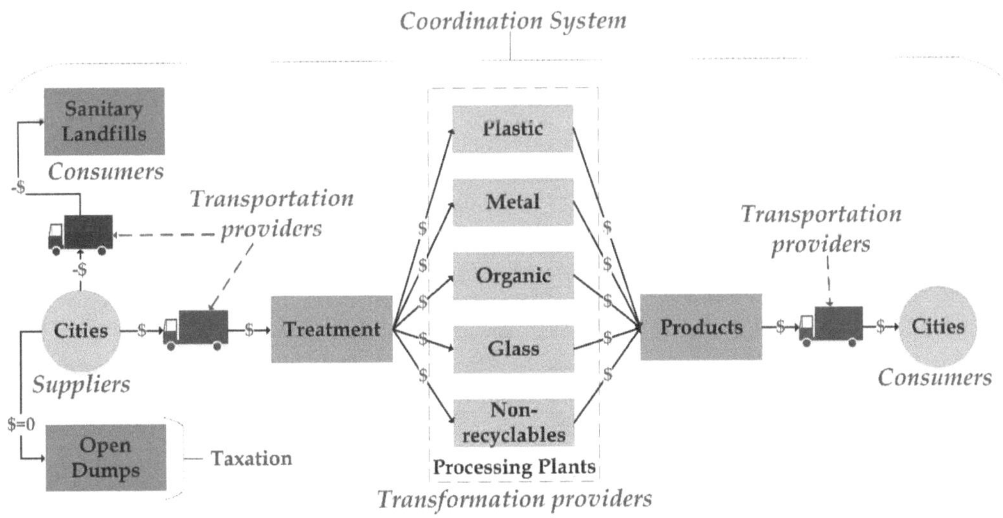

Fig. 1. Proposed superstructure for the coordinated MSW system. [Printed (Adapted) with permission from Munguia-López, A.C., Zavala, V.M., Santibañez-Aguilar, J.E., and Ponce-Ortega, J.M. 2020. Optimization of municipal solid waste management using a coordinated framework. Waste Management, *115*, 15–24].

a coordinator, known as an independent system operator (ISO), uses bid information to determine allocations and prices to clear the market. With the exception of landfills, which submit negative bids, all parties submit positive offers. On the one hand, landfill suppliers' negative offer represents the amount they are prepared to pay the market to have their garbage removed. The landfill will only accept the waste if it is compensated for doing so, such as by a tipping fee or disposal cost, according to the negative bid of a landfill demand. With this knowledge, the coordinator solves an optimization problem to optimize the overall welfare (collective profit of all stakeholders), clearing the market.. The optimization formulation solved by the ISO includes the sales and costs of the market stakeholders. The cleared stakeholders are paid based on their product allocations and associated clearing prices. Transportation providers are paid based on differences in clearing prices at the source and destination locations, while the processing providers are paid based on the clearing prices or their input and output products. When a stakeholder is not cleared (it is allocated no product), this means that this player does not participate in the market. All these considerations are used for the formulation of the model, as shown in section 8.3. Furthermore, two scenarios are analyzed: I) the base case (without taxation) and II) a case with a tax for waste that ends up in open dumps. This tax represents the service that the environment provides and may prevent open dump disposal. The solutions to the ISO problem satisfy a set of fundamental economic properties of a competitive market (Sampat et al. 2019). Specifically, the clearing process guarantees that no cleared player loses money and that there is revenue adequacy (total payments collected equal total payments made) (Munguia-López et al. 2020).

8.3 Formulation of the Coordination Problem

The proposed model is deterministic, and it is based on the superstructure shown in Fig. 1 as well as on the considerations mentioned above. In the following, the equations of the model for the MSW management system using the coordinated framework are presented. Additionally, complementary equations, based on the formulation proposed by Santibañez-Aguilar et al. (2013), are shown in Chapter 4. As stated in section 8.2, we consider a framework that is composed of a set of geographical locations *N*, products *P*, consumers *D*, suppliers *S*, transportation providers *K*, and transformation (technology) providers *M*.

In the addressed MSW system, previous elements are identified as (i) the geographical locations refer to where waste is generated, where the products are consumed, and where sanitary landfills and processing plants are placed; (ii) the products represent the different types of waste and derived products obtained from the processing facilities; (iii) the suppliers represent the urban centers that generate waste, while the consumers are the urban centers that demand waste (sanitary landfills) and useful products (from processing plants); (iv) the transportation providers refer to the service of transport to move the waste from urban centers to landfills and plants, and from plants to urban centers; and (v) the processing providers refer to the different technologies to treat waste.

The bidding information $(\alpha_d, \beta_s, \gamma_k, \delta_m)$ as well as the maximum capacities $(c_d^*, g_s^*, q_k^*, f_m^*)$ for the demand (c_d), supply (g_s), transportation (q_k), and transformation (f_m) are given (as previously indicated in section 8.2). The objective function in Equation (8.1) maximizes the total welfare, which refers to the difference between the demand served and the costs of supply, transportation, and transformation. The solution to the problem includes finding the corresponding allocations (c_d, g_s, q_k, f_m) and prices that clear the market and maximize the collective profit of all the stakeholders (total welfare). These allocations satisfy the physical conservation laws in Equation (8.2), and capacity constraints in Equations (8.3-8.6). The dual variables $(\pi_{n,p})$, $(n,p) \in N \times P$ act as market clearing prices by setting values for products at different locations. Here, $\varsigma_{m,p}$ refers to the conversion factor for each technology and product.

$$\max \sum_{d \in D} \alpha_d c_d - \sum_{s \in S} \beta_s g_s - \sum_{k \in K} \gamma_k q_k - \sum_{m \in M} \delta_m f_m \tag{8.1}$$

$$\text{s.t.} \sum_{s\in S_{n,p}} g_s - \sum_{d\in D_{n,p}} c_d + \sum_{k\in K_{n,p}^{in}} q_k - \sum_{k\in K_{n,p}^{out}} q_k + \sum_{m\in M_n} \varsigma_{m,p} f_m = 0, \quad (n,p)\in N\times P, \ (\pi_{n,p}) \tag{8.2}$$

$$0 \le c_d \le c_d^*, \quad d\in D \tag{8.3}$$

$$0 \le g_s \le g_s^*, \quad s\in S \tag{8.4}$$

$$0 \le q_k \le q_k^*, \quad k\in K \tag{8.5}$$

$$0 \le f_m \le f_m^*, \quad m\in M \tag{8.6}$$

The computed allocations and prices from the optimization problem are used to remunerate providers and charge consumers. This leads to revenue adequacy, which means that the revenue collected is equal to the payments made. We use the notation π_d, π_s, π_k, π_m to refer to the locational marginal prices (clearing prices) at the corresponding locations of each stakeholder. For consumers, $\alpha_d c_d$ refers to the monetary value of the allocated demand, while $\pi_d c_d$ is the payment made to the market. Therefore, the profit for consumers (ϕ_d^D) is the difference between these values, as shown in Equation (8.7). For suppliers, $\pi_s g_s$ represents their revenue and $\beta_s g_s$ refers to their operating cost. The profit for suppliers is thus the difference between these values (Equation (8.8)). The profit for transportation providers is estimated by Equation (8.9). Here, π_k are the transportation prices that are estimated by the difference between the prices of the destination nodes and the prices of the origin nodes. The quantity $\pi_k q_k$ is the payment made to the transportation providers and $\gamma_k q_k$ is their operating cost. The transformation prices π_m are calculated as a weighted sum of marginal prices (weighted by conversion factors) for the products involved in the processing step. Note that the conversion factors are given parameters. The profit of these providers is computed as shown in Equation (8.10), $\pi_m f_m$ represents their revenue while $\delta_m f_m$ is their operating cost.

$$\phi_d^D = (\alpha_d - \pi_d)c_d, \quad d\in D \tag{8.7}$$

$$\phi_s^S = (\pi_s - \beta_s)g_s, \quad s\in S \tag{8.8}$$

$$\phi_k^K = (\pi_k - \gamma_k)q_k, \quad k\in K \tag{8.9}$$

$$\phi_m^M = (\pi_m - \delta_m)f_m, \quad m\in M \tag{8.10}$$

The clearing process guarantees that these profits are non-negative (no stakeholder loses money), this is a key benefit of the coordinated market.

8.4 Results and Discussion

We apply our framework to a case study that seeks to analyze how an MSW system would operate in the central-west region of Mexico. As shown in Fig. 2, five urban centers (Morelia, Celaya, Apatzingan, Lazaro Cardenas, and Leon) act as suppliers and consumers in the coordination system. Here, the different participants are identified as follows: *S* for suppliers, *D* for consumers, *K* for transportation providers, and *M* for processing providers. The numbers *1–5* denote where each stakeholder is situated. For instance, the technology provider *M1* is situated in the city of Morelia. Note that each processing provider has equal technologies locally available to treat each type of waste (plastic, metal, organic, glass, and non-recyclables). The different types of plastic are denoted by *R1*, *R2*, *R3*, *R4*, and *R5*. The different types of glass are denoted by *G1*, *G2*, and *G3*. Representative parameters of this application, such as the generation rate of waste in each city, are shown in Chapter 4. The complete information of the case study, as well as the corresponding parameters, have been previously reported and can be

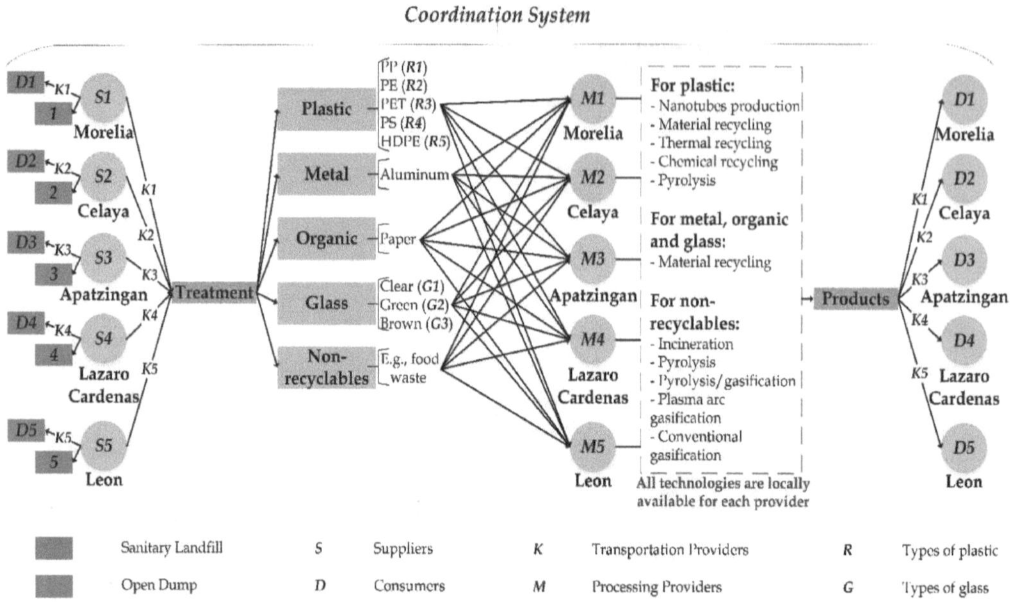

Fig. 2. Schematic representation of the addressed case study for the MSW system. [Printed (Adapted) with permission from Munguia-López, A.C., Zavala, V.M., Santibañez-Aguilar, J.E., and Ponce-Ortega, J.M. 2020. Optimization of municipal solid waste management using a coordinated framework. *Waste Management*, 115, 15–24].

found in Santibañez-Aguilar et al. (2013). Note that Fig. 2 is a schematic representation of the case study to facilitate the visualization of all the participants and the possibilities of the system.

Associated with each of the MSW system participants there is a specific flow, product type, capacity, location, and bidding cost. For all types of waste, the same geographical locations (nodes) are considered, along with two possible pathways: sending the waste to a processing facility for treatment and sending the waste to a sanitary landfill. It also included the possibility of sending waste to open dumps. To account for this environmental impact, two scenarios are analyzed: I) base case in which we ignore the impact of dumps, and II) a tax is applied to any quantity of waste disposed of at open dumps. After solving the problem, we find that the total welfare (collective profit) of the system is 871,744 USD and 784,061 USD for Scenarios I and II, respectively. As expected, accounting for the environmental impact incurs a penalty in the total welfare. Note that previous approaches have included taxation schemes, but to foster recycling. For instance, a tax applied to landfill disposal to increase the level of recycling has been reported by Lavee (2007). Furthermore, a tax collection scenario to install a recycling facility was evaluated by Ko et al. (2020). However, we focus on the taxation for open dump disposal to monetize this environmental impact. Our approach also considers processing facilities to treat waste and sanitary landfill disposal, but without applying direct taxation to these activities.

A tax of 5.1 USD/tonne was used, this value was identified as the minimum penalization that avoids diverting waste to open dumps. To obtain this value, we first used a tax equal to 12.35 USD/tonne (which corresponds to the cost of sending waste to the landfill). Then, the obtained lowest prices (marginal values) for the landfill supply of all types of waste were identified and evaluated as tax values. Distinct values from these prices were evaluated until the reported minimum tax (5.1 USD/tonne) was found. Comparing the allocations (profits) in the following results to those obtained in Appendix A (found in the Appendix section of this book) with a tax equal to 12.35 USD/tonne, we can see that the minimum tax yields similar profit levels for all stakeholders. These results illustrate how our methodology can be used to systematically identify policies that incentivize different stakeholders in MSW systems.

The impact of the variability of the distinct prices involved in MSW systems has been previously addressed. For instance, the prices of waste recycling in an industrial park were analyzed by He et al. (2020). Mohammadi et al. (2019) found that the profit of an MSW system, involving the efficient use of waste, is affected by changes in electricity prices. Lavee et al. (2009) reported that the uncertainty in recycling prices may make decision-makers prefer landfill disposal. On the other hand, our approach provides the clearing prices for all the involved stakeholders in the MSW system, including suppliers, consumers, and providers (of transportation and transformation). It should be noted that depending on the scenario (with or without taxation for open dump disposal), the prices vary. Nevertheless, the prices related to the transformation providers do not change with the taxation since the tax does not have an impact on the recycled waste. The profits for each participant are estimated as well. Note that all the profits are non-negative. This occurs because participants do not lose money within a coordinated system.

8.5 Analysis of Plastic Waste

First, they analyzed the solutions for the evaluated suppliers (*S1*-*S5*), consumers (*D1*-*D5*), and transportation providers (*K1*-*K5*) of plastic. No transformation providers participate in the optimal solution for this type of waste. Thus, supply, demand, and transport refer to the plastic sent to the landfill. For both scenarios, the supply and demand offer negative bid costs (Table 1). These costs are equal for all types of plastic. The landfill supply with a negative bid is a payment that the suppliers are willing to pay the market to take away the waste. On the other hand, it was reported that disposal facilities can act as consumers (Sampat et al. 2019). Specifically, the landfill demand with a negative bid indicates that the landfill will take the waste only if it is paid for doing so (such as a disposal cost or a penalty cost). This behavior is included in the model because such incentives (suppliers are willing to pay) and penalties (disposal cost) are common in this type of market to promote appropriate waste management. Paut Kusturica et al. (2020) reported that residents are willing to pay for a pharmaceutical disposal program. Ko et al. (2020) found that people agree with a tax to be able to discard their recyclable waste. All the obtained clearing prices for the landfill supply (Table 8.1) are such that the profit is non-negative for all players (no player loses money in the MSW system). This is a key property of coordinated markets (Sampat et al. 2019). In Scenario I, there is no plastic sent to the sanitary landfill and, thus, the profit is zero. The landfill suppliers do not participate in this scenario since all the waste is disposed of at open dumps. Contrary to Scenario II, where all the waste is disposed of at sanitary landfills due to the involved tax. On the one hand, the prices only vary depending on the stakeholder and not on the type of plastic; on the other hand, the profit varies by type of plastic due to the distinct flows of each material sent to the sanitary landfill (Fig. 3).

Table 1 shows the clearing prices for the landfill demand. In Scenario I all the clearing prices are higher than the bids; thus, the consumers are not allocated any product (they do not participate in the market). Therefore, all waste is allocated to open dump systems. In Scenario II (with taxation), the obtained prices are such that all profits are non-negative, and all waste is sent to sanitary landfills. This illustrates how, regardless of the tax used, participants do not lose money in a coordinated MSW system. Regarding the transportation providers of plastic entering the landfill, their prices are shown in Table 1 as well. The obtained prices for Scenario I are lower than the bids, while for Scenario II all the clearing prices are greater. Therefore, only the prices from Scenario II allow finding positive profits, since they satisfy the constraints required for the transportation providers to have non-negative profits. Therefore, the waste is transported to sanitary landfills in this scenario. For the transportation providers' profits only there are considered the operational costs and not the optimal number of collection trucks as in Lee et al. (2016). Alternatively, the proposed approach includes the optimal number of transportation providers that participate according to the clearing process. Within their operational costs, the expenses related to the needed trucks are included. The profits are categorized in Fig. 3 by type of plastic. Here, there are observed some interesting trends, specifically, stakeholders 1 and 5 always make the largest profits followed by stakeholders 2 and 4; while stakeholder 3 makes

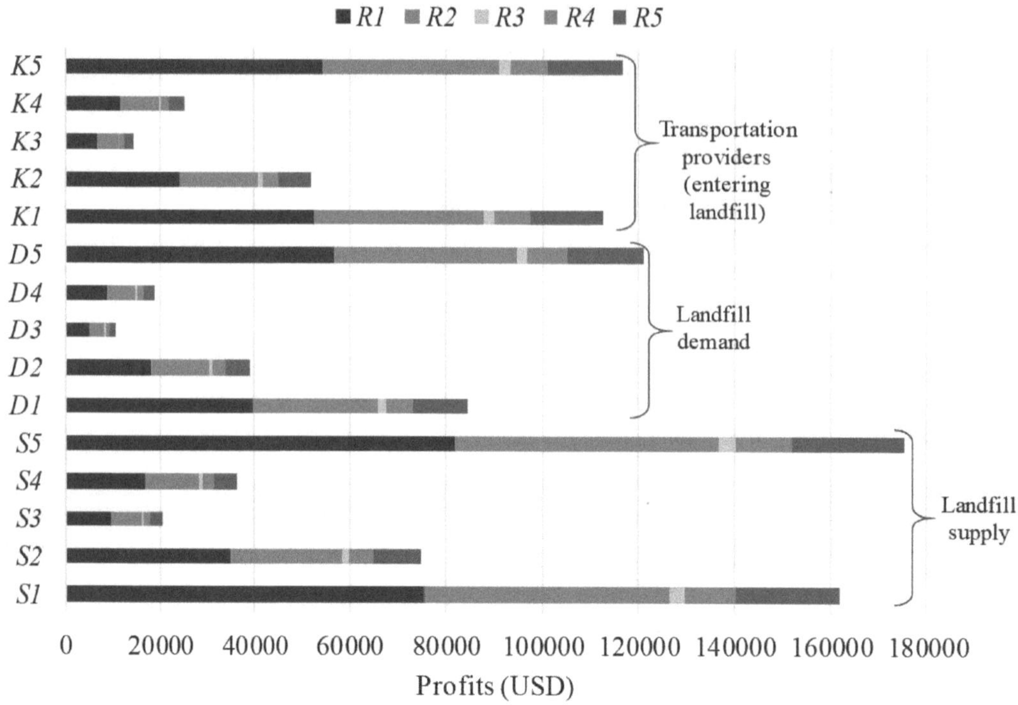

Fig. 3. Profits for landfill supply, landfill demand, and transportation providers by types of plastic in Scenario II. [Printed (Adapted) with permission from Munguia-López, A.C., Zavala, V.M., Santibañez-Aguilar, J.E., and Ponce-Ortega, J.M. 2020. Optimization of municipal solid waste management using a coordinated framework. Waste Management, *115*, 15–24].

Table 1. Bids and clearing prices for landfill supply, landfill demand, and transportation providers for plastic waste. [Printed (Adapted) with permission from Munguia-López, A.C., Zavala, V.M., Santibañez-Aguilar, J.E., and Ponce-Ortega, J.M. 2020. Optimization of municipal solid waste management using a coordinated framework. Waste Management, *115*, 15–24].

	Bids (USD/tonne)	**Prices (USD/tonne)**				
Landfill supply	S1-S5	S1	S2	S3	S4	S5
Scenario I	–12.35	0	0	0	0	0
Scenario II (with tax)	–12.35	–4.27	–4.26	–4.28	–4.29	–5.1
Landfill demand	D1-D5	D1	D2	D3	D4	D5
Scenario I	–12.35	–12.29	–12.31	–12.28	–12.28	–12.26
Scenario II (with tax)	–12.35	–16.57	–16.57	–16.57	–16.57	–17.36
Transportation providers (entering landfill)	K1-K5	K1	K2	K3	K4	K5
Scenario I	0.0165	–12.29	–12.31	–12.28	–12.28	–12.26
Scenario II (with tax)	0.0165	5.64	5.64	5.64	5.64	4.84

the smallest profit. Regarding the types of plastic, *R1* and *R2* (corresponding to PP and PE) represent most of the total profit. Note that landfill suppliers attain the highest profits. Overall, the taxation scenario in the coordinated system avoids diverting waste to open dumps and simultaneously allows landfill providers to attain profits.

8.6 Analysis of Metal Waste

The following results refer to the solutions for metal waste. Table 2 and Fig. 4 present the information for the different suppliers (*S1-S5*), consumers (*D1-D5*), transportation providers (*K1-K5*), and technology providers (*M1-M5*). Only one type of metal (aluminum) was evaluated. For the landfill supply (Table 2), all the prices are greater than the bids and, thus, the profits are all non-negative. Positive profits are only achieved in Scenario II (Fig. 4). This occurs because the involved taxation prevents open dump disposal by involving the participation of landfill suppliers. Consequently, the waste is disposed of at sanitary landfills. Contrary to Scenario I, where metal waste is sent to open dumps since the prices are not lower than the bids for the landfill demand. Therefore, these consumers are not cleared (there is a lack of demand for waste) according to the requirements for the consumers to have non-negative profits. On the other hand, in Scenario II, the clearing prices are equal to the bids, so these consumers participate in the market (there is a landfill demand for waste), but their profit is zero. The optimal solution for metal waste includes transformation providers to process the waste in a processing facility for sale. Regarding the processing supply, all the clearing prices are greater than the bids and equal for both scenarios. Thus, positive and equal profits for both scenarios (Fig. 4) are observed. For the processing plant demand, in both scenarios, only the prices of the consumer *D1* are lower than the bids. Therefore, *D1* attains a profit greater than zero. He et al. (2020) reported that industries tend to cooperate when the marginal cost of recycling is lower than the cost of raw materials. Note that the other consumers (*D2-D5*) could be cleared obtaining a profit equal to zero, however, only *D4* participates in the market. As expected, the taxation scheme does not impact the

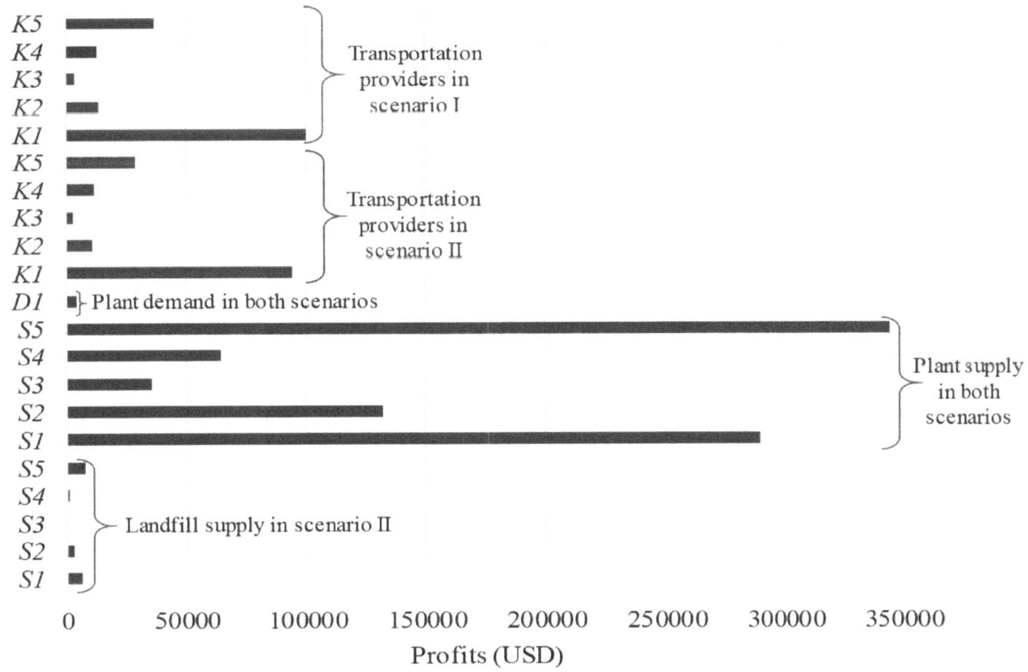

Fig. 4. Profits for supply, demand, and transportation providers of metal for Scenarios I and II. [Printed (Adapted) with permission from Munguia-López, A.C., Zavala, V.M., Santibañez-Aguilar, J.E., and Ponce-Ortega, J.M. (2020). Optimization of municipal solid waste management using a coordinated framework. Waste Management, *115*, 15–24].

processing plant supply and demand, since the tax is applied only to the waste at open dump systems. The results for the transportation providers of the metal entering the landfill and processing plant, and of the products leaving the processing plant are also presented in Table 8.2. For the transportation providers of the metal entering the processing plant, all the prices are greater than the bids, which allows the profits to be positive. These profits are equal for both scenarios since the clearing prices are equal as well. The prices for the transportation providers of the metal leaving the processing plant are equal to those obtained for the transportation providers of the metal entering the processing plant. However, only the providers *K1* and *K4* are cleared here. Again, the profits are equal for both scenarios. The total profits for the transportation providers of each scenario are presented in Fig. 4. These total profits are the sum of the profits of the providers of the metal entering the sanitary landfill and processing plant and of the providers of the products leaving the processing plant. The profits are positive for both scenarios, but in Scenario I the profits are greater since no metal is sent to the

Table 2. Bids and clearing prices for supply, demand, transportation, and technology providers for metal waste. [Printed (Adapted) with permission from Munguia-López, A.C., Zavala, V.M., Santibañez-Aguilar, J.E., and Ponce-Ortega, J.M. 2020. Optimization of municipal solid waste management using a coordinated framework. Waste Management, *115*, 15–24].

	Bids (USD/tonne)	**Prices (USD/tonne)**				
Landfill supply	S1-S5	S1	S2	S3	S4	S5
Scenario I	–12.35	0	0	0	0	0
Scenario II (with tax)	–12.35	–0.14	–0.11	–0.16	–0.17	–0.21
Landfill demand	D1-D5	D1	D2	D3	D4	D5
Scenario I	–12.35	–12.24	–12.19	–12.18	–12.14	–12.21
Scenario II (with tax)	–12.35	–12.35	–12.35	–12.35	–12.35	–12.35
Plant supply	S1-S5	S1	S2	S3	S4	S5
Scenario I	223.5	1204.2	1198.4	1196.6	1198.5	1196
Scenario II (with tax)	223.5	1204.2	1198.4	1196.6	1198.5	1196
Plant demand	D1-D5	D1	D2	D3	D4	D5
Scenario I	1300	1290	1300	1300	1300	1300
Scenario II (with tax)	1300	1290	1300	1300	1300	1300
Transportation providers (entering landfill)	K1-K5	K1	K2	K3	K4	K5
Scenario I	0.04	–12.21	–12.24	–12.19	–12.18	–12.14
Scenario II (with tax)	0.04	–11.56	–11.56	–11.56	–11.56	–11.56
Transportation providers (entering plant / leaving plant)	K1-K5	K1	K2	K3	K4	K5
Scenario I	0.04	90.1	101.64	103.44	101.52	104
Scenario II (with tax)	0.04	90.1	101.64	103.44	101.52	104
Technology providers	M1-M5	M1	M2	M3	M4	M5
Scenario I	90	1294.2	1300	1300	1300	1300
Scenario II (with tax)	90	1294.2	1300	1300	1300	1300

landfill (instead, open dump disposal occurs). For the technology providers, all the clearing prices are greater than the bids, but only the provider *M1* participates in the market for both scenarios (the tax does not impact these stakeholders). Here, material recycling is used to process the metal and the profit for provider *M1* is 1,070,035 USD. Note that this provider attains the highest level of profit followed by the processing plant suppliers, the transportation providers, the landfill suppliers, and the processing plant demand. Furthermore, it should be noted that including taxes fosters sanitary landfill disposal, despite only landfill suppliers (and not landfill consumers) attaining profits greater than zero. This occurs since even with the profit of landfill consumers being zero, they do not lose money and, thus, participate in the market.

8.7 Analysis of Glass Waste

The solutions for the different types of glass (*G1*, *G2*, and *G3*) are presented in Table 3 and Fig. 5. Here no processing players participate (as in the case of plastic). The results reveal that taxation incentivizes

Table 3. Bids and clearing prices for landfill supply, landfill demand, and transportation providers for glass. [Printed (Adapted) with permission from Munguia-López, A.C., Zavala, V.M., Santibañez-Aguilar, J.E., and Ponce-Ortega, J.M. 2020. Optimization of municipal solid waste management using a coordinated framework. Waste Management, *115*, 15–24].

	Bids (USD/tonne)	Prices (USD/tonne)				
Landfill supply	S1-S5	S1	S2	S3	S4	S5
Scenario I	−12.35	0	0	0	0	0
Scenario II (with tax)	−12.35	−3.09	−3.07	−3.12	−3.13	-5.1
Landfill demand	D1-D5	D1	D2	D3	D4	D5
Scenario I	−12.35	−12.21	−12.24	−12.19	−12.18	−12.14
Scenario II (with tax)	−12.35	−15.31	−15.31	-15.31	-15.31	-17.24
Transportation providers (entering landfill)	K1-K5	K1	K2	K3	K4	K5
Scenario I	0.04	12.21	−12.24	−12.19	−12.18	-12.14
Scenario II (with tax)	0.04	2.2	2.2	2.2	2.2	0.27

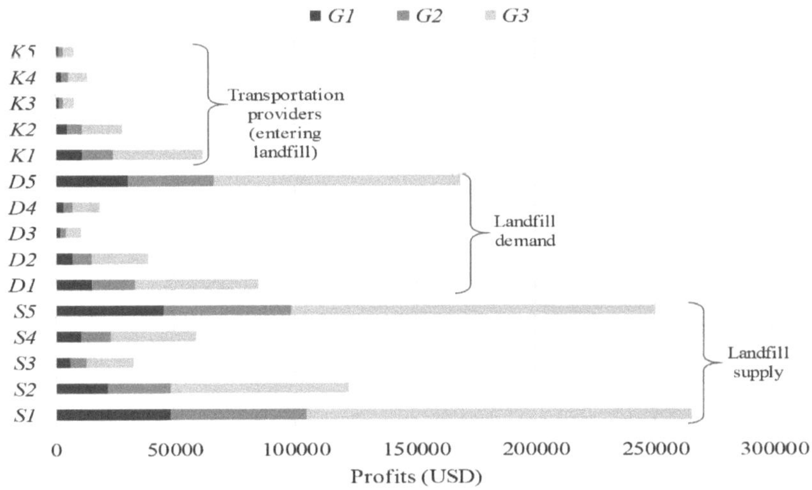

Fig. 5. Profits for landfill supply, landfill demand, and transportation providers by types of glass in Scenario II. [Printed (Adapted) with permission from Munguia-López, A.C., Zavala, V.M., Santibañez-Aguilar, J.E., and Ponce-Ortega, J.M. 2020. Optimization of municipal solid waste management using a coordinated framework. *Waste Management*, *115*, 15–24]

the provision of landfill services and avoids open dump systems for every type of glass. Specifically, for landfill supply, all the prices are greater than the bids; however, positive profits are attained only for Scenario II because in Scenario I there is no glass sent to the sanitary landfill. Therefore, all waste is diverted to open dumps. For the landfill demand, every consumer in Scenario II has positive profits as well. Note that the taxation scenario allows the generation of landfill demand to prevent open dump disposal of glass waste. Interestingly, only the transportation providers of glass entering the landfill participate. For these providers, the prices in Scenario II allow finding positive profits because they are greater than the bids. Therefore, glass waste is transported to the landfill only when the tax is involved. Otherwise, the waste is sent to open dump systems (scenario I). As in the case of plastic, the prices vary depending on the stakeholder and not on the type of glass since the bids are equal for all types. The transportation providers make the smallest profits while the landfill suppliers attain the highest benefits. Regarding the type of glass, *G3* (corresponding to brown glass) represents most of the total profit. Comparing these results to those obtained for plastic, it is observed that the profits for providers of glass are greater than the profits for providers of plastic. This is because of the different bids and flows of waste (more glass waste is generated).

8.8 Analysis of Organic Waste and Non-recyclables

Figure 6 and 7 show the profits for organic waste and non-recyclables. Only one type of waste was evaluated (Fig. 6). For organic waste, there were considered all types of paper-based products (such as cardboard). While for non-recyclables, food waste is considered. No transformation providers participate in the optimal solution for these types of waste. For both organic waste and non-recyclables, their bids and clearing prices for the landfill supply are equal to those obtained for plastic. The same occurs with their bids and prices for the landfill demand (Table A8.1 in Appendix A). However, their profits are different due to the involved flows. Regarding the transportation providers of organic and

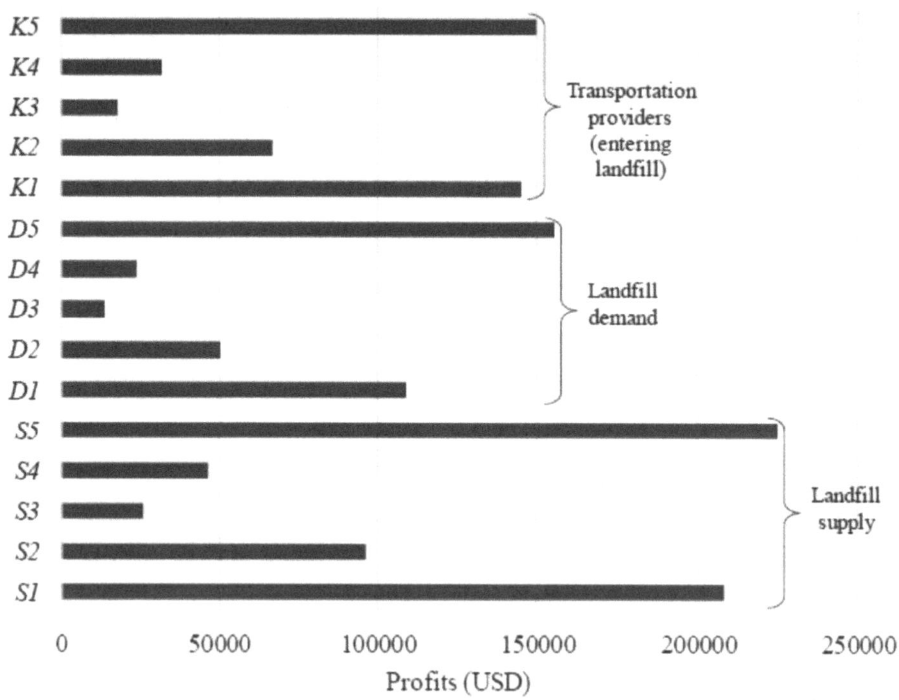

Fig. 6. Profits for landfill supply, landfill demand and transportation providers of organic waste in Scenario II. [Printed (Adapted) with permission from Munguia-López, A.C., Zavala, V.M., Santibañez-Aguilar, J.E., and Ponce-Ortega, J.M. 2020. Optimization of municipal solid waste management using a coordinated framework. Waste Management, *115*, 15–24].

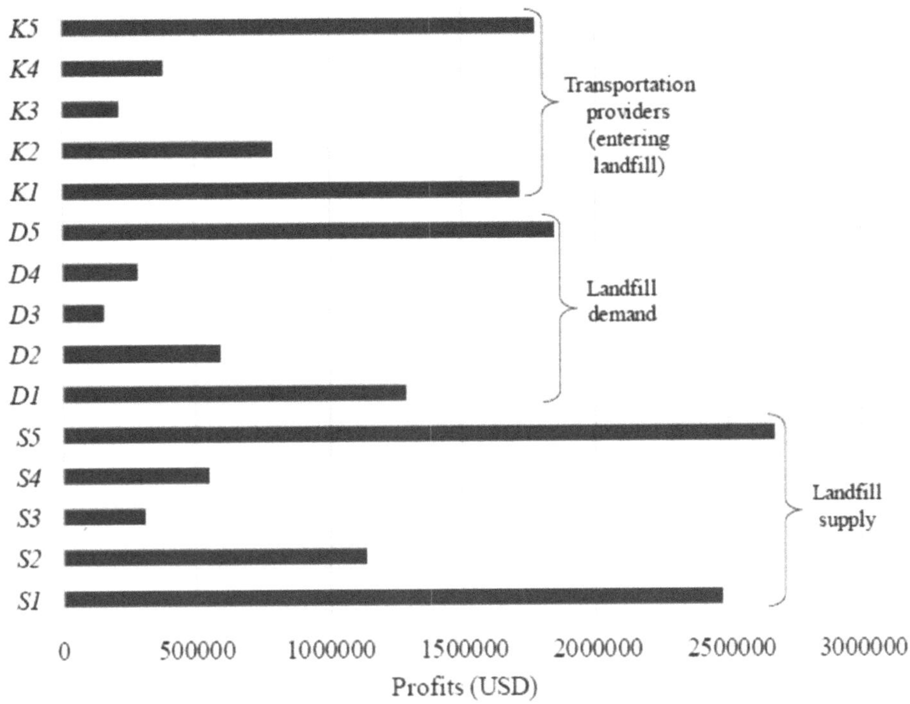

Fig. 7. Profits for landfill supply, landfill demand and transportation providers of non- recyclables in Scenario II. [Printed (Adapted) with permission from Munguia-López, A.C., Zavala, V.M., Santibañez-Aguilar, J.E., and Ponce-Ortega, J.M. 2020. Optimization of municipal solid waste management using a coordinated framework. Waste Management, *115*, 15–24].

non-recyclable waste entering the sanitary landfill, their bids and prices are also equal to those obtained for plastic (Table A8.1 in Appendix A) and, thus, the waste is transported to sanitary landfills. It is observed positive profits for all landfill stakeholders only in Scenario II. This indicates that taxation is necessary to activate the market and foster the participation of landfill providers to avoid open dump disposal. On the other hand, in Scenario I, all organic waste and non-recyclables are disposed of at open dumps and no participant makes a profit. For the profits of these types of waste in scenario II, the same trends found for plastic are observed. For instance, stakeholders 1 and 5 of every service (supply, demand, and transport) obtain greater profits than the other providers. Furthermore, all landfill suppliers attain the highest profits. It is observed that the profits for the participants of non-recyclables are slightly higher than those obtained for the stakeholders of organic waste.

It is important to note that in Scenario II, for every type of waste (plastic, metal, organic, glass, and non-recyclables), nothing is disposed of at open dumps. This occurs due to the involved taxation (5.1 USD/tonne). This tax was identified as the minimum penalization that avoids diverting waste (of all types) to open dumps. Lower taxes allow sending only some types of waste to sanitary landfills. However, this partial improvement is not of interest since the aim is to eliminate open dump systems and promote appropriate waste management.

Note that metal is the only type of waste that involves transformation providers (for both scenarios). Regarding the metal that is not recycled, in scenario I, it is disposed of at open dumps, while in scenario II it is disposed of at sanitary landfills (equally to the other types of waste). Specifically, 36% of the generated metal is processed in a treatment facility for sale. As mentioned above, the collective profit of the system is positive in all cases. Contrary to the study reported by Santibañez-Aguilar et al. (2013), where the recycled metal is maximized (100% is recycled), the total profit is negative (this is not a coordinated approach).

8.9 Concluding Remarks

This chapter has presented a coordinated market framework for an MSW system. Here, different stakeholders were identified, including suppliers, consumers, as well as transportation and transformation providers. ach of these makes a distinct bid, and their clearing prices and profits are determined by solving the problem. Additionally, the option of treating waste in sanitary plants or landfills is taken into account. Additionally implicated is the garbage that ends up in open dumps. To monetize this environmental impact, a tax is incorporated into the formulation. There has been presented the effect of activating the market and preventing open dump disposal. An MSW system in Mexico was assessed as a case study to demonstrate the applicability. However, the presented model is general and can be applied to any case study. The allocations and pricing that maximize the total profit for all parties involved and the types of waste involved in various scenarios (with and without taxation) are displayed in the results. The clearing process ensures that individual earnings are non-negative by balancing supply and demand for waste and products, which is a significant advantage of the suggested structure. The marginal values were used to determine the minimum tax needed to prevent waste from ending up in open dumps. Additionally, it was discovered that metal is the only waste type that can be recycled profitably. The proposed system in this approach results in special interest in regions where MSW collection is not efficient and MSW management needs to be greatly improved. Therefore, in order to find a successful method, starting points like taxing open dumps and taking into account the earnings of all stakeholders were covered here. An analysis that incorporates different recyclable bids based on clearing prices would be intriguing for future research. Furthermore, although the impact of open dump systems was the main emphasis of this chapter, other environmental issues might be covered using the same methodology.

8.10 Nomenclature

8.10.1 Parameters

c_d^* Maximum capacities for the consumers

f_m^* Maximum capacities for the technology providers

g_s^* Maximum capacities for the suppliers

q_k^* Maximum capacities for the transportation providers

α_d Bidding information for the consumers

β_s Bidding information for the suppliers

δ_m Bidding information for the technology providers

γ_k Bidding information for the transportation providers

$\varsigma_{m,p}$ Conversion factor for each technology and product

8.10.2 Variables

c_d Allocations for the consumers

f_m Allocations for the technology providers

g_s Allocations for the suppliers

q_k Allocations for the transportation providers

π_d Clearing prices for the consumers

π_s Clearing prices for the suppliers

π_k Clearing prices for the transportation providers

π_m Clearing prices for the technology providers

ϕ_d^D Profits for the consumers

ϕ_s^S Profits for the suppliers

ϕ_k^K Profits for the transportation providers

ϕ_m^M Profits for the technology providers

8.11 References

Botello-Álvarez, J.E., Rivas-García, P., Fausto-Castro, L., Estrada-Baltazar, A., and Gomez-Gonzalez, R. 2018. Informal collection, recycling and export of valuable waste as transcendent factor in the municipal solid waste management: A Latin-American reality. Journal of Cleaner Production, *182*, 485–495. https://doi.org/10.1016/j.jclepro.2018.02.065

Das, S., Lee, S.H., Kumar, P., Kim, K.-H., Lee, S.S., and Bhattacharya, S.S. 2019. Solid waste management: Scope and the challenge of sustainability Journal of Cleaner Production, *228*, 658–678. https://doi.org/10.1016/j.jclepro.2019.04.323

Garibay-Rodriguez, J., Laguna-Martinez, M.G., Rico-Ramirez, V., and Botello-Alvarez, J.E. 2018. Optimal municipal solid waste energy recovery and management: A mathematical programming approach. Computers and Chemical Engineering, *119*, 394–405. https://doi.org/10.1016/j.compchemeng.2018.09.025

Guerrero, L.A., Maas, G., and Hogland, W. 2013. Solid waste management challenges for cities in developing countries. Waste Management, *33*, 220–232. https://doi.org/10.1016/j.wasman.2012.09.008

Gutberlet, J. 2015. Cooperative urban mining in Brazil: Collective practices in selective household waste collection and recycling. Waste Management, *45*, 22–31. https://doi.org/10.1016/j.wasman.2015.06.023

He, M., Jin, Y., Zeng, H., and Cao, J. 2020. Pricing decisions about waste recycling from the perspective of industrial symbiosis in an industrial park: A game model and its application. Journal of Cleaner Production, *251*, 119417. https://doi.org/10.1016/j.jclepro.2019.119417

Hoornweg, D., and Bhada-Tata, P. 2012. What a waste: a global review of solid waste management. https://siteresources.worldbank.org/INTURBANDEVELOPMENT/Resources/336387-1334852610766/What_a_Waste2012_Final.pdf (accessed February 2020).

Jiang, P., Fan, Y. Van, Zhou, J., Zheng, M., Liu, X., and Klemeš, J.J. 2020. Data-driven analytical framework for waste-dumping behaviour analysis to facilitate policy regulations. Waste Management, *103*, 285–295. https://doi.org/10.1016/j.wasman.2019.12.041

Ko, S., Kim, W., Shin, S.C., and Shin, J. 2020. The economic value of sustainable recycling and waste management policies: The case of a waste management crisis in South Korea. Waste Management, *104*, 220–227. https://doi.org/10.1016/j.wasman.2020.01.020

Lavee, D. 2007. Is municipal solid waste recycling economically efficient?. Environmental Management, *40*, 926-943. https://doi.org/10.1007/s00267-007-9000-7

Lavee, D., Regev, U. and Zemel, A. 2004. Optimal intervention in uncertain waste recycling markets, Monaster Center for Economic Research, Ben-Gurion University of the Negev, Israel.

Lavee, D., Regev, U. and Zemel, A. 2009. The effect of recycling price uncertainty on municipal waste management choices. Journal of Environmental Management, *90*, 3599–3606. https://doi.org/10.1016/j.jenvman.2009.06.012

Lee, C.K.M., Yeung, C.L., Xiong, Z.R., and Chung, S.H. 2016. A mathematical model for municipal solid waste management –A case study in Hong Kong. Waste Management, *58*, 430–441. https://doi.org/10.1016/j.wasman.2016.06.017

Lohri, C.R., Camenzind, E.J., and Zurbrügg, C. 2014. Financial sustainability in municipal solid waste management–Costs and revenues in Bahir Dar, Ethiopia. Waste Management, *34*, 542–552. https://doi.org/10.1016/j.wasman.2013.10.014

Medina, M. 2010. Solid wastes, poverty and the environment in developing country cities: Challenges and opportunities. (No. 2010/23). WIDER Working Paper. UNU-WIDER, Helsinki. https://doi.org/10.1093/acprof:oso/9780199590148.003.0015

Medina, M. 2007. Waste picker cooperatives in developing countries, in: Chen, M., Jhabvala, R., Kanbur, R., Richards, C., Membership Based Organizations of the Poor. Routledge, London, pp. 125–141. https://doi.org/10.4324/9780203934074

Minoglou, M., and Komilis, D. 2013. Optimizing the treatment and disposal of municipal solid wastes using mathematical programming—A case study in a Greek region. Resources, Conservation and Recycling, *80*, 46–57. https://doi.org/10.1016/j.resconrec.2013.08.004

Mohammadi, M., Jämsä-Jounela, S., and Harjunkoski, I. 2019. Sustainable supply chain network design for the optimal utilization of municipal solid waste. AIChE Journal, *65*, e16464. https://doi.org/10.1002/aic.16464

Munguia-López, A.C., Zavala, V.M., Santibañez-Aguilar, J.E., and Ponce-Ortega, J.M. 2020. Optimization of municipal solid waste management using a coordinated framework. Waste Management, *115*, 15–24.

Ojeda-Benitez, S., and Beraud-Lozano, J. 2003. The municipal solid waste cycle in Mexico: final disposal. Resources, Conservation and Recycling, *39*, 239–250. https://doi.org/10.1016/S0921-3449(03)00030-2

Paul, K., Chattopadhyay, S., Dutta, A., Krishna, A.P., and Ray, S. 2018. A comprehensive optimization model for integrated solid waste management system: A case study. Environmental and Engineering Research, *24*, 220–237. https://doi.org/10.4491/eer.2018.132

Paut Kusturica, M., Golocorbin-Kon, S., Ostojic, T., Kresoja, M., Milovic, M., Horvat, O., Dugandzija, T., Davidovac, N., Vasic, A., and Tomas, A. 2020. Consumer willingness to pay for a pharmaceutical disposal program in Serbia: A double hurdle modeling approach. Waste Management, *104*, 246–253. https://doi.org/10.1016/j.wasman.2020.01.029

Sampat, A.M., Hu, Y., Sharara, M., Aguirre-Villegas, H., Ruiz-Mercado, G., Larson, R.A., and Zavala, V.M. 2019. Coordinated management of organic waste and derived products. Computers and Chemical Engineering, *128*, 352–363. https://doi.org/10.1016/j.compchemeng.2019.06.008

Santibañez-Aguilar, J.E., Ponce-Ortega, J.M., Betzabe González-Campos, J., Serna-González, M., and El-Halwagi, M.M. 2013. Optimal planning for the sustainable utilization of municipal solid waste. Waste Management, *33*, 2607–2622. https://doi.org/10.1016/j.wasman.2013.08.010

Semarnat 2012. Informe de la Situación del Medio Ambiente en México. Compendio de Estadísticas Ambientales. Indicadores Clave y de Desempeño Ambiental. México. https://apps1.semarnat.gob.mx:8443/dgeia/informe15/tema/pdf/Informe15_completo.pdf (accessed January 2020).

Yousefloo, A., and Babazadeh, R. 2020. Designing an integrated municipal solid waste management network: A case study. Journal of Cleaner Production, *244*, 118824. https://doi.org/10.1016/j.jclepro.2019.118824

Section III

Challenges and Innovations in Waste Management

9

Optimal Waste Management during the COVID-19 Pandemic

Numerous issues have been brought forward by the COVID-19 epidemic. One of these is the alarming rise in medical waste production and as a result of that the serious health risks they provide. This chapter presents a mathematical model for optimal solid waste management incorporating a circular value chain where all types of waste are treated in an intensified industrial park. The model selects the processing technologies and their production capacity. The problem was formulated as a mixed-integer linear programming problem to maximize profits and the waste processed and minimize environmental impact. The proposed strategy was applied to the case study of the city of New York, where the increase in the generation of medical waste has been very significant. To promote recycling, different tax rates are proposed, depending on the amount of waste sent to the landfill. The results are presented on a Pareto curve showing the trade-off between profits and processed waste. The taxes promote recycling, even of those wastes that are not very convenient to recycle (from an economic point of view), favoring profits, reducing the environmental impact, and the risk to health inherent to the medical waste.

9.1 Introduction

The rise in the production of certain waste kinds, particularly medical waste, is one of the primary issues brought on by the pandemic. This kind of waste poses a health danger to patients, healthcare professionals, garbage collectors, and the general public. (Das et al. 2021). Solid waste management was already a problem even before the pandemic, especially in developing countries, due to the lack of infrastructure for waste processing. The existing risk from poor waste management is already so steep that it is estimated that around 9.2 million people die annually as a result (Star 2021).

There are many alternatives and strategies for the management of municipal solid waste (MSW), one of the main trends currently is the conversion of waste-to-energy (WTE) (Serrano-Ruiz 2020, Sharma et al. 2022), using technologies such as gasification, pyrolysis, and incineration. However, not all types of waste are convenient to be converted into energy. Metallic waste, for example, has great economic potential, due to the high price at which its derivatives can be sold (Santibañez-Aguilar et al. 2013). In addition, new technologies for waste treatment have been developed in particular for the treatment of waste from COVID-19, such as the processing system for waste from respirators proposed by Zhao and You (2020). Despite the different technologies currently available to MSW, the recycling percentages remain low (AP 2021, EPA 2021). Different strategies have recently been developed for waste management seeking to increase the amount of waste that is processed, considering different objectives, whether social, environmental, or economic.

This chapter presents a multi-objective optimization strategy for MSW treatment. The approach presented involves the evaluation of the construction of an industrial park that promotes the intensification of recycling processes, reduces transport costs, and considers different alternatives for

MSW, such as recycling, WTE, and landfill disposal. The model consists of a mixed-integer linear problem (MILP) that includes a variable tax rate to increase recycling rates, especially for those wastes with low economic potential, such as the treatment of medical waste, which has increased significantly because of the pandemic (Munguia-López et al. 2022).

This chapter is structured as follows. Section 9.2 presents the related literature, Section 9.3 presents the problem statement, Section 9.4 shows the proposed model, Section 9.5 discusses the applicability of the proposed model to a case study from New York. Section 9.6 shows the discussion of the obtained results, and Section 9.6 presents the conclusion and recommendations.

9.2 Literature Review

This section presents the literature review, which is segregated as follows.

9.2.1 Waste Management in General

To increase the amount of waste that is recycled, various political and financial strategies have been proposed to promote recycling while increasing the economic and environmental benefits. Zhao and You (2021) reported an optimization system for the f ood-energy-water-waste nexus, to reduce the social and environmental impact generated by the COVID-19 pandemic due to the increase in food waste generated. Ko et al. (2020) developed a strategy for a hypothetical case in South Korea, where recycling is encouraged through the collection of taxes. These taxes that promote the preservation of the environment are called "Green taxes" (Oates 1 995, Paut-Kusturica et al. 2020). The main challenge is to find the tax rate that best promotes recycling since a very high tax will generate great social disapprobation, while a very small one would not be enough to promote recycling since the inhabitants could be willing to pay this tax and the recycling rates would remain the same. Munguía-López et al. (Munguia-López et al. 2020a, 2020b) reported a coordinated structure for the recycling of MSW, driven by the implementation of taxes on all waste destined for the open landfill. They also included a sensibility analysis with different tax rates, which allowed finding the minimum tax rate that avoids sending large amounts of waste to the open landfill. Santibañez-Aguilar et al. (2013) reported an optimization strategy for the sustainable use of MSW. However, the results, although favorable, showed a certain tendency not to recycle certain types of waste, such as glass, this is due to the high costs of separation and transportation inherent in the glass recycling process and the long distances between the treatment plants. Munguía-López et al. (2020) presented an optimization strategy for the optimal design of residential complexes, which involves the satisfaction of energy and is associated with the consumption of freshwater demands, considering alternatives such as rainwater harvesting, water treatment, recycling, cogeneration, gasification of solid wastes, and algae cultivation. Chen et al. (2016) developed a model that considers uncertainty for long-term planning of waste management, which provides information for possible locations of waste processing facilities. The aforementioned strategies coincide with the process intensification approach, where natural or residual raw materials allow for maximizing the yield and indirectly the economic benefits (Ponce-Ortega et al. 2012, Meyer et al. 2021). From a massive integration framework, this could be especially useful for those wastes with low economic potential and the construction of an industrial park destined for the processing of MSW would be able to promote this type of approach (Rivas et al. 2018).

9.2.2 Post COVID-19 Waste Management

A waste catastrophe has resulted from the rise in medical and organic waste production since the start of the COVID-19 epidemic (Ives 2021) and this has changed the way waste is managed. Govindan et al. (2021) developed a MILP model that considers simultaneous economic and social objective functions. This model considers vehicle scheduling, vehicle failure, split delivery, population risk, and fuel consumption to select the optimal transportation routes for waste that minimize the costs and risks for the population. Valizadeh and Mozafari (2022) developed a mathematical model using four cooperative game theory methods, such as the Shapley value, the *t* value, the central center, and the minimum kernel, employing genetic algorithms for its solution. Incentives are provided to municipal garbage collector contractors under this technique, and the solutions demonstrate that collaboration among collectors lowers overall costs, strengthening the collectors' alliance. Homayouni and Pishvaee (2020) developed a robust multi-objective model for the management of medical waste, using the epsilon constraint method and generating a distribution network that assigns to existing processing centers different amounts of medical waste that minimize cost and health risks. Although good results are observed, the distances between each processing center can be large, thus transportation costs can be relatively high, depending on the region where it is applied.

There are many opportunities for innovation in the field of process intensification since during the last decade the development of processes in this area has been oriented almost exclusively toward large-scale chemical processes (Becht et al. 2009, Strube et al. 2018). Therefore, it is necessary to highlight the relevance of including process intensification in other fields such as urban solid waste management, since the reuse and revaluation of waste are a form of intensification (Cabrera and Labatut 2021) applied to a different supply chain. None of the previous works have considered a strategy that contemplates a circular framework composed of an intensified industrial park with a wide range of available technologies and a variable rate of green taxes that encourages the processing of hazardous waste and minimizes negative impacts on the environment.

9.3 Problem Statement

The proposed system comprises different types and subtypes of waste, and different technologies for each type of waste. Urban centers are considered the suppliers of waste and the consumers of new materials and energy resources generated in the process, seeking to generate a change in the supply chain from linear to circular, improving the efficiency of the use of waste, coinciding with the concept of circular economy (Monsiváis-Alonso et al. 2020) (Fig. 1). The processing and installation costs of

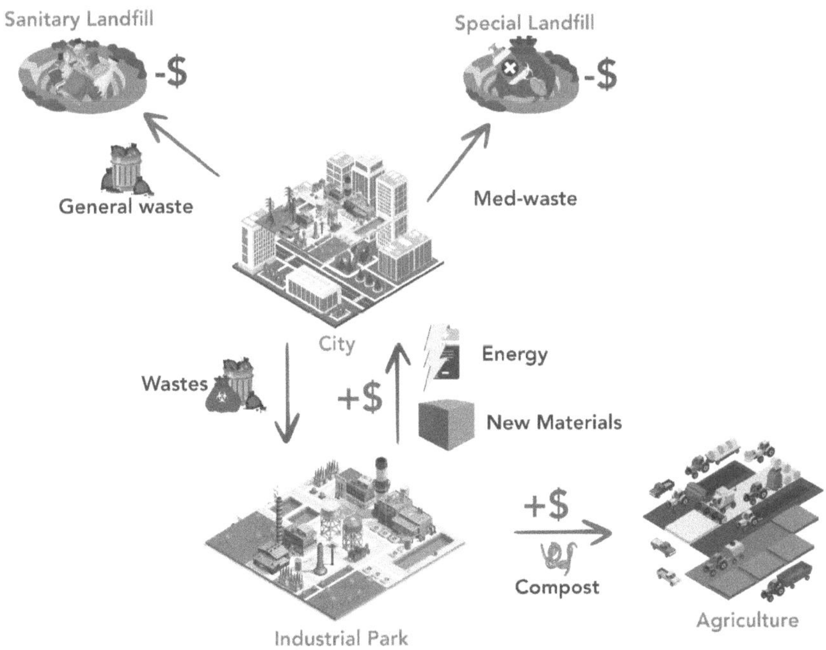

Fig. 1. Proposed strategy in Chapter 9. [Prineted (adapted) with permission from Munguia-López, A.C., Ochoa-Barragán, R., and Ponce-Ortega. J.M. 2022. Optimal waste management during the COVID-19 pandemic. Chemical Engineering and Processing: Process Intensification, *176*, 108942].

the various recycling equipment are calculated using a ten-year useful life and amortization period. Transport providers are taken into account when processing waste and new materials. The increased industrial park is situated in a particular area, close to the trash source. One of the biggest expenses of the entire process is waste separation, which is taken into account. (Table A.9.6 in the Appendix). The proposed system has two sanitary landfills. In the first one, waste of any type that cannot be processed is sent (due to the processing capabilities of the available technologies), except medical waste that must be sent to a safe landfill, due to the high risk to health that this class of waste represents because of the COVID-19 pandemic (Hantoko et al. 2021). Disposing of waste in landfills is cheap economically, but has a high environmental cost. Thus, to avoid sending large amounts of waste to these landfills, a variable tax rate is proposed depending on the amounts of waste sent to the landfills.

In the proposed superstructure (Fig. 2), two main types of waste processes are handled, those that are destined to produce new materials (e-waste, metal, glass, plastic, paper and, as a special case, organic waste that is converted to compost) and processes that convert waste into energy (med-waste, organic waste). The mathematical formulation of the model is general and is not limited by the technologies proposed in this chapter. The technologies selected for the proposed analysis are discussed below.

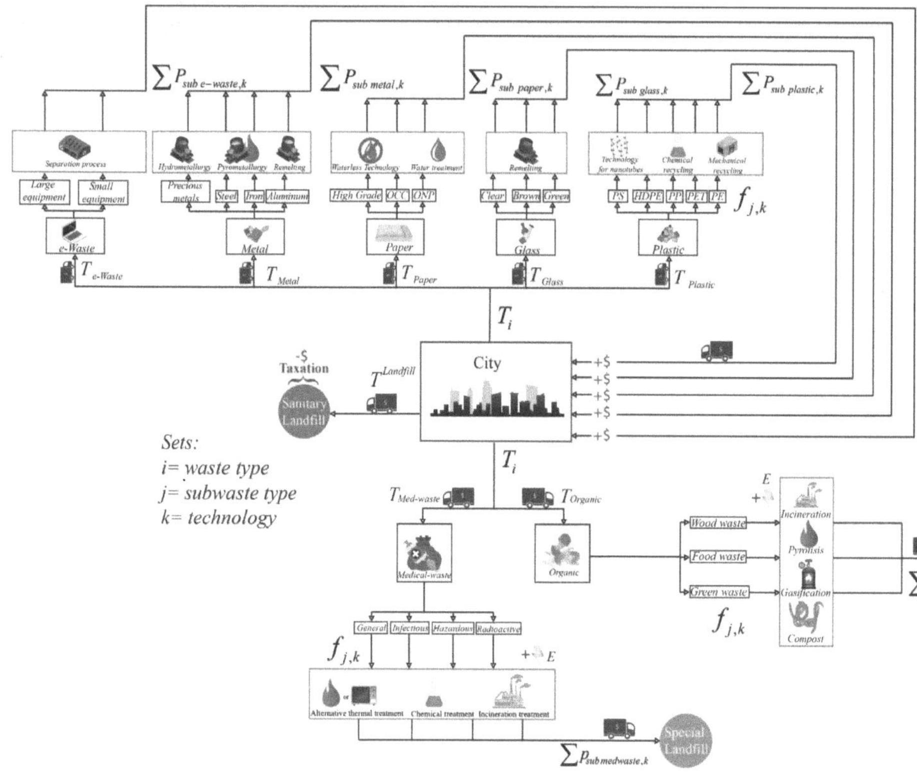

Fig. 2. Superstructure for solid waste management for the model of Chapter 9. [Prineted (adapted) with permission from Munguia-López, A.C., Ochoa-Barragán, R., Ponce-Ortega. J.M. 2022. Optimal waste management during the COVID-19 pandemic. Chemical Engineering and Processing: Process Intensification, *176*, 108942].

9.4 Mathematical Model

The formulation of the mathematical model for Chapter 9 is based on the superstructure presented in Fig. 2, using parameters that allow selecting one process over another when there is more than one alternative for a type of waste. The values for the different parameters can be found in the Appendix section..

9.4.1 Generated Waste in the City

The total waste (TW) produced by the city per year must be separated and classified (Equation (9.1)) into the different types of waste *i*: e-waste, metal, paper, glass, plastic, medical waste, and organic.

$$T_i = (1 - wf)\mathrm{TW} \cdot \alpha_i, \ \forall i \in I \tag{9.1}$$

$$T^{Landfill} = wf \cdot \mathrm{TW} \tag{9.2}$$

The above equations allow calculating the quantities of waste to be processed, α_i is the existing fraction of each type of waste *i* in the selected city (Table A.9.1). $T^{Landfill}$ represents the amount of total waste that cannot be sent to processing (either for reasons of investment or processing capacity), and *wf* represents the fraction of the total waste that cannot be sent to processing (Equation (9.2)). T_i represents the waste that is processed.

Each type of waste must be separated to be processed by the different available technologies for e-waste subtypes, metal subtypes, paper subtypes, glass subtypes, plastic subtypes, medical waste subtypes, and organic waste subtypes (Equation (9.3)).

$$cw_{i,j} = \beta_{i,j} \cdot T_i, \forall i \in I, j \in J \tag{9.3}$$

where $\beta_{i\ j}$ are the fractions of each subtype of waste *j* (Table A.9.2).

The classified waste is sent to different facilities to be processed (Equation (9.4)).

$$cw_{i,j} = \sum_{k \in K} f_{j,k}, \forall i \in I, j \in J \tag{9.4}$$

where $f_{j,k}$ represents the classified subtypes of waste that will be processed by the different technologies *k* available and compatible with each subtype of waste. The technologies used are presented in Table A.9.5.

9.4.2 Disjunctions to Choose the Treatment or Processing Technology

The selection of one technology or another depends on the objective function. For the optimal selection of the technologies proposed in the model, the following disjunction is established.

$$\begin{bmatrix} H_{j,k} \\ f_{j,k} \geq \delta_{j,k}^{\min} \\ f_{j,k} \leq \delta_{j,k}^{\max} \\ cost_{j,k} = \mathrm{A}_{j,k} + \mathrm{B}_{j,k} \cdot f_{j,k} \end{bmatrix} \vee \begin{bmatrix} \neg H_{j,k} \\ f_{j,k} = 0 \\ cost_{j,k} = 0 \end{bmatrix}, \forall j \in J, k \in K \tag{9.5}$$

The decision of what type of technology is used depends on the Boolean variable $H_{j,k}$. If it is true, then, that processing technology is used, while if it is false it is discarded. In the previous disjunction, $\mathrm{A}_{j,k}$ and $\mathrm{B}_{j,k}$ are parameters used to consider fixed and variable costs. This disjunction can be algebraically reformulated as follows:

$$f_{j,k} \geq \delta_{j,k}^{\min} \cdot h_{j,k}, \forall j \in J, k \in K \tag{9.6}$$

$$f_{j,k} \leq \delta_{j,k}^{\max} \cdot h_{j,k}, \forall j \in J, k \in K \tag{9.7}$$

$$cost_{j,k} = kf\left(\mathrm{A}_{j,k} \cdot h_{j,k} + \mathrm{B}_{j,k} \cdot f_{j,k}\right), \forall j \in J, k \in K \tag{9.8}$$

where $h_{j,k}$ is a binary variable, that when its value is 0, the technology is discarded, while when it is 1, it is decided to use that technology. δ represent the search limits for the quantities of the material to be processed, these values are established with a lower limit close to 0 and an upper limit depending on the proposed objectives, *kf* represents the annualization factor applied to fixed and variable costs.

An example of using the previous disjunction for metal waste is described as follows. The types of proposed technologies compatible with metals are hydrometallurgy, pyrometallurgy, and remelting. Then, to decide if the technology of remelting for iron is used, the decision rests on the value of the Boolean variable $H_{iron,remelting}$, where if its value is *true*, that processing technology is used, while if it is *false* it is discarded.

$$\begin{bmatrix} H_{iron,remelting} \\ f_{iron,remelting} \geq \delta^{\min}_{iron,remelting} \\ f_{iron,remelting} \leq \delta^{\max}_{iron,remelting} \\ cost_{iron,remelting} = \mathrm{A}_{iron,remelting} + \mathrm{B}_{iron,remelting} \cdot f_{iron,remelting} \end{bmatrix} \vee \begin{bmatrix} \neg H_{iron,remelting} \\ f_{iron,remelting} = 0 \\ cost_{iron,remelting} = 0 \end{bmatrix} \quad (9.9)$$

Similarly, $\mathrm{A}_{iron,remelting}$ and $\mathrm{B}_{iron,remelting}$ are parameters used to consider fixed and variable costs. The above disjunction can be algebraically reformulated as follows:

$$f_{iron,remelting} \geq \delta^{\min}_{iron,remelting} \cdot h_{iron,remelting} \quad (9.10)$$

$$f_{iron,remelting} \leq \delta^{\max}_{iron,remelting} \cdot h_{iron,remelting} \quad (9.11)$$

$$cost_{iron,remelting} = kf\left(\mathrm{A}_{iron,remelting} \cdot h_{iron,remelting} + \mathrm{B}_{iron,remelting} \cdot f_{iron,remelting}\right) \quad (9.12)$$

where $h_{iron,remelting}$ is a binary variable that when its value is 0, the technology is discarded, while when it is *1*, it is decided to use that technology. This applies analogously to the other technologies.

The technologies evaluated in the model have been widely studied technologies. Therefore, they can be included in the model through different parameters, such as coefficients for the fixed and variable costs. The different values for the parameters of the fixed and variable costs are shown in Tables A.9.3 and A.9.4.

All the processed inorganic waste becomes products that can be sold to the city of origin, regardless of the type of technology used for processing:

$$p_{j,k} = \gamma_{j,k} \cdot f_{j,k}, \forall j \in J, k \in K \quad (9.13)$$

where $p_{j,k}$ is the product obtained from the treatment of the different types of solid waste, $\gamma_{j,k}$ are the fractions that according to their respective and compatible technology can be converted into products (Table A.9.5). The processed waste such as electronic, metal, plastic, paper, and glass are sent in the form of new materials to the city of origin, while the processed medical waste must be sent to a special sanitary landfill mitigating the risk inherent to this kind of waste. It is feasible to create agriculturally usable materials from the residual organic waste, therefore the untreated organic waste serves as an extra revenue stream.

9.4.3 Costs and Sales

The treatment of solid waste requires different levels of investment depending on the type of technology to be implemented or the desired processing capacity, in addition to the expense associated with the transportation of the raw material. Total costs are calculated as follows:

$$TC = Transport\ cost + Treatment\ cost + Taxes \quad (9.14)$$

$$Transport\ cost = RMTC + PTC \tag{9.15}$$

$$RMTC = \mathrm{TCPT}\left(\sum_{i\in I} T_i\right) + \sum_{i\in I}\left(\mathrm{SC}_i \cdot T_i\right) \tag{9.16}$$

$$PTC = \mathrm{TCPT}\left(\sum_{j\in J}\sum_{k\in K} p_{j,k}\right) \tag{9.17}$$

$$Treatment\ cost = \sum_{j\in J}\sum_{k\in K} cost_{j,k} \tag{9.18}$$

where *RMTC* is the raw material transport cost (for solid waste), which also includes the separation cost for each type of waste represented by SC_i, it is worth noting that by only processing waste from a single city, it is possible to establish a strategic location and leave transportation costs fixed. *PTC* is the transportation cost for products, TCPT is the total cost per ton transported. Table A.9.6 shows the different parameters associated with costs of separation and transport.

For taxes, the first two new variables are considered to separate the waste that will be taxed with a lower and a higher penalization. represents the waste that will be taxed with a lower penalization and T^{landH} the waste that will be taxed with a higher penalization.

$$T^{Landfill} = T^{\mathrm{landH}} + T^{\mathrm{landL}} \tag{9.19}$$

Then, the following disjunction is proposed.

$$\begin{bmatrix} H^{Landfill} \\ T^{\mathrm{landL}} < T^{II} \end{bmatrix} \vee \begin{bmatrix} \neg H^{Landfill} \\ T^{\mathrm{landH}} \geq T^{II} \\ T^{\mathrm{landH}} \leq T^{III} \end{bmatrix} \tag{9.20}$$

where is the upper bound (for this case, 80,000) for the generated waste for the lower tax, but also it is the lower bound for the higher tax. is the upper bound (for this case, 250,000) for the highest tax. The previous disjunction can be algebraically reformulated as follows:

$$T^{landL} < T^{II} \cdot h^{Landfill} \tag{9.21}$$

$$T^{landL} \geq T^{II}\left(1 \quad h^{Landfill}\right) \tag{9.22}$$

$$T^{landH} \leq T^{III}\left(1 - h^{Landfill}\right) \tag{9.23}$$

The taxes can be obtained with the following e quations:

$$Taxes^{L} = TPT^{L} \cdot T^{\mathrm{landL}} \tag{9.24}$$

$$Taxes^{H} = TPT^{H} \cdot T^{\mathrm{landH}} \tag{9.25}$$

where is the lower unitary tax and is the higher unitary tax.

$$Taxes = Taxes^{L} + Taxes^{H} \tag{9.26}$$

The income varies depending on the type of material processed and the quantity, and the production of energy. Therefore, the income can be calculated as follows:

$$Revenue = \sum_{j\in J}\sum_{k\in K}\left(\omega_{j,k} \cdot p_{j,k}\right) \tag{9.27}$$

where $\omega_{j,k}$ is the sale price per ton of waste processed (Table A.9.7). The total profit can be calculated as follows:

$$Total\ profit = Revenue - TC \tag{9.28}$$

9.4.4 Objective Functions

For the model presented in this chapter, it is possible to propose different objectives depending on the case, favoring the economic benefit, reducing the impact on the environment, or favoring social development are some examples. To minimize the environmental impact, the amount of untreated waste must be reduced (Equation (9.8)).

Sanitary landfills present an enormous environmental risk, not only because of the nature of the waste deposited (which also represents a health risk), but also because of toxic chemical seepage into the soil and water supply. These leaks are produced by water precipitation, generating high-risk leachates that may contain volatile organics, chloride, nitrogen, solvents, phenols, and heavy metals. No matter how many barriers, liners, and pipes are installed to try to mitigate risk, landfills will always leach toxic chemicals into the soil and water (Pecci, 2018).

A certain quantity of solid waste must be dumped in a landfill because waste processing facilities are limited. Reducing the amount of waste that is sent would benefit the environment, in addition to reducing the amount of taxes (Equation (9.29)).

$$\min T^{Landfill} \tag{9.29}$$

Also to maximize the economic benefit can be considered as an objective function (Equation (9.30)):

$$\max Total\ profit \tag{9.30}$$

9.4.5 Solution Procedure

The constrained method was selected for this chapter (Diwekar 2008). This method was implemented initially maximizing the environmental objective function, that is, solving the model considering an unlimited maximum production capacity, which leads to a solution that requires huge areas and investments for the construction of an industrial park capable of processing 100% of the waste generated. Then, the same model is solved seeking to maximize income, which leads to the minimization of the processing capacities and the selection of the cheapest technologies. Here, in addition to sending waste to the landfill, the maximum values obtained from the economic solution, and are added in the form of a restriction to the model that minimizes the environmental impact, now the multi-objective optimization model becomes a single objective, where each economic restriction gives a point on the Pareto curve (Fig. 3), where trade-offs between environmental and economic objectives are observed.

9.5 Case Study

The case study selected for the application of the proposed mathematical model is the city of New York, where the construction of the intensified industrial park in the Staten Island region is proposed. New York City is one of the cities that generates the most waste in the United States (DSNY 2020), in addition to being one of the cities that has been affected by COVID-19 and the generation of medical waste since the outbreak, on March 3, 2020 (Zhang et al. 2020). In the United States, 2.5 kilograms of medical waste is generated per bed every day (Janagi and Shah 2015). Currently, the waste management system is primarily in the charge of the New York City Department of Sanitation (DSNY). Waste collection is carried out through public and private contractors where waste is separated into 3 categories: paper, metal/plastic/glass, and mixed solid waste (non-recyclable garbage) (Nagle 2014). Regarding recycling, the main facility is the Sunset Park Material Recovery Facility in Brooklyn, where mainly about 75% of the discarded paper, glass, metal, and plastic is recycled, however, 80% of the mixed solid waste ends up in a landfill 600 miles from the city, and the remaining 20% is converted to energy using incineration (Galka 2016). During the COVID-19 pandemic, the increased demand for medical goods such as masks, gowns, and goggles, along with the fact that domestic trash generated per patient bed is also highly infectious, resulted in a major increase in medical waste. The volume of medical waste is increasing at an exponential rate due to the COVID-19 pandemic, and hospital staging room storage capacity cannot keep up (Chad et al. 2021). In New York City, there are nine medical waste disposal centers with an estimated disposal capacity of 12,000 tons per year, with 60% of the capacity being used to treat highly contagious medical waste under COVID-19 (Mei, 2021). Tables A.91, A.9.2, and A.9.6 show the required parameters of the case study for its implementation to the model.

9.6 Results

The formulation of the mathematical model related to the selected case study (New York City) consists of 317 continuous variables, 63 binary variables, and 341 equations. The resulting model is a mixed-integer linear programming problem (MILP) that is solved using the epsilon-constraint method varying the maximum processing capabilities for each of the scenarios. It was solved through the GAMS modeling environment with a solution time of 0.015 s using the CPLEX solver and an AMD 5600x processor at 4.5 GHz with 16 GB of RAM at 4000 MHz.

Of all the possible solutions, 4 of them are selected for further analysis (Fig. 3). Starting with solution A which is a hypothetical case where all the MSW is sent to processing, eliminating the expenses and taxes associated with sending the waste to the sanitary landfill. In this solution, the processing capacity of each one of the recycling facilities would have to be immensely large, which is quite difficult to implement. On the other hand, there is solution B (which remains in the region of the minimum tax rate of 2 $US/ton). In this solution, less than half of the waste is sent to the sanitary landfill, and the processing capacities are not excessively large as it turns out to be in solution A. Solution C is found in the region with the highest tax rate that promotes recycling (5.1 $ US/ton) proposed by Munguía-López et al. (2020). In this solution, more than 100,000 tons of garbage are being

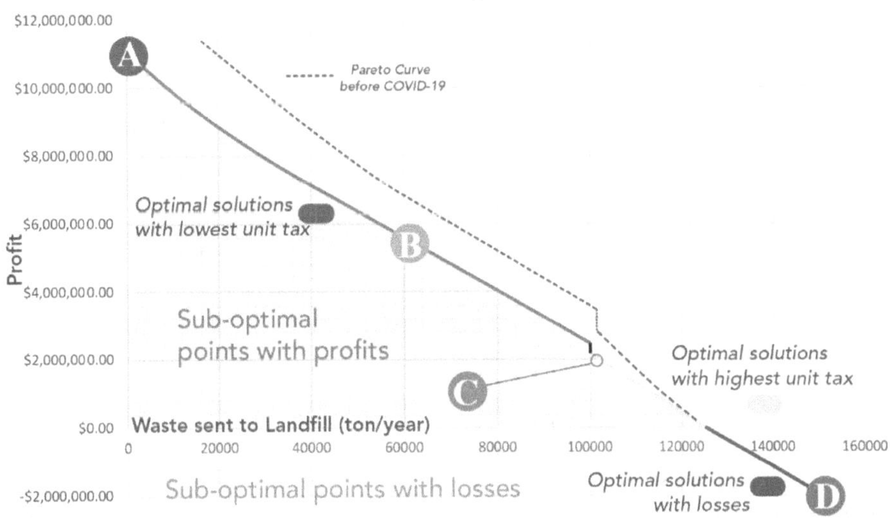

Fig. 3. Pareto curve for a system of processing facilities based on municipal solid waste during COVID-19. [Prineted (adapted) with permission from Munguia-López, A.C., Ochoa-Barragán, R., and Ponce-Ortega. J.M. 2022. Optimal waste management during the COVID-19 pandemic. Chemical Engineering and Processing: Process Intensification, *176*, 108942].

sent to the landfill, however, it is economically viable, since it continues to generate profits, although these can be small. In solution D, which produces annual economic losses, an even greater quantity of garbage is sent to the landfill, making the investment costs necessary for the implementation of the processing plants exceed the profits that could be obtained (Fig. 4). On the other hand, the model is solved using data before the pandemic (Themelis et al. 2002, Klemez et al. 2020, Jiang et al. 2021), in which, as expected, a better income is obtained, mainly due to a lower generation of waste with low economic potential, such as organic and medical waste (Roy et al. 2021).

Figure 4 shows a sample of the distribution of expenses and sales for different solutions. In all scenarios, the highest costs result from the transport and separation of the waste, thus the distance at which the waste source is from the processing center directly affects the viability of the strategy. Santibañez-Aguilar et al. (2013) proposed an optimization strategy where the different processing facilities were located in different cities. Therefore, the distances between facilities in some cases were especially large, which meant that some waste such as glass was not convenient to recycle, since the costs of separation and transport exceeded the possible income to be obtained by recycling it. The taxes in the first three cases influence significantly, while for the rest of the scenarios, it becomes insignificant compared to the rest of the costs. Probably the reduction in the tax rate (solution B) allows for increasing the rate of organic and medical waste destined for energy production, which is visible when solution B is compared to solution C (which receives the highest tax rate because it sends large amounts of waste to landfills). The main income comes from the recycling of inorganic waste, so, from an economic point of view, it might be better to discard the waste destined to produce energy and compost, to reduce the costs associated with the separation and treatment of this waste. However, from an environmental point of view, the amount of waste sent to the sanitary landfill would increase dramatically, which would cause a negative environmental and social impact, since within this waste are those that are generated because of the fight against COVID-19 (Celis et al. 2021, Liang et al. 2021).

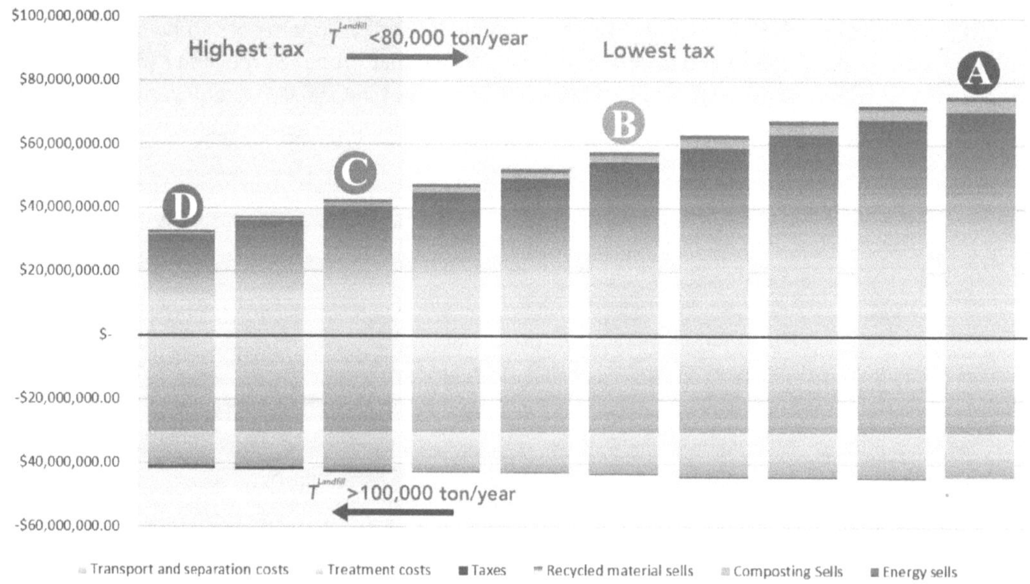

Fig. 4. Distribution for costs and sales of the supply chain based on solid waste. [Prineted (adapted) with permission from Munguia-López, A.C., Ochoa-Barragán, R., and Ponce-Ortega. J.M. 2022. Optimal waste management during the COVID-19 pandemic. Chemical Engineering and Processing: Process Intensification, *176*, 108942].

9.6.1 Analysis of e-waste

In this section, the results for e-waste residues are analyzed. Figures 5 and 8 present the flow diagrams for solid waste processing, and Figs. 7 and 10 compare the different costs and sales with the other types of waste. For solutions B and C, electronic waste is the one that produces the least tax expenses

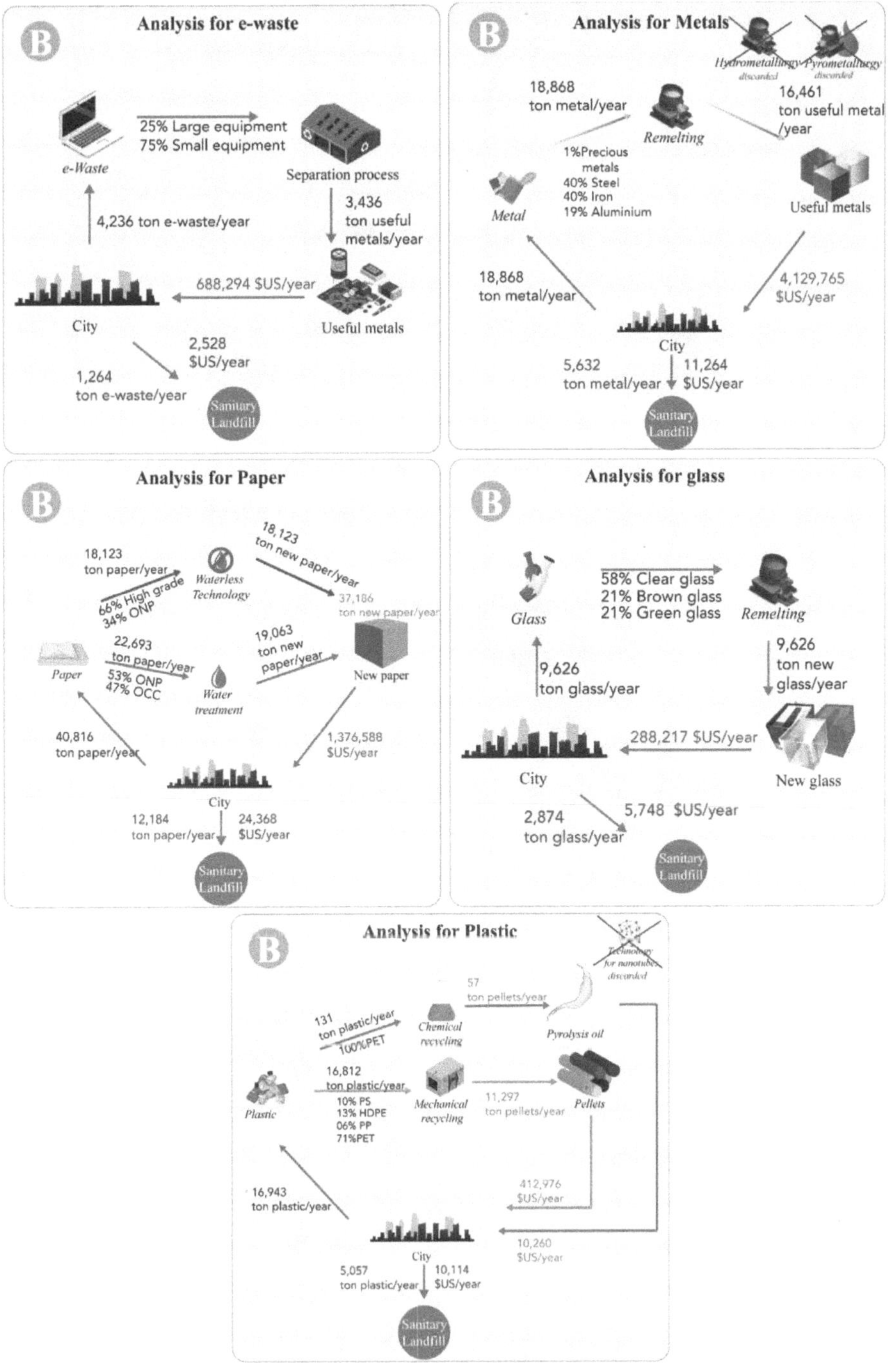

Fig. 5. Solution B: Analysis of inorganic wastes. [Prineted (adapted) with permission from Munguia-López, A.C., Ochoa-Barragán, R., and Ponce-Ortega. J.M. 2022. Optimal waste management during the COVID-19 pandemic. Chemical Engineering and Processing: Process Intensification, *176*, 108942].

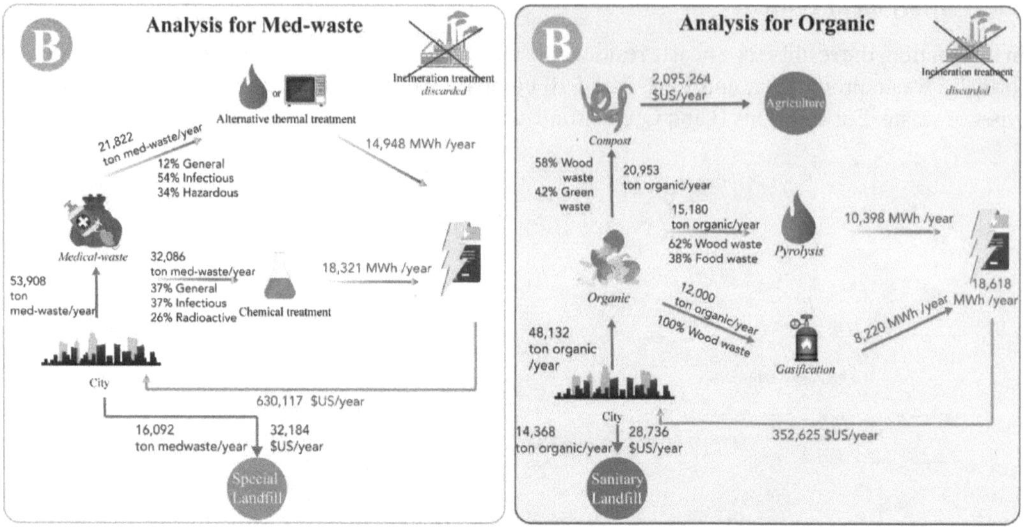

Fig. 6. Solution B: Analysis of medical waste and organic waste. [Prineted (adapted) with permission from Munguia-López, A.C., Ochoa-Barragán, R., and Ponce-Ortega. J.M. 2022. Optimal waste management during the COVID-19 pandemic. Chemical Engineering and Processing: Process Intensification, *176*, 108942].

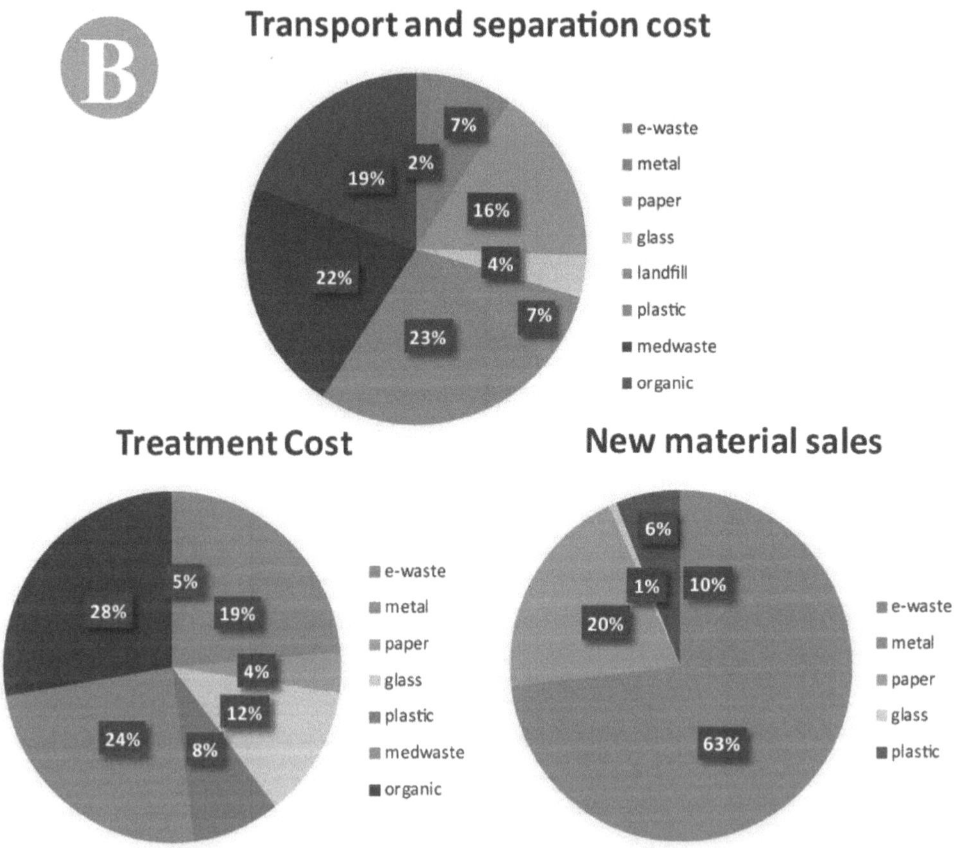

Fig. 7. Solution B: Costs and sales distribution for different types of waste. [Prineted (adapted) with permission from Munguia-López, A.C., Ochoa-Barragán, R., and Ponce-Ortega. J.M. 2022. Optimal waste management during the COVID-19 pandemic. Chemical Engineering and Processing: Process Intensification, *176*, 108942].

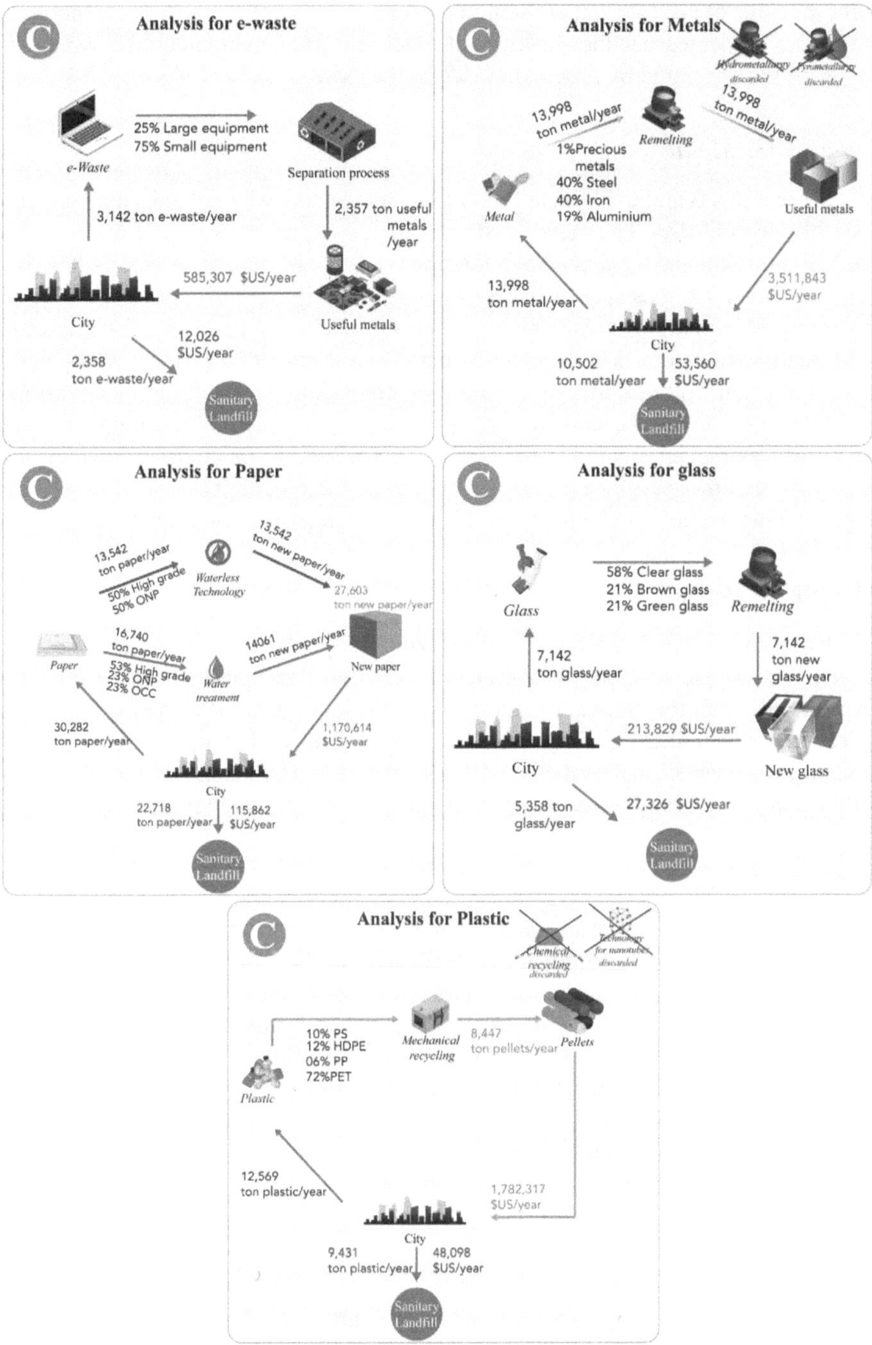

Fig. 8. Solution C: Analysis of inorganic wastes. [Prineted (adapted) with permission from Munguia-López, A.C., Ochoa-Barragán, R., and Ponce-Ortega. J.M. 2022. Optimal waste management during the COVID-19 pandemic. Chemical Engineering and Processing: Process Intensification, *176*, 108942].

due to the low amount of waste generated annually compared to the other types. However, it is not the one that produces the least profit, mainly because of how expensive the new raw materials generated can be. Therefore, e-waste is an important part of the income generated. It is observed that the change in the tax rate does not affect the amount of e-waste that is sent to be processed. But, in scenario B, although the amounts of e-waste that are sent to the landfill are similar, the tax rate that is applied is the lowest (due to the decrease in the quantities of waste sent from other materials), which drastically increases the profits generated by processing electronic waste.

9.6.2 Analysis for Metals

The results in scenarios B and C for metallic waste are presented, where thanks to metal processing, the strategy increases profits. The metal waste alone produces more than half of the profits (Figs. 7 and 10). Metals are the most expensive in the prices per ton of new materials. Solutions B and C rule out the use of technologies such as hydrometallurgy and pyrometallurgy, possibly because they have a lower processing efficiency factor than remelting, in addition to the fact that Tables A.9.3 and A.9.4 show that they are significantly more expensive in terms of investment. However, variable costs are even less than half as expensive as for remelting. Compared to electronic waste, although its generated products obtain the same benefits per ton, in this case implementing a lower tax rate, significantly reduces the amounts of metal waste that are sent to the landfill, reducing it by 50%. In the same way, the lower tax rate greatly favors the possible benefits to be obtained from metal processing, reducing the taxes to be paid by a factor of 5 comparing solutions B and C.

9.6.3 Analysis of Paper

The use of two technologies is observed in a nalyzing the solutions for paper processing, also the reduction of the tax rate promotes an increase in the production of recycled paper by one-third (solution B) which reduces by half the amount of waste sent to the landfill. On the one hand, it is feasible to obtain an economic benefit from using water-based treatments and an environmental benefit from using water-free treatments. On the other hand, it is observed that by increasing the amounts of waste, the use of recycling technologies that use water is favored, to maximize profits.

9.6.4 Analysis of Glass

Regarding the recycling of glass, for any of the solutions, the expenses inherent to transport, separation, and treatment are more expensive, compared to the possible profits from sales of new glass. Therefore, the rational solution could be to send all the glass waste to the landfill. However, the variability in the tax rate causes a trend toward the reduction of waste sent to the landfill. Thus, it is convenient to recycle a part of this glass waste, to remain below the 100,000 tons per year of waste sent to the landfill. If solutions B and C are compared, solution C pays about 3 times more taxes than solution B since the processing increases by about 20%. Due to the high costs of transport and separation, a possible way to reduce expenses and increase profits could be by placing the processing facilities in a different location and closer to the source of waste.

9.6.5 Analysis of Plastic

Solutions B and C for plastic are the ones that differ the most. In solution C, only plastic is processed using mechanical recycling for the production of pellets, while for solution B, chemical recycling is incorporated, the production of nanotubes is ruled out in both scenarios due to the associated high investment costs. Compared with glass waste solutions, something similar happens with plastic, the profits are low, compared to the processing of other waste such as metal or even paper, in addition to being the waste that generates the most expenses in separation and transport (Figs. 7 and 10). The lower tax rate significantly affects the amount of plastic waste sent to the landfill since from solution C to solution B there is a 46% decrease. In solution B, a small amount of pyrolysis oil is produced, which represents a minimum income. However, if more plastic waste is required to be processed, the production of pyrolysis oil would increase due to the inherent production costs of the process, which are lower than those for pellet production (Tables A.9.3 and A.9.4).

9.6.6 Analysis of Medical Waste

Medical waste requires special attention due to the COVID-19 pandemic , when medical waste gained prominence and volume. For solution C (Fig. 9), chemical treatment and alternative thermal treatment are used as medical waste disinfection technologies, which alternatively produce energy, although, comparing the gains from energy sales, they are very small (Fig. 4). Table A.9.1 shows that medical waste makes up most of the waste generated by the city; this fraction has increased due to the current COVID-19 pandemic. Therefore, the strategy presented implies that this waste increases the tax rate. The amount of waste generated is so large that it can hardly be processed completely, thus, a special sanitary landfill for this type of waste is required. In solution B, the waste sent to the special sanitary landfill is reduced by about half. However, the profits do not increase significantly, but it is worth noting that the main benefit of processing this waste lies in the social benefit. In both solutions, it is suggested that the incineration is discarded, a technology that produces more emissions of toxic chemicals for humans (Liang et al. 2021).

9.6.7 Analysis of Organic Waste

For organic waste, the process generates two types of income: those associated with the production of material for composition and those related to the production of electrical energy. For any of the solutions, the gains inherent to the agricultural part are always greater. Compared to the results obtained by Zhao and You (2020), particularly for food waste (in solutions B and C), the model uses only Pyrolysis for this type of waste, processing only 10% of the food waste generated (Figs. 6 and 9), however, although in this case food waste is not prioritized, there is a significant reduction in organic waste in general. In scenarios where the higher tax rate is applied, the amount of organic waste processed is minimal, while when a lower tax rate is applied, it takes on greater relevance,

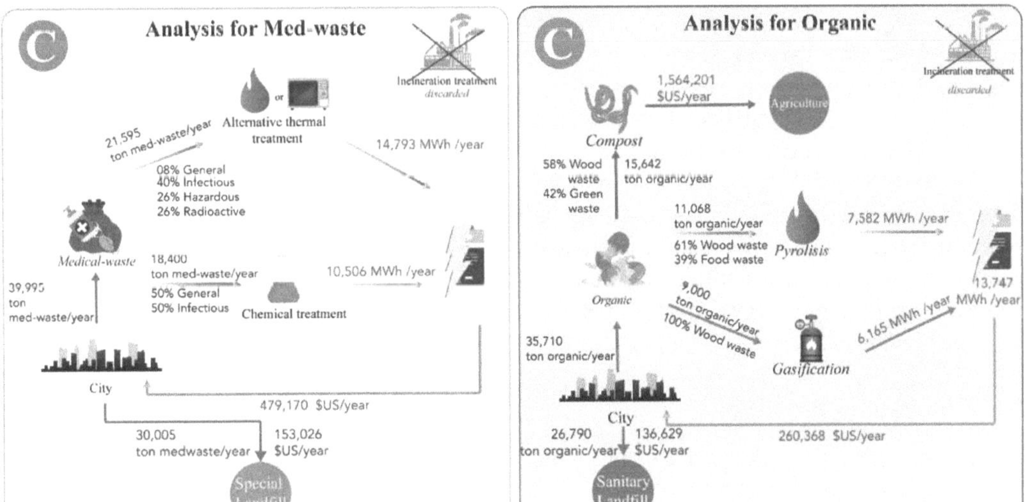

Fig. 9. Solution C: Analysis of medical waste and organic waste. [Prineted (adapted) with permission from Munguia-López, A.C., Ochoa-Barragán, R., and Ponce-Ortega. J.M. 2022. Optimal waste management during the COVID-19 pandemic. Chemical Engineering and Processing: Process Intensification, 176, 108942].

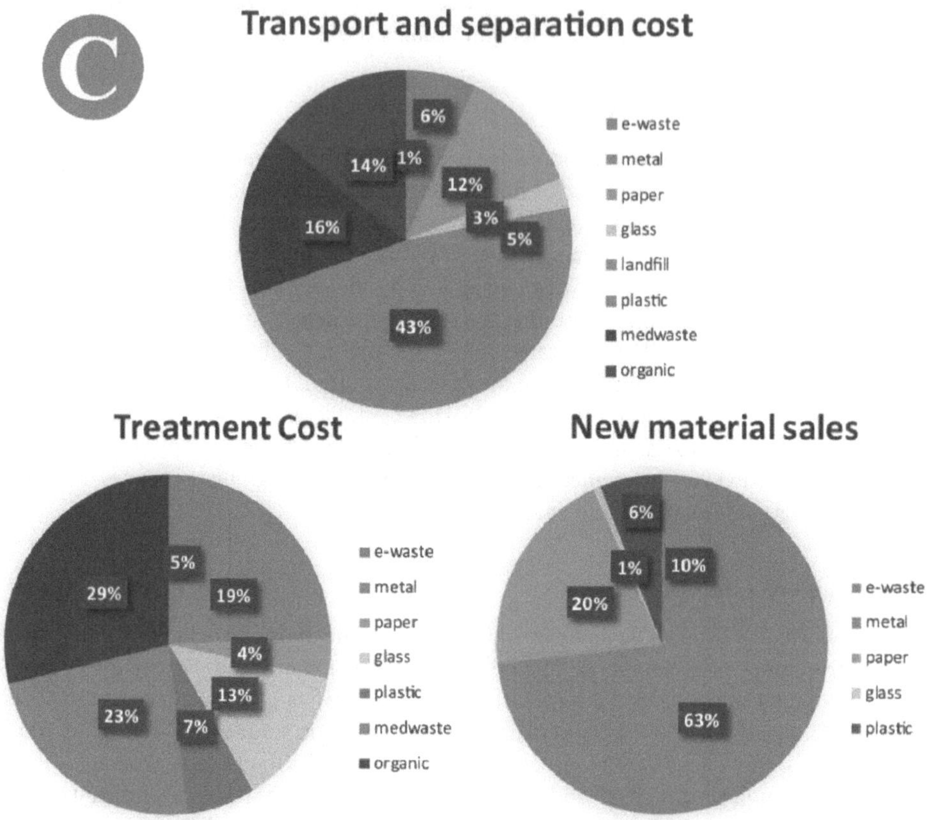

Fig. 10. Solution C: Costs and sales distribution for different types of waste. [Prineted (adapted) with permission from Munguia-López, A.C., Ochoa-Barragán, R., and Ponce-Ortega. J.M. 2022. Optimal waste management during the COVID-19 pandemic. Chemical Engineering and Processing: Process Intensification, 176, 108942].

and the profits from the processing of this waste become important (Fig. 4). Similar to the previous analysis, incineration is discarded while pyrolysis and gasification are selected as the technologies for solutions B and C, as for medical waste, organic waste is one of the most generated, so its processing is essential to achieve a lower tax rate.

The results show that the application of the strategy generates greater processing of previously unprocessed waste, such as those classified as non-recyclable garbage, of which only 20% were processed to be converted to energy, compared to the strategy shown, around 70% of the waste is processed even though it does not represent a significant source of income. This becomes more relevant when a large part of this waste represents a health risk, as is the case of medical waste produced by the pandemic.

9.7 Concluding Remarks

A circular framework to address the management of the waste generated by the COVID-19 pandemic employing strategies based on process intensification is presented in this chapter. This approach incorporates variable green tax rates, which promote the processing of MSW for the production of energy and new materials and minimize environmental and social impacts. The involved objectives include maximizing the economic benefit and minimizing the waste sent to landfills while simultaneously minimizing the social impact generated by the COVID-19 pandemic related to the high risk that the medical waste represents for health. The problem in this chapter was formulated as a multi-objective mixed-integer linear programming problem. The model considers a wide range of options to process the different types of waste, in addition, it selects and assigns the production capacities for the processing plants of the selected technology. The application of the model was illustrated, selecting New York City in the United States as a case study. The results show that it is feasible to implement an intensified industrial park that processes all the waste generated. Furthermore, the variable tax rate promotes waste recycling, maximizing profits even for those wastes with low economic potential. The different Pareto solutions obtained allow the decision-maker to have a wide variety of options to choose the solution that best suits the proposed objectives. The location of the intensified industrial park is key to minimizing transportation costs, which, for waste with less economic potential, can be even higher than the possible sale prices of its associated products. This model can be useful in regions where the municipal solid waste management is not adequate and there is a high risk of exposure to the waste.

9.8 Nomenclature

9.8.1 Binary Variables

$h_{j,k}$	Binary decision variable for which type of technology k to use for subtype of waste j
$h^{Lanfill}$	Binary decision variable for what tax rate to use for waste sent to landfill

9.8.2 Continuous Variables

AS	Sales associated with agriculture
$cost_{j,k}$	Costs for the implementation of technology k for subtype of waste j
$cw_{i,j}$	Subwaste classified
ES	Sales associated with energy production
$f_{j,k}$	Classified subtype of waste j that will be processed by technology k
MS	Sales associated with the processed materials
$p_{j,k}$	Product obtained from the subtype of waste j using technology k
PTC	Transportation cost for products
$RMTC$	Raw material transport cost
T_i	Classified waste i
$T^{Landfill}$	Amount of total waste sent to landfill
T^{landL}	Waste that will be taxed with a lower penalization
T^{landH}	Waste that will be taxed with a higher penalization
T^{II}	Lower bound for the lower tax
T^{III}	Upper bound for the higher tax
$Taxes^L$	Taxes with the lower tax rate
$Taxes^H$	Taxes with the higher tax rate
TC	Total cost
wf	Fraction of the total waste sent to processing

9.8.3 Parameters

$A_{j,k}$	Fixed cost for using technology k for subtype of waste *j*
$B_{j,k}$	Variable cost for using technology k for subtype of waste *j*
EMPT	Sales per ton associated with electronic material
kf	Annualization factor
SCi	Separation cost
TCPTR	Total cost per ton of waste transported
α_i	Existing fraction of each waste *i* classified
$\beta_{i,j}$	Waste separation fraction i for each type of sub-waste *j*
$^{\min}_{j,k}$	Limit for the minimum processing capacity
$\delta^{\max}_{j,k}$	Limit for the maximum processing capacity
$\gamma_{j,k}$	Efficiency factor for processing subtype of waste j using technology *k*
$\omega_{j,k}$	Sale price per ton of subtype of waste j processed using technology *k*

9.10 Appendix

9.10.1 Data for the Mathematical Model

Table A.9.1. Separation fractions for each type of waste for New York City. [Prineted (adapted) with permission from Munguia-López, A.C., Ochoa-Barragán, R., Ponce-Ortega. J.M. (2022). Optimal waste management during the COVID-19 pandemic. Chemical Engineering and Processing: Process Intensification, 176, 108942].

Waste type	α_i	Description
e-waste	0.022	Fraction of e-waste present in total solid waste
Glass	0.05	Fraction of glass present in total solid waste
Med-waste	0.28	Fraction of medical waste present in total solid waste
Metal	0.098	Fraction of metal present in total solid waste
Organic	0.25	Fraction of organic present in total solid waste
Paper	0.212	Fraction of paper present in total solid waste
Plastic	0.088	Fraction of plastic present in total solid waste

Table A.9.2. Separation fractions for the different sub-wastes in New York City. [Prineted (adapted) with permission from Munguia-López, A.C., Ochoa-Barragán, R., Ponce-Ortega. J.M. (2022). Optimal waste management during the COVID-19 pandemic. Chemical Engineering and Processing: Process Intensification, *176*, 108942].

Waste-type (*i*)	Description	Sub-waste (*j*)	$\beta_{i,j}$
e-waste	Represents the fraction of each type of waste *i* for e-waste	Large equipment	0.25
		Small equipment	0.75
Glass	Represents the fraction of each type of waste *l* for glass	Clear	0.58
		Brown	0.21
		Green	0.21

Table A.9.2. contd. ...

Table A.9.2. contd. ...

Waste-type (*i*)	Description	Sub-waste (*j*)	$\beta_{i,j}$
Med-waste	Represents the fraction of each type of waste *n* for medical waste	General	0.27
		Infectious	0.44
		Hazardous	0.14
		Radioactive	0.15
Metal	Represents the fraction of each type of waste *j* for metal	Precious metals	0.01
		Steel	0.4
		Iron	0.4
		Aluminum	0.19
Organic	Represents the fraction of each type of waste *o* for organic waste	Food waste	0.694
		Wood waste	0.12
		Green waste	0.186
Paper	Represents the fraction of each type of waste *k* for paper	High grade	0.294
		OCC[1]	0.15
		ONP[1]	0.556
Plastic	Represents the fraction of each type of waste *m* for plastic	PS[1]	0.102
		HDPE[2]	0.125
		PET[2]	0.057
		PE[2]	0.716

[1] OCC: corrugated cardboard boxes; ONP: Old Newsprint paper

[2] PS: Polystyrene; HDPE: High-density polyethylene; PET: polyethylene terephthalate; PE: Polyethylene

Table A.9.3. Fixed costs for processing technologies. [Prineted (adapted) with permission from Munguia-López, A.C., Ochoa-Barragán, R., Ponce-Ortega. J.M. (2022). Optimal waste management during the COVID-19 pandemic. Chemical Engineering and Processing: Process Intensification, *176*, 108942].

Technology (*k*)	Description and compatibility (*j*)	$A_{j,k}$	Units
Chemical recycling	Fixed cost for employing chemical recycling for all **plastic subwastes**	2,090,000	$US
Composting	Fixed cost for employing composting for all **organic subwastes**	9,573,000	$US
Gasification	Fixed cost for employing gasification for all **organic subwastes**	6,412,410	$US
Hydrometallurgy	Fixed cost for hydrometallurgy for all **metal subwastes**	5,384,000	$US
Incineration	Fixed cost for employing incineration for all **medical subwastes**	4,000,000	$US
Incineration	Fixed cost for employing incineration for all **organic subwastes**	4,000,000	$US
Mechanical recycling	Fixed cost for employing mechanical recycling for all **plastic subwastes**	1,590,000	$US
Nanotube's technology	Fixed cost for employing nanotubes technology for all **plastic subwastes**	200,000,000	$US
Pyrolysis	Fixed cost for employing pyrolysis for all **organic subwastes**	2,000,000	$US
Pyrometallurgy	Fixed cost for employing pyrometallurgy for all metal subwastes	6,044,000	$US
Remelting	Fixed cost for employing Remelting for all **glass subwastes**	5,000,000	$US
Remelting	Fixed cost for employing remelting for all **metal subwastes**	4,440,000	$US

Table A.9.3. contd. ...

Table A.9.3. contd. ...

Technology (*k*)	Description and compatibility (*j*)	$A_{j,k}$	Units
Separation process	Fixed cost for employing separation process for all **e-subwastes**	3,137,030	$US
Thermal treatment	Fixed cost for employing Thermal treatment for all **medical subwastes**	2,000,000	$US
Waterless technology	Fixed cost for employing waterless technology for **high grade paper**	8,700,000	$US
Waterless technology	Fixed cost for employing waterless technology for all **paper subwastes**	1,880,000	$US

Table A.9.4. Variable costs for processing technologies. [Prineted (adapted) with permission from Munguia-López, A.C., Ochoa-Barragán, R., Ponce-Ortega. J.M. (2022). Optimal waste management during the COVID-19 pandemic. Chemical Engineering and Processing: Process Intensification, *176*, 108942].

Technology (*k*)	Description and compatibility (*j*)	$B_{j,k}$	Units
Chemical recycling	Variable cost for employing chemical recycling for all **plastic subwastes**	106.6	$US/Ton
	Variable cost for employing chemical recycling for all **plastic subwastes**	144	$US/Ton
	Variable cost for employing chemical treatment for all **medical subwastes**	200	$US/Ton
Composting	Variable cost for employing composting for all **organic subwastes**	20.15	$US/Ton
Gasification	Variable cost for employing gasification for all **organic subwastes**	36	$US/Ton
Hydrometallurgy	Variable cost for hydrometallurgy for all **metal subwastes**	112	$US/Ton
Incineration	Variable cost for employing incineration for all **medical subwastes**	380	$US/Ton
	Variable cost for employing incineration for all **organic subwastes**	400	$US/Ton
kf	Annualization factor	0.3	$Year^{-1}$
Nanotubes technology	Variable cost for employing nanotubes technology for all **plastic subwastes**	55,000	$US/Ton
Pyrolysis	Variable cost for employing pyrolysis for all **organic subwastes**	354	$US/Ton
Pyrometallurgy	Variable cost for employing pyrometallurgy for all **metal subwastes**	217	$US/Ton
Remelting	Variable cost for employing remelting for **clear glass waste**	70	$US/Ton
	Variable cost for employing remelting for **Brown and green glass waste**	90	$US/Year
Separation process	Variable cost for employing separation process for all **e-subwastes**	2.82	$US/Ton
Thermal treatment	Variable cost for employing thermal treatment for all **medical subwastes**	400	$US/Ton
Water treatment technology	Variable cost for water treatment technology for **high grade paper waste**	56	$US/Ton
Water treatment technology	Variable cost for water treatment technology for **OCC paper waste**	31	$US/Ton
Water treatment technology	Variable cost for water treatment technology for **ONP paper waste**	32	$US/Ton
Waterless technology	Variable cost for employing waterless technology for **high grade paper**	1,800	$US/Ton

Table A.9.5. Efficiency factors for the processing technologies. [Prineted (adapted) with permission from Munguia-López, A.C., Ochoa-Barragán, R., Ponce-Ortega. J.M. (2022). Optimal waste management during the COVID-19 pandemic. Chemical Engineering and Processing: Process Intensification, *176*, 108942].

Waste type	Technology (*k*)	Product	$\gamma_{j,k}$ (ton product/ton waste) $\forall j$
E-waste	Separation process	Useful metals	0.75
Glass	Remelting	New Glass	1
Med-waste	Alternative thermal treatment	Energy	0.685[A]
	Chemical treatment	Energy	0.571[A]
	Incineration treatment	Energy	0.685[A]
Metal	Hydrometallurgy	Useful metals	0.990
	Pyrometallurgy	Useful metals	0.990
	Remelting	Useful metals	1
Organic waste	Incineration treatment	Energy	0.544[A]
	Pyrolysis	Energy	0.685[A]
	Gasification	Energy	0.685[A]
	Compost	Organic fertilizer	1
Paper	Water treatment	New paper	0.84
	Waterless tech	New paper	1
Plastic	For nanotubes	Nanotubes	0.320
	Chemical recycling	Pyrolysis oil	0.432
	Mechanical recycling	Pellet	0.672

[A] This factor has units of MWH Energy/Ton Waste.

Table A.9.6. Parameters assigned for the costs of separation and processing of waste and products for the case of New York City. [Prineted (adapted) with permission from Munguia-López, A.C., Ochoa-Barragán, R., Ponce-Ortega. J.M. (2022). Optimal waste management during the COVID-19 pandemic. Chemical Engineering and Processing: Process Intensification, 176, 108942].

Parameter	Value	Units	Description
$SC_i, \forall_i$	235	$US/Ton	Estimated separation cost
TCPT	0.686	$US/Ton	Total cost per ton of waste transported
TW	250,000	Ton/Year	Total waste

Table A.9.7. Processing costs, [Prineted (adapted) with permission from Munguia-López, A.C., Ochoa-Barragán, R., Ponce-Ortega. J.M. (2022). Optimal waste management during the COVID-19 pandemic. Chemical Engineering and Processing: Process Intensification, 176, 108942].

Waste-type	Subwaste(*j*)	Technology (*k*)	$\omega_{j,k}$	Units
e-waste	For **all electronic subwastes**	Separation process	1300	$US/Ton
Glass	For **all glass subwastes**	Remelting	29.94	$US/Ton
Med-waste	For **all medical subwastes**	For **all medical waste technologies**	18.94	$US/MWh
Metal	For **all metal subwastes**	For **all metal technologies**	1300	$US/Ton
Organic	For **all organic subwastes**	Compost	100	$US/Ton
		Pyrolysis, gasification, incineration	18.94	$US/MWh
Paper	For **all paper subwastes**	For **all paper technologies**	210	$US/Ton
Plastic	For **all plastic subwastes**	Nanotube's production	150,000	$US/Ton
		Chemical recycling	180	$US/Ton
		Mechanical recycling	211	$US/Ton

9.11 References

AP. 2021. Tasa de reciclaje en México es del 9.6%. Ambiente Plastico, https://www.ambienteplastico.com/tasa-de-reciclaje-en-mexico-es-del-9-6/, 2021 (Accessed 11 November 2021).

Becht, S., Franke, F., Geißelmann, A., and Hahn, H. 2009. An industrial view of process intensification. Chemical Engineering and Processing - Process Intensification, *48*(1), 329–332. Doi: 10.1016/j.cep.2008.04.012

Cabrera, D.V., and Labatut, R.A. 2021. Outlook and challenges for recovering energy and water from complex organic waste using hydrothermal liquefaction. Sustainable Energy and Fuels, *5*(8), 2201–2227. DOI: 10.1039/d0se01857k

Celis, J.E., Espejo, W., Paredes-Osses, E., Contreras, S.A., Chiang, G., and Bahamonde, P. 2021. Plastic residues produced with confirmatory testing for COVID-19: Classification, quantification, fate, and impacts on human health. Science of The Total Environment, *760*, 144167. DOI: 10.1016/j.scitotenv.2020.1441

Chand, S., Shastry, C.S., Hiremath, S., Joel, J.J., Krishnabhat, C.H., and Mateti, U.V. 2021. Updates on biomedical waste management during COVID-19: The Indian scenario. Clinical Epidemiology and Global Health, *11*, 100715. DOI:10.1016/j.cegh.2021.100715

Chen, X., Huang, G., Zhu, H., Suo, M., and Dong, C. 2016. Inexact inventory theory–based waste management planning model for the City of Xiamen, China. Journal of Environmental Engineering, *142*(5), 04016013. DOI:10.1061/(asce)ee.1943-7870.0001069

Das, A.K., Islam,M.N., Billah, M.M., and Sarker, A. 2021. COVID-19 pandemic and healthcare solid waste management strategy—A mini-review. Science of The Total Environment, *778*, 146220. DOI: 10.1016/j.scitotenv.2021.1462.

Diwekar, U. 2008. Introduction to applied optimization. Springer Optimization and Its Applications. Springer Science and Business Media: Clarendon Hills. 22 (2008) DOI:10.1007/978-0-387-76635-5

DSNY 2020. Annual Report: New York City curbside and containerized municipal refuse and recycling statistics. The City of New York Department of Sanitation. https://dsny.cityofnewyork.us/wp-content/uploads/2020/08/about_dsny-collections-annual-2020.pdf (Accessed 29 September 2021).

EPA 2021. National Overview: Facts and Figures on Materials, Wastes and Recycling. U.S. Environmental Protection Agency, https://www.epa.gov/facts-and-figures-about-materials-waste-and-recycling/national-overview-facts-and-figures-materials#NationalPicture, 2021 (Accessed 11 November 2021).

Galka, M. 2016. What does New York do with all its trash? One city's waste –in numbers. The Guardian. https://www.theguardian.com/cities/2016/oct/27/new-york-rubbish-all-that-trash-city-waste-in-numbers (Accessed 24 February 2021).

Govindan, K., Nasr, A.K., Mostafazadeh, P., and Mina, H. 2021. Medical waste management during coronavirus disease 2019 (COVID-19) outbreak: A mathematical programming model. Computers and Industrial Engineering, *162*, 107668. DOI: 10.1016/j.cie.2021.107668

Hantoko, D., Li, X., Pariatamby, A., Yoshikawa, K., Horttanainen, M., and Yan, M. 2021. Challenges and practices on waste management and disposal during COVID-19 pandemic. Journal of Environmental Management, *286*, 112140. DOI: 10.1016/j.jenvman.2021.112140

Homayouni, Z., and Pishvaee, M.S. 2020. A bi-objective robust optimization model for hazardous hospital waste collection and disposal network design problem. Journal of Material Cycles and Waste Management, *22*, 1965–1984. DOI:10.1007/s10163-020-01081-8

Ives, M. 2021. The Covid Crisis Is Now a Garbage Crisis, Too, The New York Times https://www.nytimes.com/2021/09/18/world/covid-trash-recycling.html, 2021 (Accessed 24 February 2022).

Janagi, R., Shah, J., and Maheshwari, D. 2015. Scenario of management of medical waste in US and UK: a review. Journal of Global Trends in Pharmaceutical Sciences, *6*(1), 2328–2339. https://www.jgtps.com/admin/uploads/ERmDV8.pdf, 2015 (Accessed 16 October 2021).

Jiang, P., Fan, Y.V., and Klemeš, J.J. 2021. Impacts of COVID-19 on energy demand and consumption: Challenges, lessons and emerging opportunities. Applied Energy. *285*, 116441. DOI: 10.1016/j.apenergy.2021.116441

Klemeš, J.J., Fan, Y.V., and Jiang, P. 2020. The energy and environmental footprints of COVID-19 fighting measures –PPE, disinfection, supply chains. Energy, *211*, 118701. DOI: 10.1016/j.energy.2020.118701

Ko, S., Kim, W., Shin, S.C., and Shin, J. 2020. The economic value of sustainable recycling and waste management policies: The case of a waste management crisis in South Korea. Waste Management, *104*, 220–227. DOI: 10.1016/j.wasman.2020.01.020.

Liang, Y., Song, Q., Wu, N., Li J., Zhong, Y., and Zeng, W. 2021. Repercussions of COVID-19 pandemic on solid waste generation and management strategies. Frontiers of Environmental Science and Engineering, *15*(115), 1767 DOI:10.1007/s11783-021-1407-5

Mei, X., Hao, H., Sun, Y., Wang, X., and Zhou U. 2021. Optimization of medical waste recycling network considering disposal capacity bottlenecks under a novel coronavirus pneumonia outbreak. Environmental Science and Pollution Research, *29*, 79669–79687, DOI: 10.1007/s11356-021-16027-2

Meyer, F., Johannsen, J., Liese, A., Fieg, G., Bubenheim, P., and Waluga, T. 2021. Evaluation of process integration for the intensification of a biotechnological process. Chemical Engineering and Processing - Process Intensification, *167*, 108506. DOI: 10.1016/j.cep.2021.108506

Monsiváis-Alonso, R., Mansouri, S.S., and Román-Martínez, A. 2020. Life cycle assessment of intensified processes towards circular economy: Omega-3 production from waste fish oil. Chemical Engineering and Processing-Process Intensification, 108171. DOI: 10.1016/j.cep.2020.108171

Munguía-López, A.C., Zavala, V.M., Santibañez-Aguilar, J.E., and Ponce-Ortega, J.M. 2020a. Optimization of municipal solid waste management using a coordinated framework. Waste Management, *115*, 15–24. DOI: 10.1016/j.wasman.2020.07.006

Munguía-López, A.C., Núñez-López, J.M. and Ponce-Ortega, J.M. 2020b. Identifying fair solutions in the optimal design of integrated residential complexes. Chemical Engineering and Processing - Process Intensification, 108116, DOI: 10.1016/j.cep.2020.108116

Munguia-López, A.C., Ochoa-Barragán, R., and Ponce-Ortega. J.M. 2022. Optimal waste management during the COVID-19 pandemic. Chemical Engineering and Processing: Process Intensification, *176*, 108942.

Nagle, R. 2014. Picking up: on the streets and behind the trucks with the sanitation workers of New York City. 1st edition. New York: Farrar, Straus and Giroux. ISBN:978-0-374-29929-3.

Oates, W.E. 1995. Green Taxes: Can We Protect the Environment and Improve the Tax System at the Same Time? Southern Economic Journal, *61*(4), 915. DOI: 10.2307/1060731

Paut-Kusturica, M., Golocorbin-Kon, S., Ostojic, T., Kresoja, M. Milovic, M., Horvat, O., Dugandzija, T., Davidovac, N., Vasic, A., and Tomas, A. 2020. Consumer willingness to pay for a pharmaceutical disposal program in Serbia: A double hurdle modeling approach. Waste Management, *104*, 246–253. DOI: 10.1016/j.wasman.2020.01.029.

Pecci, K. 2018. All Landfills Leak, and our health and environment pay the toxic price. Conservation Law Foundation. https://www.clf.org/blog/all-landfills-leak-and-our-health-and-environment-pay-the-toxic-price/ (Accessed 24 February 2021).

Ponce-Ortega, J.M., Al-Thubaiti, M.M. and El-Halwagi, M.M. 2012. Process intensification: New understanding and systematic approach. Chemical Engineering and Processing: Process Intensification, *53*, 63–75. DOI: 10.1016/j.cep.2011.12.010

Rivas, D.F., Castro-Hernández, E., Villanueva Perales, A.L., and van der Meer, W. 2018. Evaluation method for process intensification alternatives. Chemical Engineering and Processing: Process Intensification, *123*, 221–232. DOI: 10.1016/j.cep.2017.08.013

Roy, P., Mohanty, A.K., Wagner, A., Sharif, S., Khalil, H., and Misra, M. 2021. Impacts of COVID-19 Outbreak on the Municipal Solid Waste Management: Now and beyond the Pandemic. ACS Environmental Au, *1*, 32–45 DOI:10.1021/acsenvironau.1c00005

Santibañez-Aguilar, J.E., Ponce-Ortega, J.M., González-Campos, J.B., Serna-González, M., and El-Halwagi, M.M. 2013. Optimal planning for the sustainable utilization of municipal solid waste. Waste Management, *33*(2), 2607–2622. DOI: 10.1016/j.wasman.2013.08.010.

Serrano-Ruiz, J.C. 2020. Biomass: A renewable source of fuels, chemicals and carbon materials. Molecules, *25*(21), 5217. DOI: 10.3390/molecules25215217.

Sharma, R., Kodamana, H., and Ramteke, H. 2022. Multi-objective dynamic optimization of hybrid renewable energy systems. Engineering and Processing - Process Intensification, *170*, 108663, DOI: 10.1016/j.cep.2021.108663.

Star, 2021. Poor medical waste management will increase infections. The Daily Star, https://www.thedailystar.net/editorial/news/poor-medical-waste-management-will-increase-infections-1909561, 2021 (Accessed 11 November 2021).

Strube, J., Ditz, R., Kornecki, M. Huter, M., Schmidt, A., Thiess, H., and Zobel-Roos, S. 2018. Process intensification in biologics manufacturing. Chemical Engineering and Processing - Process Intensification, *133*, 278–293. DOI: 10.1016/j.cep.2018.09.022

Themelis, N.J., Kim, Y.H., and Brady, M.H. 2002. Energy recovery from New York City municipal solid wastes. Waste Management and Research, *20*(3), 223–233. DOI:10.1177/0734242X0202000303

Valizadeh, J., and Mozafari, P. 2022. A novel cooperative model in the collection of infectious waste in COVID-19 pandemic, Journal of Modelling in Management, *17*, 363-401. DOI:10.1108/JM2 07 2020-0189

Zhao, X., and You, F. 2020. Waste respirator processing system for public health protection and climate change mitigation under COVID-19 pandemic: Novel process design and energy, environmental, and techno-economic perspectives. Applied Energy, 116129. DOI:10.1016/j.apenergy.2020.11612

Zhang, L., Li, D., and Ren, J. 2020. COVID-19 propagation analysis of a multistage dynamic time lag dynamics model. Geomatics and Information Science of Wuhan University, *45*, 658–666. DOI: 10.13203/j.whugis20200206

Zhao, N., and You, F. 2021. Food-energy-water-waste nexus systems optimization for New York State under the COVID-19 pandemic to alleviate health and environmental concerns. Applied Energy, *282*, 116181. DOI:10.1016/j.apenergy.2020.11618

10

Hybrid Machine Learning-Mathematical Programming Approach

This chapter presents a mathematical optimization strategy to address problems related to municipal waste management. This approach incorporates two strategies: deterministic optimization and machine learning. First, the optimization model determines the optimal supply chain for the municipal waste management system. Then, machine learning prediction models estimate the required parameters over time, which helps generate future projections. The optimization model was coded in the General Algebraic Modeling System, while the prediction model was coded in the Python programming environment. The proposed strategy was applied to the case study of New York City, which includes extensive socioeconomic data sets to train the machine learning model. The predicted waste collection over time is based on socioeconomic data. The results show trade-offs between the economic (profit) and environmental (waste sent to landfill) objectives for future scenarios, which can be helpful for possible pandemic scenarios in the following years.

10.1 Introduction

The COVID-19 pandemic has created issues in the majority of human development efforts. Both direct and indirect effects on the environment have resulted from the implementation of various social distancing measures, the overproduction of various hygiene goods, the restructuring of health systems, or even the way consumers plan their food purchases and the waste that this involves (Vittuari et al. 2021, Zanetti et al. 2022). Positive effects were observable from cleaner lakes, rivers, and beaches, to improvements in air quality, which offered an opportunity to formulate ecological policies that could restore environmental health (Chakraborty et al. 2021). The negative effects caused by this pandemic on the environment are related to the large increase in the production of municipal solid waste (MSW) and how the improper handling of it would increase the likelihood of the virus spreading as well as cause it to contaminate regular municipal waste (Nabavi-Pelesaraei et al. 2022), in addition to the low recycling or treatment capacity that existed at the time of the pandemic (Mohamed et al. 2022). The COVID-19 pandemic changed the rules of waste management, since inadequate management practices, poor waste management, and inadequate final disposal can increase the susceptibility to the spread of the virus, standards for the disposal of medical waste have been developed by several environmental and public health groups, but it is challenging to execute these guidelines effectively in the public sectors of developing nations (Bhar et al. 2022), especially in those regions where waste management was already unsustainable even before the pandemic due to ineffective treatment methods, the capacity limitations of healthcare facilities, and insufficient waste disposal techniques (El-Ramady et al. 2021).

The pandemic-related waste management issues have persisted for at least five years (Tripathi et al. 2020); for this reason alone, the implementation of different waste management strategies is necessary. I solation and changes in the lifestyles of the population have caused various problems in waste management since the consumption of single-use plastics, and healthcare products has increased greatly, in addition to the large increases in the generation of medical waste, representing a risk to society if not managed correctly (Sharma et al. 2020).

In recent studies, it has not been possible to determine if infectious waste plays a critical role in the distribution of the virus, however, the deposition of different medical residues in the water and soil ecosystem can cause microbial transformations, which represents a potential threat to the public health system (Ekhaise et al. 2023). The enormous amounts of waste generated during the pandemic make processing this waste a priority (Andeobu et al., 2022, Dharmaraj et al. 2021). To minimize the environmental and social impact in the short and long term, implementing efficient waste management strategies is necessary to reduce the long-term effects that the pandemic may generate, solving this challenge and creating technological techniques and knowledge will enable the safe treatment of municipal waste in the current pandemic, and above all to generate prevention strategies in the face of the possibility of future pandemics (Skrzyniarz et al. 2022).

In the search for possible solutions to the management of MSW, various strategies have been proposed considering different methodologies. Using an optimization perspective, Govidan et al. (2019) developed a multi-objective optimization model to consider social and economic objectives. The scheduling of the collection vehicles, possibilities of failures, and fuel consumption allow for establishing the optimal waste transport route. Valizadeh and Mozafari (2021) presented a mathematical model that uses four methods of cooperative game theory using genetic algorithms. This strategy offers different incentives to municipal waste collector contractors, allowing for solutions that show cooperation between collectors that reduce total costs and promote coalition between collectors. Homayouni and Pishvaee (2020) reported a multi-objective optimization model for medical waste management, where the epsilon constraint method was used to generate a distribution network that assigns different amounts of medical waste to the existing processing centers, this allows for minimizing the cost and health risk inherent in medical waste generated during the pandemic. Despite the positive outcomes, this approach ignores the distances between processing centers, which, depending on the area, may result in comparatively high transportation expenses..

Artificial intelligence-based (AI-based) methodologies have also been reported to solve MSW management problems (Andeobu et al. 2022). Smart waste management is a strategy that makes use of advanced technologies to dispose of waste in a profitable, and efficient manner. Artificial intelligence provides several methods that can be used to build intelligent waste management systems (Sinthiya et al. 2022). AI-based systems are used to solve challenging issues, deal with uncertainty, and demonstrate the effectiveness of intelligent systems (Ihsanullah et al. 2022). Integrated solid waste management strategies seek to generate methodologies that allow reducing the amounts of waste that are sent to landfills, where waste management is divided into stages (waste separation, collection and transportation, sub-waste separation, waste treatment, and final disposal) (Antonopoulos et al. 2012). Within these stages, it is possible to incorporate Machine Learning (ML) models with different objectives; Rubab et al. (2022) proposed different alternatives for the use of AI-based models that allow for improving integrated solid waste management, the use of AI for the optimization of collection and transport routes, assignment of locations, and optimal numbers of incinerators and sanitary landfills are some examples of the proposed alternatives.

Understanding the COVID-19 pandemic and utilizing all of the information produced will highlight the present and upcoming waste management challenges, where a new set of tools will be employed. When applied creatively, these tools will enable better planning in the event of future emergency situations.

The primary contributions of this chapter are based on the implementation of ML models to a mathematical model of multi-objective optimization that proposes the construction of an industrial park in charge of treating the different MSW generated. This hybridization of the model will allow the prediction of different parameters over time to evaluate the long-term viability of installing different industrial parks under different scenarios. This will provide alternatives to the decision maker that, based on the observable benefits, will facilitate the decision of whether or not to install an industrial park for the treatment of the different kinds MSW under a strategy based on the intensification of processes and the circular economy.

10.2 Problem Statement

The increase in the generation of MSW during the pandemic carries different social and environmental risks (Luhar et al. 2022). Face masks, syringes, and surgical gloves were the first line of defense against COVID-19, and once used, they represent a great risk to anyone who may come into contact with these residues of medical waste (Singh et al. 2022), at the same time, it reduces the possibilities of reuse or treatment thanks to established medical waste treatment guidelines (Hou et al. 2022), since the disposal and utilization of medical waste has not been explicitly planned (Le et al. 2022).

Social distancing and the changes in lifestyle during the pandemic were reflected in an increase in the use of disposable plastics (Haque and Fan 2022, Igalavithana et al. 2022), as well as in food waste (Iranmanesh et al. 2022, Çavuş et al. 2022). In developed countries, different strategies were implemented to deal with these new amounts of waste, new waste treatment facilities were installed, and policies were implemented in favor of prioritizing the treatment of medical waste (Andeobu et al. 2022). However, in underdeveloped countries there was no planning that could mitigate the problem represented by these new rates of generation of toxic waste, thus, a large part of the waste generated ended up in open-air landfills, which represents tremendous health and environmental risks (Masud et al. 2023).

Although different alternatives have been explored to address waste management, the conditions that arose during the COVID-19 pandemic changed the dynamics with which waste must be treated (Ali and Parvin 2022). Currently, it is possible to take advantage of the information generated during a health emergency that allows to generate better waste management strategies (Chen 2022), to be better prepared for the possibility of a future pandemic. The supply chain in waste management consists mainly of classification, collection and transportation, separation of sub-waste, treatment, and final disposal stages in which high levels of uncertainty are associated (Fathollahi-Fard et al. 2020, Santibañez-Aguilar et al. 2013). Therefore, a deterministic approach may not be enough to generate robust waste management strategies. To address this problem, the use of hybrid mathematical models (mathematical optimization and machine learning) can help optimize an MSW management system during sanitary emergencies (Valizadeh et al. 2021).

10.3 Implemented Approach

The implemented approach is based on a set of tools that allows for solving a specific part of the problem. The general solution approach is shown in Fig. 1, which consists of the iterative solution of a mathematical optimization model that describes the supply chain, assigns the processing technologies for each sub-residue, and generates a green tax rate that seeks to promote the recycling even of those residues with low economic potential, this model is coded in the General Algebraic Modeling System (GAMS) while storing its results in GAMS Data Exchange (GDX) files to later be fed to an ML model programmed in Python that will be responsible for the prediction of the useful parameters for future iterations of the optimization model. Among these parameters, the waste generation rates are estimated based on different socio demographic attributes such as unemployment, average family income, and the population over 45 years of age, characteristics of the case study. This optimization strategy based on "closed loop" data allows an organic integration of ML and mathematical programming, this «hybrid» system generates information on both sides (ML and mathematical programming) offering an efficient alternative for optimization problems under uncertainty (Ning and You 2019). Below is a detailed description of the parts of the optimization model.

10.3.1 Data-based system

First, the use of prediction tools is explored, particularly ML models that have shown excellent results in predicting the generation of MSW in different regions. According to Jahandideh et al. (2009), waste prediction using ML algorithms presents better behavior in complex non-linear type problems than traditional regression methods. Figure 2 shows the graphic scheme of the methodology used for the data-based system. In the following section, the development of the ML model proposed for this work is discussed in detail.

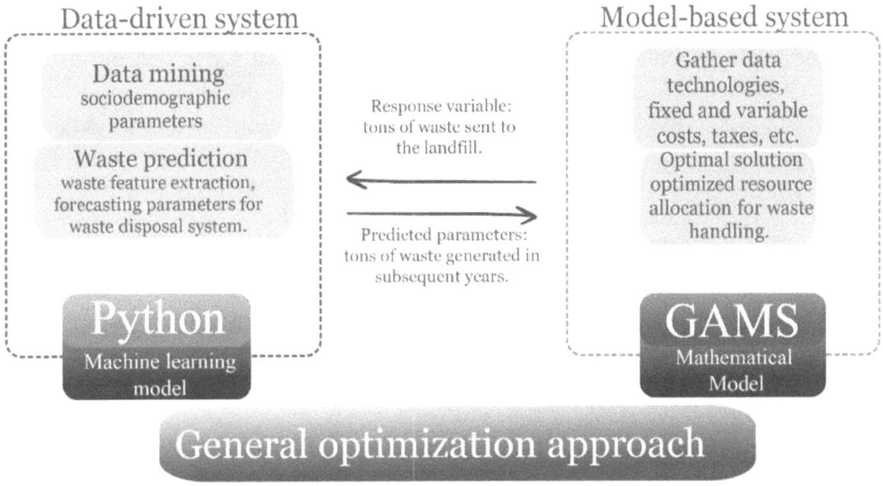

Fig. 1. Optimization approach based on hybrid modeling and data systems. [Printed (adapted) with permission from Munguia-López, A.C., Ochoa-Barragán, R., and Ponce-Ortega, J.M. 2022. Optimal waste management during the COVID-19 pandemic. Chemical Engineering and Processing: Process Intensification, *176*, 108942].

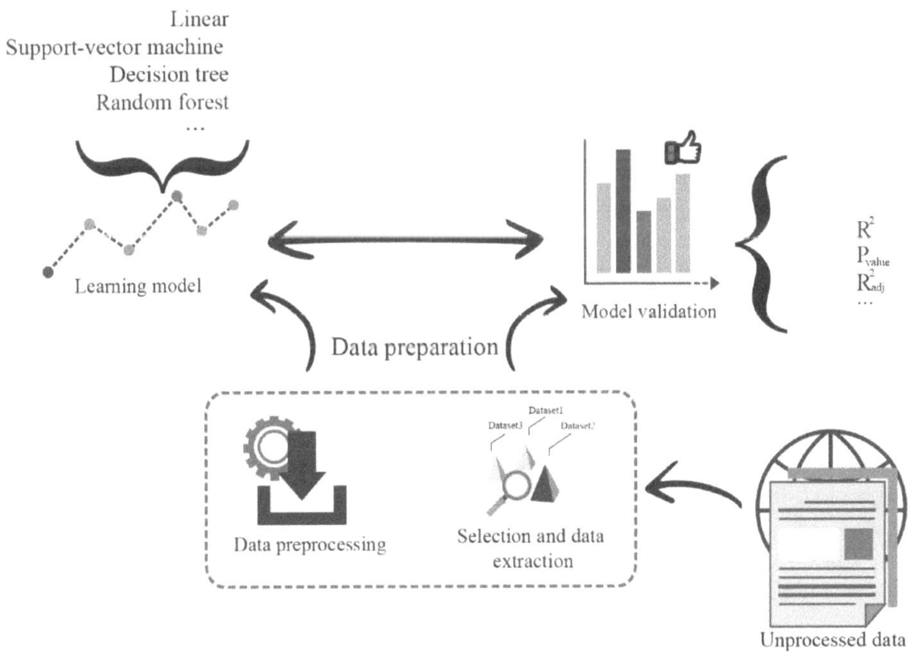

Fig. 2. Scheme of the machine learning model workflow. [Printed (adapted) with permission from Munguia-López, A.C., Ochoa-Barragán, R., and Ponce-Ortega, J.M. 2022. Optimal waste management during the COVID-19 pandemic. Chemical Engineering and Processing: Process Intensification, 176, 108942].

10.3.1.1 Selection of significant socio-economic parameters

Databases have become an essential tool in modern times, they allow access to a large amount of data over time, however, data doesn't initially express information, or there might be so much data that it can be overwhelming to obtain useful information , for this reason, data mining must be carried out to allow for extracting a delimited set of useful data from which it is possible gain relevant information. According to Ghinea et al. (2016), age plays a fundamental role in the generation of waste, this is mainly because of the different social behaviors such as expenses, savings, or changes in the increasingly sedentary lifestyle as the population approaches adulthood. Among other relevant socio-economic parameters for waste generation is the quality of life, since a better quality of life also implies a greater waste generation (Monavari et al. 2012). Estimating the quality of life of a region is possible through different characteristic data of the region such as unemployment, education level, and life expectancy, among others. Knowing these data allows for finding a correlation between various social behaviors and waste generation, however, one of the main problems is precisely the availability of data, for this investigation different data sets available in the City of New York Data Database (NYC open data, 2022) were used, the selected socioeconomic data were then annualized to homogenize the available data. Table 1 shows the list of socioeconomic indicators selected for further analysis and the possible selection of parameters for the ML model.

Table 1. Selected socioeconomic indicators and their acronyms. [Printed (adapted) with permission from Munguia-López, A.C., Ochoa-Barragán, R., and Ponce-Ortega, J.M. 2022. Optimal waste management during the COVID-19 pandemic. Chemical Engineering and Processing: Process Intensification, 176, 108942].

Socioeconomic Indicator	Census Area	Abbreviation
Population over 45	Demographics	PO45
Median household income	Earnings and income	MHI
Unemployment ratio	Employment	UmR
School dropout	Education	ScD
Socioeconomic Indicator	Census Area	Abbreviation
Population over 45	Demographics	PO45
Median household income	Earnings and income	MHI
Unemployment ratio	Employment	UmR
School dropout	Education	ScD

10.3.1.2 Preprocessing of socioeconomic data

Figure 3 describes the different steps required to perform the preprocessing of the data collected in this work. First, data pre-processing must be implemented, this being one of the most important stages. Data is generally obtained in a raw format (from a data lake) and is not specifically prepared for the addressed task. Data preprocessing is carried out through a series of steps, starting with taking the collected data and generating an adequate data structure, later a normalization of the data sets that present very different scales between their components is carried out, and if necessary, a filtering of the data must be carried out whenever atypical or abnormal data are presented. Often, the collected socioeconomic data, coming from censuses, are usually taken in different temporalities, so it is necessary to carry out interpolations, and extrapolations in some cases, for this work the data sets were reduced to an annual data set. For the data preprocessing in this work, Pandas and NumPy were used, both being Python libraries.

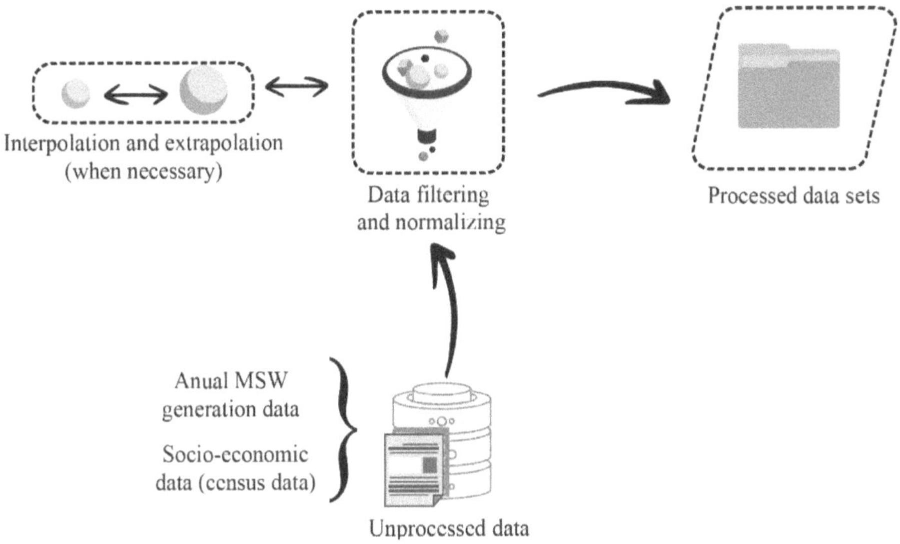

Fig. 3. Data preprocessing methodology used to generate the training and validation data sets.[Printed (adapted) with permission from Munguia-López, A.C., Ochoa-Barragán, R., and Ponce-Ortega, J.M. 2022. Optimal waste management during the COVID-19 pandemic. Chemical Engineering and Processing: Process Intensification, *176*, 108942]

10.3.1.3 Training and validation of models

Throughout the modeling stage, different models were tested for each data set seeking to find the best correlation between variables. Because ML algorithms can produce over-fit models (i.e., they seem to have a very good fit with the training set), given that this data-set division is among the most popular, it was determined to split the data sets in an 80:20 training-testing ratio (Azadi and Karimi-Jashni 2016). To determine the significant variables of the model, different modeling configurations were tested, the validation criteria used were p-value (p_{value}<0.025) and the adjusted coefficient of determination (R^2_{adj}), which are calculated as follows (Hastie et al. 2009):

$$R^2 = \frac{\sum_{n=1}^{N}\left(\hat{Y}_n - \overline{Y}\right)^2}{\sum_{n=1}^{N}\left(Y_n - \overline{Y}\right)^2} \tag{10.1}$$

$$R^2_{adj} = 1 - \left[\frac{\left(1-R^2\right)\left(n-1\right)}{n-k-1}\right] \tag{10.2}$$

Where $\hat{Y}_n$ represents the value of the variable estimated by the model, $\overline{Y}$ the average value of the variable, Y_n the real value of the variable for case *n* of the set of cases *N*, and *k* is the number of independent regressors (the number of variables in the model). The use of the adjusted coefficient of determination on the coefficient of determination (R^2) should be highlighted. This is because R^2 has a disadvantage since R^2 increases its value for each variable added to the model, which may appear to be a better fit for a model with a greater number of variables, and this is not always the case; which does not happen when using R^2_{adj} (Karch 2020).

p_{value} and R^2_{adj} were estimated on each model for each training and testing set. Where the training sets obtain better results, that is, a better fit, this is mainly because the model parameters are adjusted based on the training set. Lastly, the optimal complexity of the model (the one that best fits to evaluate unseen data) turns out to be a model composed only of 3 socioeconomic variables, PO45, MHI, and UmR, this model is selected based on the fact that it turns out to be the one that calculates the lowest

p_{values} and best R^2_{adj} values. To estimate the value of the socioeconomic parameters in years after the application of the strategy, different regressions were carried out for this set of independent data.

10.3.1.4 Machine learning strategy

Considering the data sets of the addressed case study, random forest (RF) algorithms were used. However, for other case studies (including different types of data sets), other algorithms may be attractive. In RF algorithms, each decision tree is built from samples taken from the training set, and the division of the data set is carried out through random subsets, this source of randomness allows for the reduction of the variance of the forest estimator (Genuer 2012). This randomness added to the model produces different decision tree models with different prediction errors, thus, by taking an average of all the predictions it is possible to minimize some errors present in individual decision tree models. In the particular case of the model selected, a total of 10 estimators were used, namely, the regression model is composed of 10 trees, where the depth of each tree is defined until the minimum number of samples is reached to generate a split of samples, for this work it is established as 2.

The proposed ML model is capable of generating estimates of the solid waste produced in the region of the case study, in addition to the fraction of that waste that will be sent to the landfill, this allows generating different projections for subsequent years considering different scenarios of the strategy. To implement these predictions, it is necessary to feed the solutions obtained in the mathematical optimization model to the ML (closed-loop optimization) (Ning and You 2019) model and calculate the new rates of waste sent to the landfill as shown in Equation (10.3).

$$\Delta LW_{y,s} = LW^{ML}_{y,s} - PW^{optm}_{y,s}, \forall y \in Y, s \in S \tag{10.3}$$

Where $\Delta LW_{y,s}$ represents the decrease in waste sent to the landfill, for year y and scenario s, $LW^{ML}_{y,s}$ represents the waste sent to the landfill estimated by the ML model, and $PW^{optm}_{y\ s}$ represents the processed waste estimated by the mathematical optimization model.

10.3.2 Model-based System

In the second part of the optimization approach, the mathematical model based on the superstructure shown in Fig. 10.4 is presented. In the superstructure, the different and possible configurations of the supply chain for waste management are shown, using different technologies for each type of sub-waste and penalizing the disposal of waste in sanitary landfills through the implementation of a variable rate of green taxes.

10.3.2.1 Disjunctions to select the processing or treatment technique

One of the most important parts of the mathematical model consists of its ability to select the technology and its respective processing capacity with which each type of sub-waste can be treated. To accomplish this, a series of disjunctions are proposed. It is important to highlight that the selection of one technology over another depends mainly on the proposed objective function. The assignment of technologies is not limited to those shown in this chapter, and those selected are solely for illustrative and comparative purposes. For the optimal selection of the technologies proposed in the model, the following disjunction is established.

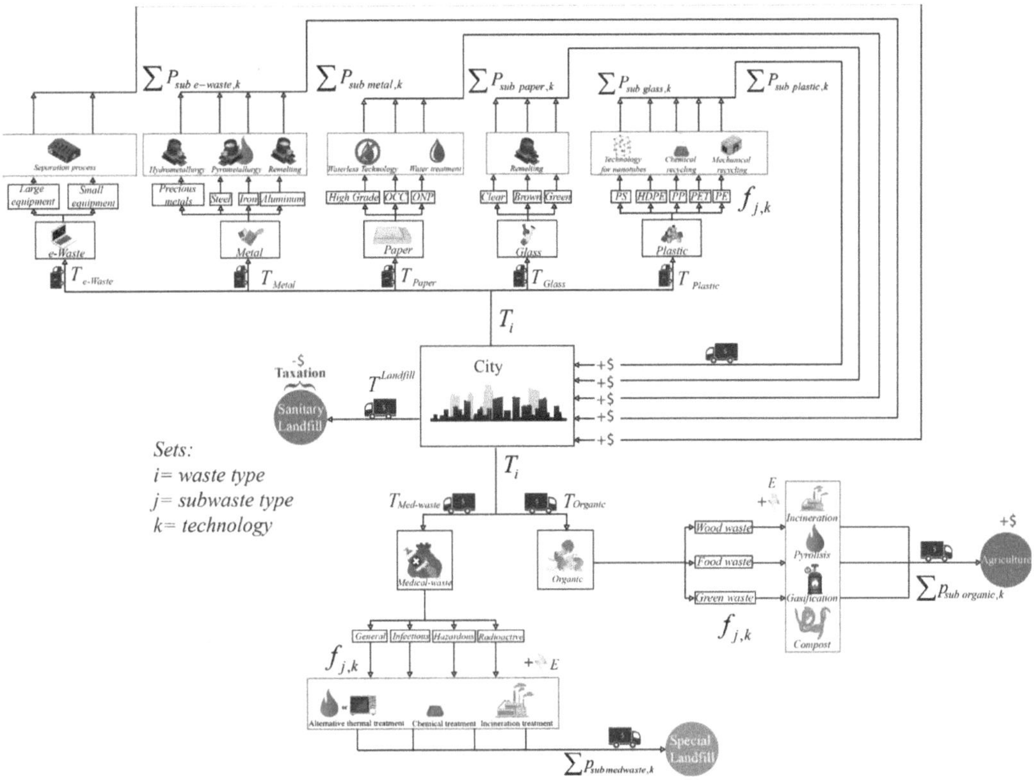

Fig. 4. Superstructure for solid waste management through different technologies and disposal in sanitary landfills. [Printed (adapted) with permission from Munguia-López, A.C., Ochoa-Barragán, R., and Ponce-Ortega, J.M. 2022. Optimal waste management during the COVID-19 pandemic. Chemical Engineering and Processing: Process Intensification, 176, 108942].

$$\begin{bmatrix} H_{j,k} \\ f_{j,k} \geq \delta_{j,k}^{\min} \\ f_{j,k} \leq \delta_{j,k}^{\max} \\ cost_{j,k} = \mathrm{A}_{j,k} + \mathrm{B}_{j,k} \cdot f_{j,k} \end{bmatrix} \vee \begin{bmatrix} \neg H_{j,k} \\ f_{j,k} = 0 \\ cost_{j,k} = 0 \end{bmatrix}, \forall j \in J, k \in K \tag{10.4}$$

The decision of which type of technology will be used resides in the Boolean variable $H_{j,k}$. If $H_{j,k}$ is true, then that processing technology is selected, while if it is false, that technology is discarded. In the previous disjunction, $\mathrm{A}_{j,k}$ and $\mathrm{B}_{j,k}$ are parameters used to consider fixed and variable costs. The reformulated disjunction results in the following set of algebraic relationships:

$$f_{j,k} \geq \delta_{j,k}^{\min} \cdot h_{j,k}, \forall j \in J, k \in K \tag{10.5}$$

$$f_{j,k} \leq \delta_{j,k}^{\max} \cdot h_{j,k}, \forall j \in J, k \in K \tag{10.6}$$

$$cost_{j,k} = kf\left(\mathrm{A}_{j,k} \cdot h_{j,k} + \mathrm{B}_{j,k} \cdot f_{j,k}\right), \forall j \in J, k \in K \tag{10.7}$$

where $h_{j,k}$ is a binary variable, that when its value is 0, the technology is discarded, while when it is 1, it is decided to use that technology. The values δ for all cases represent the search limits for the quantities of the material to be processed, these values are established with a lower limit close to 0 and an upper limit depending on the proposed objectives, *kf* represents the annualization factor applied to fixed and variable costs.

10.3.2.2 Taxes to promote recycling and waste processing

To decrease the social and environmental impacts of the COVID-19 pandemic, Ko et al. (2020) reported a strategy for a fictional situation in South Korea, where recycling is promoted through taxes. "Green taxes" are those collected to promote the preservation of the environment (Oates 1995, Paut-Kusturica et al. 2020). To establish these taxes, two new variables are considered to separate the waste that will be taxed with a low or high penalization. refers to the waste that will be taxed with the low penalization, and T^{landH} to the waste that will be taxed with the high penalization.

$$T^{Landfill} = T^{\mathrm{landH}} + T^{\mathrm{landL}} \tag{10.8}$$

Then, the following disjunction is proposed:

$$\begin{bmatrix} H^{Landfill} \\ T^{\mathrm{landL}} < T^{II} \end{bmatrix} \vee \begin{bmatrix} \neg H^{Landfill} \\ T^{\mathrm{landH}} \geq T^{II} \\ T^{\mathrm{landH}} \leq T^{III} \end{bmatrix} \tag{10.9}$$

where is the upper bound for the generated waste penalized by the low tax, but also it is the lower bound for waste penalized by the high tax. is the upper bound for the waste penalized by the high tax. For this chapter, a tax rate of 5.1 $US/ton (Munguía-López et al. 2020) is established as a high penalization to promote recycling. On the other hand, a tax rate of 2 $US/ton is evaluated as a low penalization. The previous disjunction can be reformulated as follows:

$$T^{landL} < T^{II} \cdot h^{Landfill} \tag{10.10}$$

$$T^{landL} \geq T^{II}\left(1 - h^{Landfill}\right) \tag{10.11}$$

$$T^{landH} \leq T^{III}\left(1 - h^{Landfill}\right) \tag{10.12}$$

The taxes can be obtained with the following relationships:

$$Taxes^{L} = TPT^{L} \cdot T^{\mathrm{landL}} \tag{10.13}$$

$$Taxes^{H} = TPT^{H} \cdot T^{\mathrm{landH}} \tag{10.14}$$

where is the low unitary tax and is the high unitary tax.

$$Taxes = Taxes^{L} + Taxes^{H} \tag{10.15}$$

10.3.2.3 Proposed goals to promote waste processing

For the superstructure shown above, it is possible to propose different objectives as required depending on the region where it is applied, reducing the impact on the environment, favoring economic benefit, or favoring social development are some examples. To minimize the environmental impact, the amount of untreated waste must be reduced.

On the other hand, landfills present a huge environmental risk, not only because of the nature of the waste deposited (which also poses a health risk) but also because of the leaking of toxic chemicals into the soil and water supply. These leaks are commonly caused by the precipitation of water, which generates high-risk leachates that may contain chlorides, nitrogen, solvents, volatile organic compounds, phenols, and heavy metals. No matter how many barriers, liners, and pipes are installed to try to mitigate risk, landfills will always leach toxic chemicals into the soil and water (Pecci 2018).

Since waste processing capacities are not unlimited, in large cities a certain amount of solid waste must be sent to a landfill. Therefore, reducing the amount of waste that is sent can benefit the environment, in addition to reducing the amount of green taxes assigned by the model.

$$\min T^{Landfill} \tag{10.16}$$

To maximize the economic profit the following objective function must be included:

$$\max \textit{Total profit} \tag{10.17}$$

10.3.2.5 Solution procedure

The proposed approach involves parameter prediction using ML and a deterministic optimization strategy, which is described as follows:

The prediction of the necessary parameters to run different simulations over time is carried out through an ML model that, based on different socioeconomic data, predicts the amounts of waste generated annually and the amounts of waste that cannot be treated and are sent to the landfill, to make these predictions a random forest algorithm belonging to the supervised learning branch is used, this method generates different decision trees to calculate the final output. This ML model was coded in Python and stored its results in spreadsheets to later be fed to the optimization model to calculate the objective functions of new iterations. After solving the optimization model, the data is stored in GDX files to feed the ML model that will calculate the new waste generated and sent to the landfill, which allows comparing one scenario against another.

The GAMS modeling environment is used to solve a mathematical optimization model that is based on the superstructure shown in Fig. 4. Several technologies that can be used to treat the waste are shown in this figure along with a distribution network for waste management. The input parameters of the model are determined by the ML model. The resulting model is a mixed-integer linear programming problem (MILP) and the epsilon constraint method was used (Diwekar 2008), varying the maximum production capacities of the technologies, this allows for establishing different scenarios where more or less waste is processed. Next, the mathematical model is solved to maximize the profit, which guides the solution toward the selection of those technologies that are cheaper and generate greater economic benefits. Furthermore, the model calculates the amounts of waste sent to the landfill and penalizes based on the amounts of waste sent thanks to the implementation of variable green taxes, this allows finding a balance between the economic and environmental benefits since sending large amounts of waste to landfills implies a higher tax payment.

10.4 Addressed Case Study

To apply the proposed approach, New York City was selected as a case study. This city is one which generates the largest amount of waste annually and therefore one of the cities most affected by the generation of infectious waste from COVID-19 (DSNY 2020). New York City generates a large amount of data annually (Campbell 2017), which is ideal for using supervised learning algorithms. During the COVID-19 pandemic, the increased demand for medical supplies such as masks, gloves, gowns, goggles, and disinfection supplies, together with the generation of domestic waste, became a major problem for the city (Chand et al. 2021). Waste management is carried out through public and private contractors, where waste is classified as paper, metal/plastic/glass, and mixed solid waste (non-recyclable garbage) (Nagle 2013). Regarding recycling, the primary facility is the Sunset Park Material Recovery Facility in Brooklyn; here, approximately 75% of the discarded paper, glass, plastic, and other materials are recycled; however, 80% of the mixed solid waste goes to a landfill 600 miles from the city, and only 20% of it is turned into energy through incineration (Galka 2016).

There are nine medical waste disposal facilities in New York City with an estimated disposal capacity of 12,000 tons per year. Where 60% of this capacity was used only to treat medical waste that is highly contagious due to COVID-19 (Mei et al. 2021). Thus, the construction of an industrial park is proposed in the Staten Island region, a region for which transportation cost parameters are considered.

10.5 Results and Discussion

This section presents the results of the waste management system obtained through the presented optimization strategy. First, the results of the deterministic mathematical model formulation are presented. Then, the predicted waste collection based on the socioeconomic data is described. Lastly, several selected optimal solutions over time from the projections of the machine learning model are discussed. The deterministic mathematical model formulation applied to the New York City case study consists of 63 binary variables, 317 continuous variables, and 341 equations. The solution to the resulting problem (MILP) was obtained using the restricted epsilon method, where the maximum processing capacities for each of the scenarios were employed. This model was coded in the GAMS modeling environment using the CPLEX solver. The ML model coding was implemented in the Python programming environment, the model was trained from an annualized data set collected since 2000, composed of 3 significant socio-economic variables PO45, MHI, and UmR. The RandomForestRegressor predicting tool using the Sci-kit learn library was used with several estimators of 10 using the P-value and R^2_{adj} as validation parameters.

There are multiple Pareto solutions to the established problem, however, 4 of all those possible solutions have been selected to be analyzed, these solutions are found in different regions of the Pareto curve presented in Fig. 5. Solution A shows the ideal scenario, where it is possible to build an industrial park able to process 100% of the waste generated in the region, however, this is quite complicated to achieve mainly because of the enormous size and investment that required building such capacity of an industrial park. Solution B presents a solution far from solution A, which remains in the region with the lowest rate of green taxes, which translates into a "low" level of waste sent to the landfill compared to the other scenarios, in addition to generating a good balance between economic and environmental benefits (Fig. 6 shows the technologies used to reach this solution). Solution C falls in the region belonging to a higher tax rate, although it still generates economic benefits, these can be low. Lastly, solution D is the solution where economic losses are presented and high amounts of waste are sent to the landfill.

It can be seen that for all the proposed solutions, the main income comes from the sale of recycled materials, while income from energy and compost production begins to increase as scenarios with greater recycling capacity are chosen, on the other hand, the main costs are the separation and transport of the waste, these become even higher than the total income in solution D, the treatment costs increase in scenarios with a greater processing capacity, on the contrary, the costs associated with the payment of green taxes decrease, in solution D they represent a significant expense, as the solutions approach a scenario that recycles larger amounts of waste such as solution B, this cost becomes insignificant. Figure 7 shows a comparison of solution B between the costs of waste treatment and the sales obtained for those wastes that are used for the production of new materials, it can be seen how the costs associated with the treatment of glass (where 77% of the waste generated is treated) represent a large part of the total treatment costs of this solution, however, the economic benefits associated with the sale of new glass are only 1% of the total benefits, then from an economic point of view, not processing this type of waste could be a better option, but the inclusion of green taxes would generate penalties that would be more expensive than processing this type of waste with low economic potential.

Figure 5 presents a Pareto curve solved under the conditions presented in 2018, in which better economic benefits are observed for most solutions, with the established regression models it is possible to develop an ML model that uses an algorithm of random forests that seeks to predict the amounts of MSW collected and waste sent to landfill year by year. Figure 8 presents different

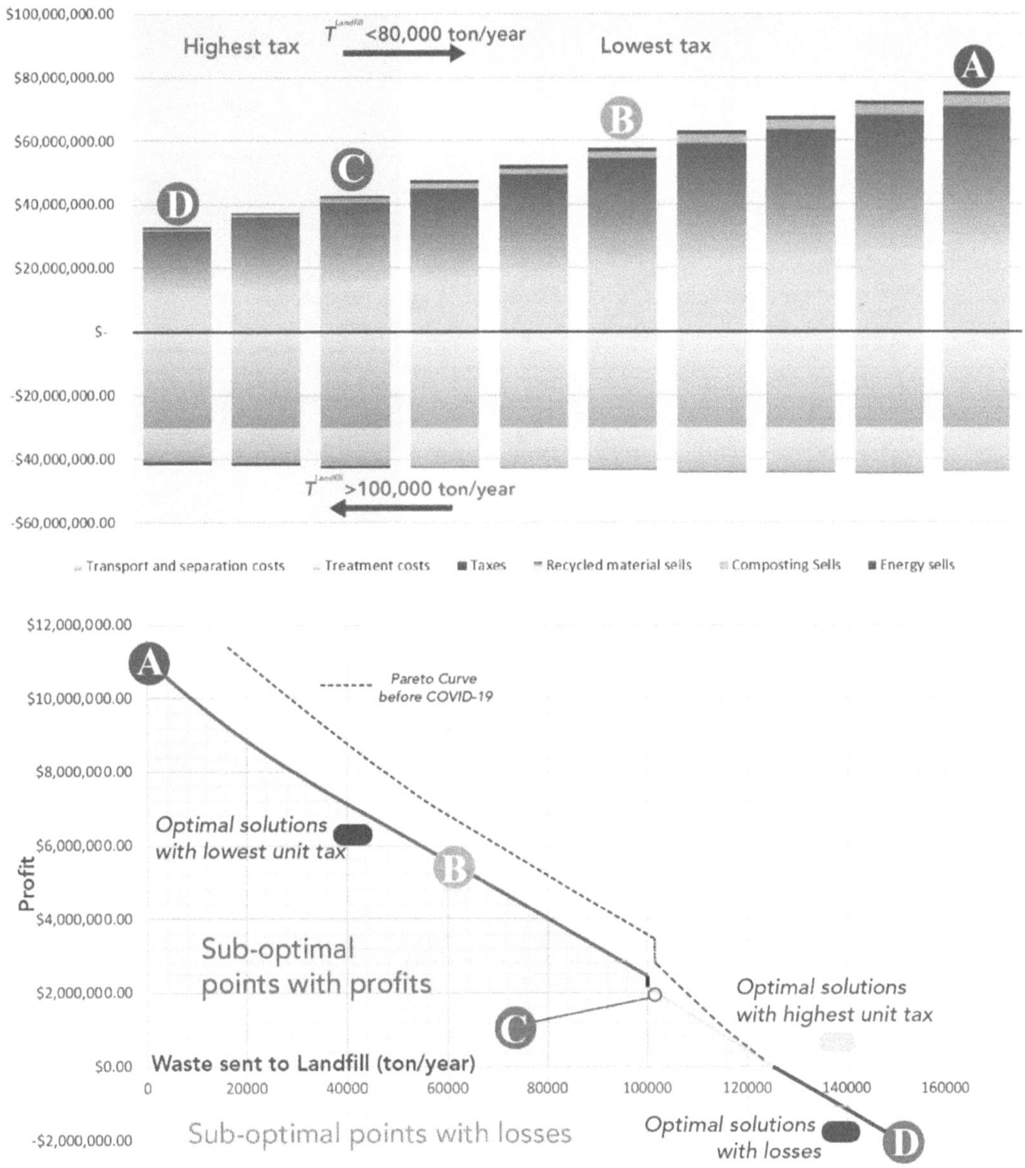

Fig. 5. Results for the deterministic model for COVID-19 conditions. [Printed (adapted) with permission from Munguia-López, A.C., Ochoa-Barragán, R., and Ponce-Ortega, J.M. 2022. Optimal waste management during the COVID-19 pandemic. Chemical Engineering and Processing: Process Intensification, *176*, 108942].

useful regressions for each set of socioeconomic data, these regressions allow for generating the complete data set that predicts the waste collected in subsequent years, these datasets show clear trends and a low statistical dispersion, which allows obtaining models and predictions with very acceptable values of R^2_{adj}, where PO45 shows a good fit with a linear regression model, MHI on the other hand, shows a better fit with a Polynomial regression model, and, finally, UmR shows its best fit with a support vector regression (SVR) type model involving data up to the year 2035. Figure 9 presents a projection for the tons of garbage that are generated from the year 2022 based on the projections generated from the aforementioned socioeconomic parameters, these estimates show a slight decrease in the production of waste for years after the pandemic, with this data it is possible to project each of the scenarios established until the year 2035 since it was the year until which the

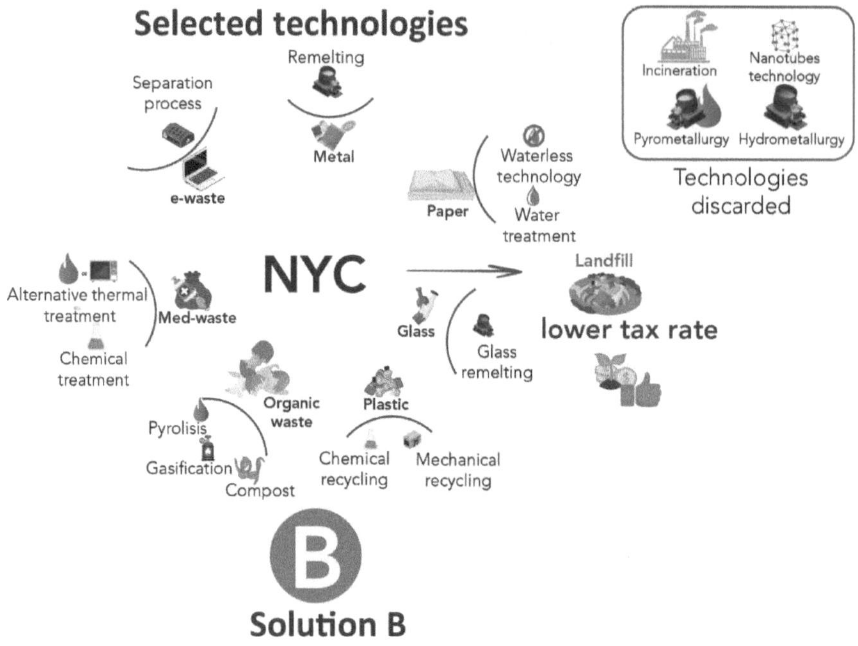

Fig. 6. Solution B: Technologies selected by the model to achieve a solution with the lowest rate of green taxes, maximizing economic benefits. [Printed (adapted) with permission from Munguia-López, A.C., Ochoa-Barragán, R., and Ponce-Ortega, J.M. 2022. Optimal waste management during the COVID-19 pandemic. Chemical Engineering and Processing: Process Intensification, *176*, 108942].

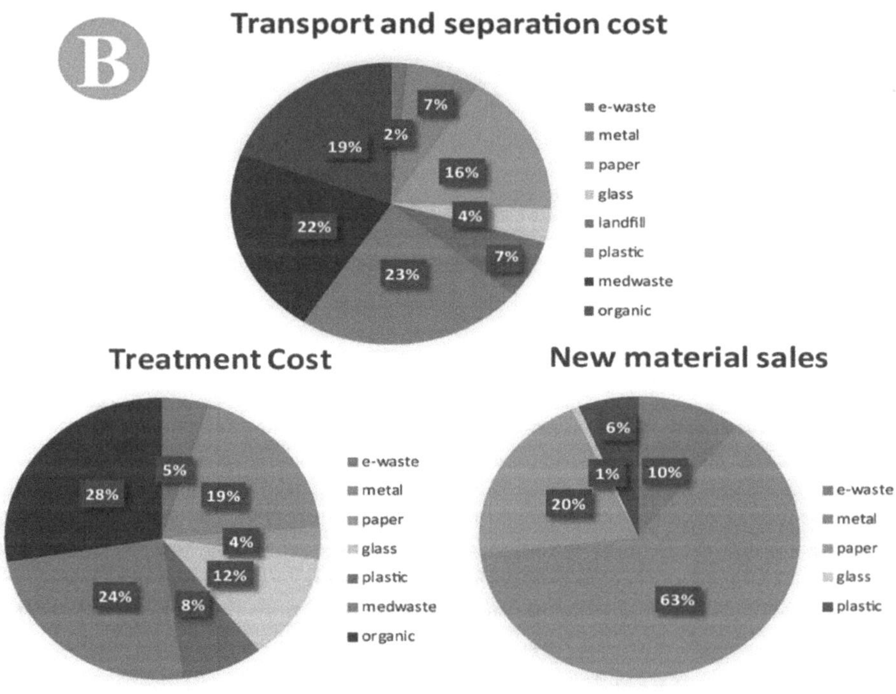

Fig. 7. Solution B: Comparison between treatment costs and economic benefits from the sale of new materials produced by the strategy. [Printed (adapted) with permission from Munguia-López, A.C., Ochoa-Barragán, R., and Ponce-Ortega, J.M. 2022. Optimal waste management during the COVID-19 pandemic. Chemical Engineering and Processing: Process Intensification, *176*, 108942].

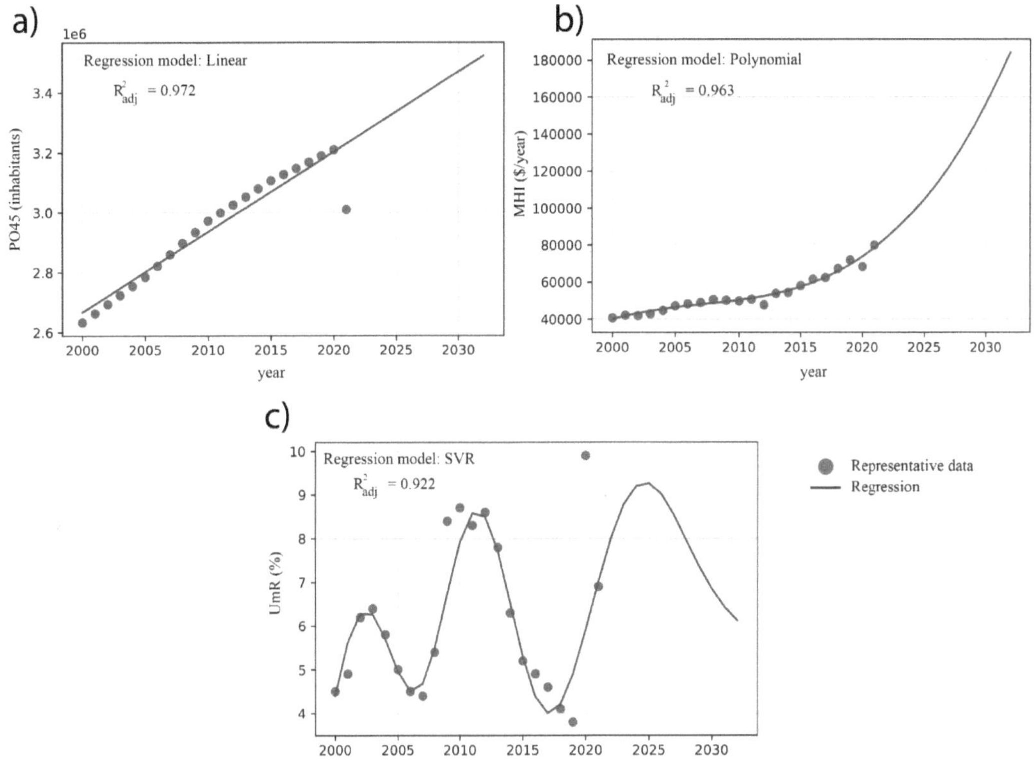

Fig. 8. a) Linear regression model for the prediction of the population over 45 years of age. b) Polynomial regression model for the prediction of median household income. c) Support vector regression model for the prediction of the unemployment ratio. [Printed (adapted) with permission from Munguia-López, A.C., Ochoa-Barragán, R., and Ponce-Ortega, J.M. 2022. Optimal waste management during the COVID-19 pandemic. Chemical Engineering and Processing: Process Intensification, *176*, 108942].

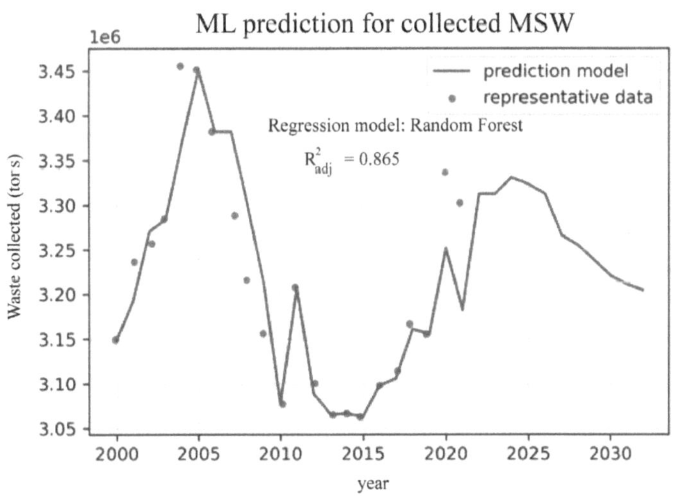

Fig. 9. Machine learning model for the prediction of municipal solid waste collected based on population over 45 years of age, median family income, and unemployment ratio. [Printed (adapted) with permission from Munguia-López, A.C., Ochoa-Barragán, R., and Ponce-Ortega, J.M. 2022. Optimal waste management during the COVID-19 pandemic. Chemical Engineering and Processing: Process Intensification, *176*, 108942]

socioeconomic parameters were projected. Figure 10.10 presents the adjustment obtained from the ML model in addition to the set of solutions B, C, and D, it is worth noting that solution A is on the time axis where all the residues are processed. For this particular case, the year 2018 is proposed for the implementation of the strategy to compare the results with a longer term, where it is observed that regardless of the proposed solution, there is an increase in the amount of waste sent to the landfill around the year 2020, this is due to the start of the COVID-19 pandemic in the New York region, however, the amounts of waste sent to the landfill decrease quite significantly even for solution D, which on the other hand, taking into account the risk associated with the waste generated during the pandemic, solution D represents the solution with the lowest social benefit, in addition to being the one that generates the most taxes, which makes solutions B and C the most viable, since there is a better balance between economic and social benefits as previously observed in Fig. 5.

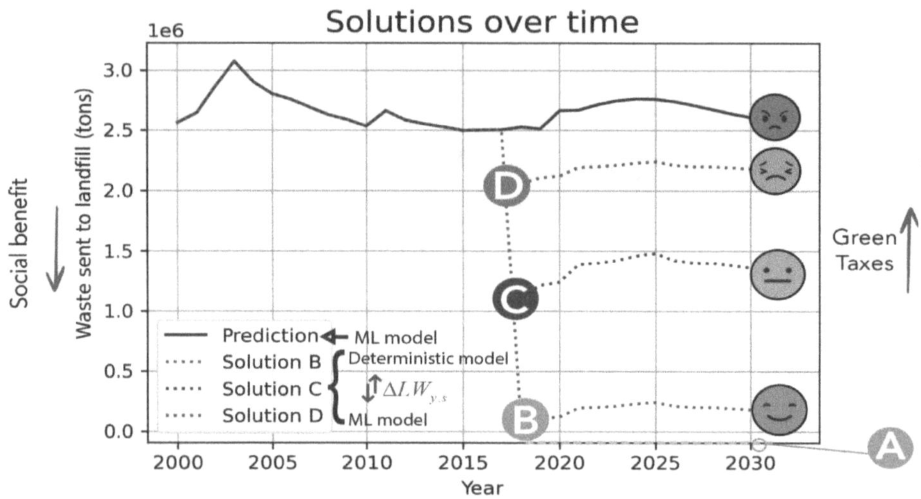

Fig. 10. Solutions obtained over time from the projections of the machine learning model. [Printed (adapted) with permission from Munguia-López, A.C., Ochoa-Barragán, R., and Ponce-Ortega, J.M. 2022. Optimal waste management during the COVID-19 pandemic. Chemical Engineering and Processing: Process Intensification, 176, 108942].

10.6 Concluding Remarks

Managing MSW effectively is crucial during pandemics. Poor MSW management can have serious negative effects on the environment and society, endangering human health. In order to encourage the recycling and treatment of garbage with little economic potential, this chapter has proposed a waste management approach that uses a variable rate of green taxes. This approach involves the use of a deterministic mathematical optimization model and an ML prediction model simultaneously. The deterministic model allows for the planning of an industrial park that is responsible for processing all types of generated waste, and thus the model can assign the processing technologies and their respective capacity that best fit the established objectives, in addition to establishing a variable green tax rate based on the waste that is sent to the landfill. Furthermore, due to the complexity of the problem, in addition to the uncertainty associated with the generation of waste, it is of paramount importance for the decision-maker to generate projections of the strategy over time that allow the selection of the option that best suits the region. In this manner, useful socio-economic data was collected to generate an ML prediction model that uses random forest algorithms.

Using New York City as a case study, the use of the suggested methodology was illustrated. The findings show that it is feasible to build an industrial park that can handle all waste generated and that, by contrasting the various scenarios, solutions that eventually strike a balance between social, environmental, and economic objectives may be found. This technique is helpful in places where there is poor urban garbage management, putting the populace at serious risk. Additionally, the prediction model aids in locating these answers for potential future situations.

10.7 References

Abdoli, M. A., Falahnezhad, M., and Behboudian, S. 2011. Multivariate econometric approach for solid waste generation modeling: Impact of climate factors. Environmental Engineering Science, *28*(9), 627–633. https://doi.org/10.1089/ees.2010.0234

Ali, S. A., and Parvin, F. 2022. Examining challenges and multi-strategic approaches in waste management during the COVID-19 pandemic: A systematic review. Waste Management and Research, *40*(9), 1356–1380. https://doi.org/10.1177/0734242X221079303

Andeobu, L., Wibowo, S., and Grandhi, S. 2022. Medical waste from COVID-19 pandemic—A systematic review of management and environmental impacts in Australia. International Journal of Environmental Research and Public Health, *19*(3), 1381. https://doi.org/10.3390/ijerph19031381

Antonopoulos, I.-S., Karagiannidis, A., Tsatsarelis, T., and Perkoulidis, G. 2012. Applying waste management scenarios in the Peloponnese region in Greece: a critical analysis in the frame of life cycle assessment. Environmental Science and Pollution Research, *20*(4), 2499–2511. https://doi.org/10.1007/s11356-012-1139-y

Azadi, S., and Karimi-Jashni, A. 2016. Verifying the performance of artificial neural network and multiple linear regression in predicting the mean seasonal municipal solid waste generation rate: A case study of Fars province, Iran. Waste Management, *48*, 14–23. https://doi.org/10.1016/j.wasman.2015.09.034

Bach, H., Mild, A., Natter, M., and Weber, A. 2004. Combining socio-demographic and logistic factors to explain the generation and collection of waste paper. Resources, Conservation and Recycling, *41*(1), 65–73. https://doi.org/10.1016/j.resconrec.2003.08.004

Bhar, A., Biswas, R. K., and Choudhury, A. K. 2022. The influence of COVID-19 pandemic on biomedical waste management, the impact beyond infection. Proceedings of the Indian National Science Academy, *88*(2), 117–128. https://doi.org/10.1007/s43538-022-00070-9

Campbell, A. 2017. New York City Open Data: A brief history. Data-smart city solutions. Retrieved September 11, 2022, from https://datasmart.ash.harvard.edu/news/article/new-york-city-open-data-a-brief-history-991

Çavuş, O., Bayhan, I., and Ismail, B. B. 2022. An overview of the effect of covid-19 on household food waste: How does the pandemic affect food waste at the household level?. *International Journal on Food System Dynamics*, 13(1), 1–16. https://doi.org/10.18461/ijfsd.v13i1.A1

Chakraborty, B., Bera, B., Adhikary, P. P., Bhattacharjee, S., Roy, S., Saha, S., and Shit, P. K. 2021. Positive effects of COVID-19 lockdown on river water quality: evidence from River Damodar, India. *Scientific Reports*, 11(1), 1—16. https://doi.org/10.1038/s41598-021-99689-9

Chand, S., Shastry, C. S., Hiremath, S., Joel, J. J., Krishnabhat, C. H., and Mateti, U. V. 2021. Updates on biomedical waste management during COVID-19: The Indian scenario. *Clinical Epidemiology and Global Health*, 11, 100715. https://doi.org/10.1016/j.cegh.2021.100715

Chen, X. 2022. Machine learning approach for a circular economy with waste recycling in smart cities. *Energy Reports*, 8, 3127- 3140. https://doi.org/10.1016/j.egyr.2022.01.193

Dharmaraj, S., Ashokkumar, V., Pandiyan, R., Halimatul Munawaroh, H. S., Chew, K. W., Chen, W.-H., and Ngamcharussrivichai, C. 2021. Pyrolysis: An effective technique for degradation of COVID-19 medical wastes. *Chemosphere*, 275, 130092. https://doi.org/10.1016/j.chemosphere.2021.130092

Diwekar, U. 2008. Introduction to Applied Optimization. *Springer Optimization and Its Applications*. https://doi.org/10.1007/978-0-387-76635-5

DSNY. 2020. Annual Report: New York City curbside and containerized municipal refuse and recycling statistics. The City of New York Department of Sanitation. Retrieved September 11, 2022, from https://dsny.cityofnewyork.us/wp-content/uploads/2020/08/about_dsny-collections-annual-2020.pdf

Ekhaise, F. O., Ikhajiagbe, B., Odozi, P. I., Erhabor, J. O., Enaigbe, A., and Enerijiofi, K. E. 2023. Effects of COVID-19 disinfectants on the environment. *Sustainability, Agri, Food and Environmental Research*, 11. http://dx.doi.org/10.7770/safer.v11i1.2813

El-Ramady, H., Brevik, E. C., Elbasiouny, H., Elbehiry, F., Amer, M., Elsakhawy, T., and Eid, Y. 2021. Planning for disposal of COVID-19 pandemic wastes in developing countries: a review of current challenges. *Environmental Monitoring and Assessment*, 193(9). https://doi.org/10.1007/s10661-021-09350-1

Fathollahi-Fard, A. M., Ahmadi, A., and Al-e-Hashem, S. M. 2020. Sustainable closed-loop supply chain network for an integrated water supply and wastewater collection system under uncertainty. *Journal of Environmental Management*, 275, 111277. https://doi.org/10.1016/j.jenvman.2020.111277

Galka, M. 2016. What does New York do with all its trash? One city's waste –in numbers. The Guardian. Retrieved September 11, 2022, from https://www.theguardian.com/cities/2016/oct/27/new-york-rubbish-all-that-trash-city-waste-in-numbers

Genuer, R. 2012. Variance reduction in purely random forests. *Journal of Nonparametric Statistics*, 24(3), 543–562. https://doi.org/10.1007/978-1-4899-0027-2f

Ghinea, C., Drăgoi, E. N., Comăniţă, E.-D., Gavrilescu, M., Câmpean, T., Curteanu, S., and Gavrilescu, M. 2016. Forecasting municipal solid waste generation using prognostic tools and regression analysis. *Journal of Environmental Management*, 182, 80–93. https://doi.org/10.1016/j.jenvman.2016.07.026

Govindan, K., Nasr, A. K., Mostafazadeh, P., and Mina, H. 2021. Medical waste management during coronavirus disease 2019 (COVID-19) outbreak: A mathematical programming model. *Computers and Industrial Engineering*, 162, 107668. https://doi.org/10.1016/j.cie.2021.107668

Haque, F., and Fan, C. 2022. Prospect of microplastic pollution control under the "New normal" concept beyond COVID-19 pandemic. *Journal of Cleaner Production*, 133027. https://doi.org/10.1016/j.jclepro.2022.133027

Hastie, T., Tibshirani, R., and Friedman, J. 2009. The elements of statistical learning: data mining, inference, and prediction. *Springer series in statistics*. Springer New York.

Homayouni, Z., and Pishvaee, M. S. 2020. A bi-objective robust optimization model for hazardous hospital waste collection and disposal network design problem. *Journal of Material Cycles and Waste Management*. https://doi.org/10.1007/s10163-020-01081-8

Hou, E. J., Hsieh, Y. Y., Hsu, T. W., Huang, C. S., Lee, Y. C., Han, Y. S., Chu, H. T. (2022). Using the concept of circular economy to reduce the environmental impact of COVID-19 face mask waste. *Sustainable Materials and Technologies*, 33, e00475. https://doi.org/10.1016/j.susmat.2022.e00475

Igalavithana, A. D., Yuan, X., Attanayake, C. P., Wang, S., You, S., Tsang, D. C., Nzihou A., and Ok, Y. S. 2022. Sustainable management of plastic wastes in COVID-19 pandemic: The biochar solution. *Environmental Research*, 212, 113495. https://doi.org/10.1016/j.envres.2022.113495

Ihsanullah, I., Alam, G., Jamal, A., and Shaik, F. 2022. Recent advances in applications of artificial intelligence in solid waste management: A review. Chemosphere, 136631. https://doi.org/10.1016/j.chemosphere.2022.136631

Iranmanesh, M., Ghobakhloo, M., Nilsashi, M., Tseng, M. L., Senali, M. G., and Abbasi, G. A. 2022. Impacts of the COVID-19 pandemic on household food waste behaviour: A systematic review. Appetite, 106127. https://doi.org/10.1016/j.appet.2022.106127

Jahandideh, S., Jahandideh, S., Asadabadi, E. B., Askarian, M., Movahedi, M. M., Hosseini, S., and Jahandideh, M. 2009. The use of artificial neural networks and multiple linear regression to predict rate of medical waste generation. Waste Management, *29*(11), 2874–2879. https://doi.org/10.1016/j.wasman.2009.06.027

Karch, J. 2020. Improving on Adjusted R-Squared. Collabra: Psychology, *6*(1). https://doi.org/10.1525/collabra.343

Ko, S., Kim, W., Shin, S.-C., and Shin, J. 2020. The economic value of sustainable recycling and waste management policies: The case of a waste management crisis in South Korea. Waste Management, *104*, 220–227. https://doi.org/10.1016/j.wasman.2020.01.020

Le, H. T., Quoc, K. L., Nguyen, T. A., Dang, K. T., Vo, H. K., Luong, H. H., Le-Van, H., Gia, K. H., Cao-Phu, L. V., Nguyen-Truong-Quoc, D., Huyen-Nguyen, T., Son, H. X., and Duong-Trung, N. 2022. Medical-waste chain: a medical waste collection, classification and treatment management by blockchain technology. Computers, *11*(7), 113. https://doi.org/10.3390/computers11070113

Luhar, I., Luhar, S., and Abdullah, M. M. A. B. 2022. Challenges and Impacts of COVID-19 Pandemic on Global Waste Management Systems: A Review. Journal of Composites Science, *6*(9), 271. https://doi.org/10.3390/jcs6090271

Masud, M. H., Mourshed, M., Hossain, M. S., Ahmed, N. U., and Dabnichki, P. 2023. Generation of waste: problem to possible solution in developing and underdeveloped nations. Waste Management and Resource Recycling in the Developing World, 21–59. https://doi.org/10.1016/B978-0-323-90463-6.00021-X

Mei, X., Hao, H., Sun, Y., Wang, X., and Zhou, Y. 2021. Optimization of medical waste recycling network considering disposal capacity bottlenecks under a novel coronavirus pneumonia outbreak. Environmental Science and Pollution Research, 1–19. https://doi.org/10.1007/s11356-021-16027-2

Mohamed, B. A., Fattah, I. M., Yousaf, B., and Periyasamy, S. 2022. Effects of the COVID-19 pandemic on the environment, waste management, and energy sectors: a deeper look into the long-term impacts. Environmental Science and Pollution Research, 1–20. https://doi.org/10.1007/s11356-022-20259-1

Monavari, S. M., Omrani, G. A., Karbassi, A., and Raof, F. F. 2011. The effects of socioeconomic parameters on household solid-waste generation and composition in developing countries (a case study: Ahvaz, Iran). Environmental Monitoring and Assessment, *184*(4), 1841–1846. https://doi.org/10.1007/s10661-011-2082-y

Munguía-López, A. C., Ochoa-Barragán, R., and Ponce-Ortega, J. M. 2022. Optimal waste management during the COVID-19 pandemic. Chemical Engineering and Processing-Process Intensification, *176*, 108942. https://doi.org/10.1016/j.cep.2022.108942

Munguía-López, A. del C., Núñez-López, J. M., and Ponce-Ortega, J. M. 2020. Identifying fair solutions in the optimal design of integrated residential complexes. Chemical Engineering and Processing - Process Intensification, 108116. https://doi.org/10.1016/j.cep.2020.108116

Nabavi-Pelesaraei, A., Mohammadkashi, N., Naderloo, L., Abbasi, M., and Chau, K. W. 2022. Principal of environmental life cycle assessment for medical waste during COVID-19 outbreak to support sustainable development goals. Science of the Total Environment, *827*, 154416. https://doi.org/10.1016/j.scitotenv.2022.154416

Nagle, R. 2013. *Picking up: on the streets and behind the trucks with the sanitation workers of New York City*. Macmillan. 1st edition. ISBN:978-0-374-29929-3.

Ning, C., You, F. 2019. Optimization under uncertainty in the era of big data and deep learning: When machine learning meets mathematical programming. Computers and Chemical Engineering, *125*, 434–448. https://doi.org/10.1016/j.compchemeng.2019.034

NYC Open Data 2022. City of New York Data. Retrieved September 11, 2022, from https://data.cityofnewyork.us/browse?category=Health&q=&sortBy=newest&utf8=%E2%9C%93

Oates, W. E. 1995. Green taxes: Can we protect the environment and improve the tax system at the same time? Southern Economic Journal, *61*(4), 915. https://doi.org/10.2307/1060731

Paut-Kusturica, M., Golocorbin-Kon, S., Ostojic, T., Kresoja, M., Milovic, M., Horvat, O. and Tomas, A. 2020. Consumer willingness to pay for a pharmaceutical disposal program in Serbia: A double hurdle modeling approach. Waste Management, *104*, 246–253. https://doi.org/10.1016/j.wasman.2020.01.029

Pecci, K. 2018. All Landfills Leak, and our health and environment pay the toxic price. Conservation Law Foundation. Retrieved September 11, 2022, from https://www.clf.org/blog/all-landfills-leak-and-our-health-and-environment-pay-the-toxic-price/

Rubab, S., Khan, M. M., Uddin, F., Abbas Bangash, Y., and Taqvi, S. A. A. 2022. A study on AI-based waste management Strategies for the COVID-19 Pandemic. ChemBioEng Reviews, *9*(2), 212–226. https://doi.org/10.1002/cben.202100044

Santibañez-Aguilar, J. E., Ponce-Ortega, J. M., González-Campos, J. B., Serna-González, M., and El-Halwagi, M. M. 2013. Optimal planning for the sustainable utilization of municipal solid waste. Waste Management, *33*(12), 2607–2622. https://doi.org/10.1016/j.wasman.2013.08.010

Sharma, H. B., Vanapalli, K. R., Cheela, V. S., Ranjan, V. P., Jaglan, A. K., Dubey, B., and Bhattacharya, J. 2020. Challenges, opportunities, and innovations for effective solid waste management during and post COVID - 19 pandemic. *Resources,* Conservation and Recycling, 105052. https://doi.org/10.1016/j.resconrec.2020.105052

Singh, E., Kumar, A., Mishra, R., and Kumar, S. 2022. Solid waste management during COVID-19 pandemic: Recovery techniques and responses. Chemosphere, *288*, 132451. https://doi.org/10.1016/j.chemosphere.2021.132451

Sinthiya, N. J., Chowdhury, T. A., and Haque, A. B. 2022. Artificial Intelligence Based Smart Waste Management—A Systematic Review. Computational Intelligence Techniques for Green Smart Cities, 67–92. https://doi.org/10.1007/978-3-030-96429-0_3

Skrzyniarz, M., Sajdak, M., Zajemska, M., Iwaszko, J., Biniek-Poskart, A., Skibiński, A., Morel, S., and Niegodajew, P. 2022. Plastic waste management towards energy recovery during the COVID-19 pandemic: the example of protective face mask pyrolysis. Energies, *15*(7), 2629. https://doi.org/10.3390/en15072629

Tripathi, A., Tyagi, V. K., Vivekanand, V., Bose, P., and Suthar, S. 2020. Challenges, opportunities and progress in solid waste management during COVID-19 pandemic. Case Studies in Chemical and Environmental Engineering, *2*, 100060. https://doi.org/10.1016/j.cscee.2020.100060

Valizadeh, J., and Mozafari, P. 2021. A novel cooperative model in the collection of infectious waste in COVID-19 pandemic. Journal of Modelling in Management, *17*(1), 363–401. https://doi.org/10.1108/JM2-07-2020-0189

Valizadeh, J., Aghdamigargari, M., Jamali, A., Aickelin, U., Mohammadi, S., Khorshidi, H. A., and Hafezalkotob, A. 2021. A hybrid mathematical modelling approach for energy generation from hazardous waste during the COVID-19 pandemic. Journal of Cleaner Production, 315, 128157. https://doi.org/10.1016/j.jclepro.2021.128157

Vittuari, M., Masotti, M., Iori, E., Falasconi, L., Gallina Toschi, T., and Segrè, A. 2021. Does the COVID-19 external shock matter on household food waste? The impact of social distancing measures during the lockdown. Resources, Conservation and Recycling, 174, 105815. https://doi.org/10.1016/j.resconrec.2021.105815

Zanetti, B., Verrascina, M., Licciardo, F., and Gargano, G. 2022. Agritourism and farms diversification in Italy: What have we learnt from COVID-19?. Land, *11*(8), 1215. https://doi.org/10.3390/land11081215

Appendix

This appendix provides supplementary material, including additional data, detailed implementation codes, and relevant resources to support the main content of this book.

A.1 Chapter 3: Using Python for waste management optimization

```
*****************************************************************************
def linear_problem(PAR):
import pyomo.environ as pyo

# --------------------------------------------------------
# Model Initialization
# --------------------------------------------------------
 model = pyo.ConcreteModel()

# --------------------------------------------------------
# Sets
# --------------------------------------------------------
model.I = pyo.RangeSet(1, 7)  # Types of subwaste
model.J = pyo.RangeSet(1, 5)  # Technologies
model.JR = pyo.Set(initialize=[1, 2, 3])  # Material Recovery Technologies
model.JE = pyo.Set(initialize=[4, 5])  # Waste-to-Energy (WtE) Technologies

# --------------------------------------------------------
# Parameters
# --------------------------------------------------------
W = {1: 12, 2: 30, 3: 40, 4: 70, 5: 55, 6: 90, 7: 20}  # (tons)
C_processing = {1: 112, 2: 2.82, 3: 144, 4: 380, 5: 354}  # (USD/
ton)
  C_transport = {1: 15, 2: 20, 3: 18, 4: 25, 5: 22}  # (USD/ton)
  EF_processing = {1: 0.05, 2: 0.10, 3: 0.08, 4: 0.15, 5: 0.12}
# (tons of CO2/ton processed)
  EF_landfill = {1: 0.20, 2: 0.30, 3: 0.25, 4: 0.40, 5: 0.35, 6:
0.45, 7: 0.50}  # (tons of CO2/ton disposed)
  CA_processing = {1: 60, 2: 50, 3: 55, 4: 80, 5: 75}  # (tons)
  eta_recov = {
 (1, 1): 0.00,
 (1, 2): 0.60,
 (1, 3): 0.85,
 (2, 1): 0.95,
 (2, 2): 0.80,
```

```
(2, 3): 0.00,
(3, 1): 0.00,
(3, 2): 0.50,
(3, 3): 0.90,
(4, 1): 0.75,
(4, 2): 0.50,
(4, 3): 0.00,
(5, 1): 0.00,
(5, 2): 0.65,
(5, 3): 0.90,
(6, 1): 0.00,
(6, 2): 0.00,
(6, 3): 0.00,
(7, 1): 0.00,
(7, 2): 0.70,
(7, 3): 0.00
  }
  eta_WtE = {
(1, 4): 750,
(1, 5): 0,
(2, 4): 0.00,
(2, 5): 0.00,
(3, 4): 0.00,
(3, 5): 0.00,
(4, 4): 0.00,
(4, 5): 0.00,
(5, 4): 0.00,
(5, 5): 8000,
(6, 4): 900,
(6, 5): 0,
(7, 4): 0,
(7, 5): 300
  } # (kWh/ton)
  C_landfill = 30 # (USD/ton)
  CΛ_truck – 15 # (tons/trip)

    # ----------------------------------------------------------
  # Variables
  # ----------------------------------------------------------
  model.f_waste = pyo.Var(model.I, model.J, domain=pyo.NonNegativeReals)
  model.f_landfill = pyo.Var(model.I, domain=pyo.NonNegativeReals)
  model.mat_recov = pyo.Var(model.JR, domain=pyo.NonNegativeReals)
  model.ene_WtE = pyo.Var(model.JE, domain=pyo.NonNegativeReals)
  model.v_number = pyo.Var(domain=pyo.NonNegativeReals)
  model.cost_total = pyo.Var(domain=pyo.NonNegativeReals)
  model.ghg_total = pyo.Var(domain=pyo.NonNegativeReals)
  model.cost_processing = pyo.Var(domain=pyo.NonNegativeReals)
  model.cost_transport = pyo.Var(domain=pyo.NonNegativeReals)
  model.ghg_processing = pyo.Var(domain=pyo.NonNegativeReals)
  model.ghg_landfill = pyo.Var(domain=pyo.NonNegativeReals)
```

```
# ------------------------------------------------------
# Constraints
# ------------------------------------------------------

# Constraint 1: Waste allocation balance
model.EQ1 = pyo.ConstraintList()
for i in model.I:
    model.EQ1.add(W[i] == sum(model.f_waste[i, j] for j in model.J) + model.f_landfill[i])

# Constraint 2: Total recovered material
model.EQ2 = pyo.ConstraintList()
for jr in model.JR:
    model.EQ2.add(model.mat_recov[jr] == sum(model.f_waste[i, jr] * eta_recov.get((i, jr), 0) for i in model.I))

# Constraint 3: Total energy generated by WtE technologies
model.EQ3 = pyo.ConstraintList()
for je in model.JE:
    model.EQ3.add(model.ene_WtE[je] == sum(model.f_waste[i, je] * eta_WtE.get((i, je), 0) for i in model.I))

# Equation 4: Total system cost
model.EQ4 = pyo.Constraint(expr=model.cost_total == model.cost_processing + model.cost_transport)

# Equation 5: Total GHG emissions
model.EQ5 = pyo.Constraint(expr=model.ghg_total == model.ghg_processing + model.ghg_landfill)

# Equation 6: Total processing cost
model.EQ6 = pyo.Constraint(expr=model.cost_processing == sum(
    model.f_waste[i, j] * C_processing[j] for i in model.I for j in model.J))

# Equation 7: Total transportation cost including landfill
model.EQ7 = pyo.Constraint(expr=model.cost_transport == sum(
    model.f_waste[i, j] * C_transport[j] for i in model.I for j in model.J) +
    sum(model.f_landfill[i] * C_landfill for i in model.I))

# Equation 8: GHG emissions from processing
model.EQ8 = pyo.Constraint(expr=model.ghg_processing == sum(
    model.f_waste[i, j] * EF_processing[j] for i in model.I for j in model.J))

# Equation 9: GHG emissions from landfill disposal
model.EQ9 = pyo.Constraint(expr=model.ghg_landfill == sum(
    model.f_landfill[i] * EF_landfill[i] for i in model.I))

# Equation 10: Number of truck trips required
model.EQ10 = pyo.Constraint(expr=model.v_number == sum(
    model.f_waste[i, j] for i in model.I for j in model.J) / CA_truck)
```

```
    # Constraint 11: Ensuring processing capacity is not exceeded
    model.EQ11 = pyo.ConstraintList()
    for j in model.J:
        model.EQ11.add(sum(model.f_waste[i, j] for i in model.I) <= CA_processing[j])

    model.epsilon = pyo.Constraint(expr = model.ghg_total <= 32.95*PAR)

    # -------------------------------------------------------
    # Objective: Minimize GHG Emissions
    # -------------------------------------------------------
    model.Objective = pyo.Objective(expr=model.cost_total, sense=pyo.minimize)

    # -------------------------------------------------------
    # Solver Execution
    # -------------------------------------------------------
    solver = pyo.SolverFactory('glpk')
    result = solver.solve(model)

        # Store results
    ghg_total_value = model.ghg_total.value
    cost_value = model.cost_total.value
    #print(obj_value)
    return cost_value, ghg_total_value


import matplotlib.pyplot as plt
import numpy as np
from scipy.signal import savgol_filter

# Initialize lists to store the results
result1 = []
result2 = []

# Loop to increase k
for i in range(70):
    k = 1 + (i * .1)
    result = linear_problem(k)
    result1.append(result[0])
    result2.append(result[1])

# Apply smoothing using the Savitzky-Golay filter
smoothed_result1 = savgol_filter(result2, window_length=7, polyorder=1)
smoothed_result2 = savgol_filter(result1, window_length=7, polyorder=1)

# Plot the original and smoothed results
plt.figure(figsize=(8, 6))
#plt.plot(result1, result2, marker='o', color='b', label='Original', alpha=0.4)  # Original
data
```

```
plt.plot(smoothed_result1, smoothed_result2 / 1000, color='r', label='Pareto Curve') #
Smoothed data
plt.title('Multi-Objective Solution: Costs [k$] vs GHG [tonCO2]')
plt.xlabel('GHG [tonCO2]')
plt.ylabel('Costs [k$]')
plt.grid()
plt.legend()
plt.show()
```

****

A.2 Chapter 6

Table A.6.1 shows the weights used in the feasible solution objective function. **Table A.6.2** shows the information for the MSW used in the case study, **Table A.6.3** presents the data for the transportation costs, and **Table A.6.4** shows the data for the separation costs. **Table A.6.5** shows the bounds for the considered scenarios. **Tables A.6.6**, A.6.7, and **A.68** show the results for Scenarios 1, 2, and 3, respectively.

Table A.6.1. Weights generated using LHS for each objective function. [Printed (adapted) with permission from Diaz-Barriga-Fernandez, A.D., Santibañez-Aguilar, J.E., Radwan, N., Nápoles-Rivera, F., El-Halwagi, M., Ponce-Ortega, J.M. (2017). Strategic planning for managing municipal solid wastes with consideration of multiple stakeholders. ACS Sustainable Chemistry and Engineering, 2017, 5(11), 10744–10762].

Case	$w^{NetProfit}$	$w^{Consume}$	$w^{Profit}_{plastics}$	w^{Profit}_{metals}	w^{Profit}_{glass}	w^{Profit}_{paper}	$w^{Profit}_{no\text{-}recyclables}$	$w^{Profit}_{government\ expenditures}$
1	0.0855	0.134	0.1024	0.2063	0.1565	0.2261	0.0494	0.0399
2	0.0396	0.1453	0.113	0.1803	0.1534	0.1869	0.0914	0.0900
3	0.0396	0.1521	0.0955	0.1818	0.2712	0.0207	0.1167	0.1224
4	0.0828	0.1607	0.1058	0.1818	0.1944	0.0241	0.1205	0.1299
5	0.2521	0.0265	0.1701	0.0167	0.1049	0.1342	0.2136	0.0819
6	0.2184	0.1335	0.1509	0.0271	0.0835	0.1241	0.1806	0.0818
7	0.0397	0.2397	0.0786	0.1223	0.204	0.1736	0.0385	0.1034
8	0.0749	0.3244	0.056	0.1331	0.1832	0.1342	0.0232	0.0709
9	0.1665	0.0225	0.1981	0.0479	0.1279	0.3015	0.0203	0.1153
10	0.1584	0.054	0.2013	0.0454	0.1573	0.2234	0.0328	0.1275

Table A.6.2. Generated waste and sub-waste for each city and efficiency for recollection. [Printed (adapted) with permission from Diaz-Barriga-Fernandez, A.D., Santibañez-Aguilar, J.E., Radwan, N., Nápoles-Rivera, F., El-Halwagi, M., Ponce-Ortega, J.M. (2017). Strategic planning for managing municipal solid wastes with consideration of multiple stakeholders. ACS Sustainable Chemistry and Engineering, 2017, 5(11), 10744–10762].

City		Morelia $(10^{-7})^a$	Celaya $(10^{-7})^a$	Apatzingan $(10^{-7})^a$	Lazaro Cardenas$(10^{-7})^a$	Leon$(10^{-7})^a$	Efficiency for recollection
MSW	**Total production**	43.50	20.0	5.49	9.720	52.30	
	Non-recyclables (70.3% MSW)	30.60	14.10	3.84	6.830	36.80	10
Plastic	**Total production**	2.520	1.160	0.316	0.564	3.030	
	HDPE(t)	0.269	0.124	0.034	6.020	0.324	6.4
	PE(t)	0.630	0.291	0.079	0.141	0.759	6.4
	PET(t)	3.680	0.017	0.004	0.001	0.004	6.4
	PP(t)	0.937	0.432	0.118	0.209	1.130	6.4
	PS(t)	0.133	0.062	0.017	29.80	0.161	6.4
Metal	**Total production**	1.350	0.062	0.169	0.301	1.620	
	Aluminum (t)	0.082	0.038	1.030	0.018	0.099	35.8
Glass	**Total production**	2.870	1.320	0.360	0.642	3.450	
	Clear glass (t)	0.521	0.240	6.540	0.117	0.627	24.7
	Green glass (t)	0.610	0.281	0.076	0.136	0.734	4.14
	Brown glass (t)	1.740	0.801	0.218	0.389	2.090	1.81
Paper	**Total production**	6.170	2.850	0.775	1.380	7.430	
	Paper useful	2.580	1.190	0.323	0.576	3.100	7.16

Table A.6.3. Distance from cities to processing facilities and unit transport costs. [Printed (adapted) with permission from Diaz-Barriga-Fernandez, A.D., Santibañez-Aguilar, J.E., Radwan, N., Nápoles-Rivera, F., El-Halwagi, M., Ponce-Ortega, J.M. (2017). Strategic planning for managing municipal solid wastes with consideration of multiple stakeholders. ACS Sustainable Chemistry and Engineering, 2017, 5(11), 10744–10762].

Cities	Distances to processing facilities located in cities (km)				
	Morelia	**Celaya**	**Apatzingan**	**Lazaro Cardenas**	**Leon**
Morelia	1	147	192	144	207
Celaya	147	1	326	447	135
Apatzingan	192	326	1	186	393
Lazaro Cardenas	144	447	186	1	514
Leon	207	135	393	514	1
Unit transportation cost ($US/kg km)	$1.65x10^{-5a}$	4.0010^{-5b}			

[a] For plastic, MSW, paper.
[b] Metal and glass.

Table A.6.4. Unitary separation cost for type of residue. [Printed (adapted) with permission from Diaz-Barriga-Fernandez, A.D., Santibañez-Aguilar, J.E., Radwan, N., Nápoles-Rivera, F., El-Halwagi, M., Ponce-Ortega, J.M. (2017). Strategic planning for managing municipal solid wastes with consideration of multiple stakeholders. ACS Sustainable Chemistry and Engineering, 2017, 5(11), 10744–10762].

Type of residue	Unitary cost for separation ($US/kg)
Plastic	0.2235
Metal	0.02
Glass	0.0008
Paper	0.04
No recyclables	0.026

Table A.6.5. Upper and Lower bonds for the different objective function for each scenario considered. [Printed (adapted) with permission from Diaz-Barriga-Fernandez, A.D., Santibañez-Aguilar, J.E., Radwan, N., Nápoles-Rivera, F., El-Halwagi, M., Ponce-Ortega, J.M. (2017). Strategic planning for managing municipal solid wastes with consideration of multiple stakeholders. ACS Sustainable Chemistry and Engineering, 2017, 5(11), 10744–10762].

Objective function	SCENARIO 1		SCENARIO 2		SCENARIO 3	
	UB	**LB**	**UB**	**LB**	**UB**	**LB**
Consumed MSW (%)	99.96	0	99.96	0	99.96	0
ONP(MM$/y)	63.18	–33.11	69.41	–19.65	74.05	–14.18
Net profit for plastics (MM$/y)	0	–6.09	0	–1.589	0	–5.196
Net profit for metals(MM$/y)	0	–17.79	1.73	0	1.730	0
Net profit for glass(MM$/y)	0	–5.70	0	–5.553	0	–5.448
Net profit for paper(MM$/y)	2.34	–17.76	9.064	0	9.065	0
Net profit for no recyclables(MM$/y)	73.53	0	75.92	0	75.92	0
Government expenditures(MM$/y)	–12.05	–14.18	–14.09	–18.48	–12.05	–14.18

Table A.6.6. Results for different solutions under Scenario 1. [Printed (adapted) with permission from Diaz-Barriga-Fernandez, A.D., Santibañez-Aguilar, J.E., Radwan, N., Nápoles-Rivera, F., El-Halwagi, M., Ponce-Ortega, J.M. (2017). Strategic planning for managing municipal solid wastes with consideration of multiple stakeholders. ACS Sustainable Chemistry and Engineering, 2017, 5(11), 10744–10762].

Objective Function/Concept	**Maximizing the ONP($MM/y)**	**Minimizing the CS ($MM/y)**	**Minimizing the AFS($MM/y)**
ONP	63.189	62.842	62.725
% Consumed MSW	85.17%	95.02%	95.02%
%Glass consumed	0%	81.5%	81.5%
%Paper consumed	100%	100%	100%
%No recyclables consumed	100%	100%	100%
Net profit for plastics	0	0	0
Net profit for metals	0	0	0
Net profit for glass	0	–0.441	–0.499
Net profit for paper	2.327	2.326	2.274
Net profit for no recyclables	73.528	73.528	73.528
Government expenditures	–12.666	–12.570	12.578
CS	0.436	0.374	0.392
AFS	0.048	0.043	0.046

Table A.6.7. Results for different solutions for Scenario 2. [Printed (adapted) with permission from Diaz-Barriga-Fernandez, A.D., Santibañez-Aguilar, J.E., Radwan, N., Nápoles-Rivera, F., El-Halwagi, M., Ponce-Ortega, J.M. (2017). Strategic planning for managing municipal solid wastes with consideration of multiple stakeholders. ACS Sustainable Chemistry and Engineering, 2017, 5(11), 10744–10762].

Objective Function/Concept	**Maximizing ONP($MM/y)**	**Compromise Solution ($MM/y)**	**Average Feasible Solution ($MM/y)**
ONP	69.407	68.158	68.991
%Consume MSW	85.685%	95.533%	87.921%
%Consumed Metals	100%	100%	100%
%Consumed Glass	100%	100%	100%
%Consumed Paper	0%	81.5%	18.5%
%Consumed No Recyclables	100%	100%	100%
Net profit for the plastics	0	0	0
Net profit for the metals	1.729	1.730	1.727
Net profit for the glass	0	–0.306	–0.319
Net profit for the paper	9.052	9.046	8.998
Net profit for the no recyclables	75.924	75.004	75.924
Government expenditures	–17.298	–17.317	–17.339
CS	0.873	0.860	0.930
AFS	0.090	0.089	0.099

Table A.6.8. Results for different solutions for Scenario 3. [Printed (adapted) with permission from Diaz-Barriga-Fernandez, A.D., Santibañez-Aguilar, J.E., Radwan, N., Nápoles-Rivera, F., El-Halwagi, M., Ponce-Ortega, J.M. (2017). Strategic planning for managing municipal solid wastes with consideration of multiple stakeholders. ACS Sustainable Chemistry and Engineering, 2017, 5(11), 10744–10762].

Objective Function/Concept	Maximizing ONP($MM/y)	Minimizing CS($MM/y)	Minimizing AFS ($MM/y)
ONP	74.047	73.861	73.819
%Consumed MSW	85.685%	95.533%	95.533%
%Consumed Metals	100%	100%	100%
%Consumed Glass	0%	81.5%	81.5%
%Consumed Paper	100%	100%	100%
%Consumed No Recyclables	100%	100%	100%
Net profit for the plastics	0	0	0
Net profit for the metals	1.729	1.729	1.724
Net profit for the glass	0	–0.304	-0.348
Net profit for the paper	9.052	9.045	9.052
Net profit for the no recyclables	75.924	75.924	75.924
Government expenditures	–12.659	–12.533	–12.533
CS	0.429	0.337	0.341
AFS	0.048	0.038	0.039

Diaz-Barriga-Fernandez, A.D., Santibañez-Aguilar, J.E., Radwan, N., Nápoles-Rivera, F., El-Halwagi, M., Ponce-Ortega, J.M. (2017). Strategic planning for managing municipal solid wastes with consideration of multiple stakeholders. ACS Sustainable Chemistry and Engineering, 2017, 5(11), 10744–10762.

A.3 Chapter 8: Complementary Results

The results of this tax are shown in the relevant section of the Appendix in the Appendix A.3 section of this book.

A.3.1 Results of plastic waste

The solutions for the suppliers (*S1–S5*), consumers (*D1–D5*), and transportation providers (*K1–K5*) of plastic are shown in **Tables A.8.1–A.8.6**. The distinct types of plastic are denoted with*R1*, *R2*, *R3*, *R4*, and *R5*. All the prices for the landfill supply (**Table A.8.1**) allow the profitto be positive. **Table A.8.2** presents the profits for the different landfill suppliers by type of plastic in scenario II. Note that the profit of Supplier 5 (*S5*) is not shown here since it is zero.This occurs because her clearing price is equal to her bidding cost (**Table A.8.1**). This supplier participates in the market since her profit is non-negative. **Tables A.8.3** and **A.8.4** present the prices and profits for the consumers (demand). Then, the clearing prices and profits for the transportation providers of the plastic entering the landfill are presented in **Tables A.8.5** and **A.8.6.**

Table A.8.1. Bids and clearing prices for the landfill supply of plastic.

Scenario	Bids (USD/tonne)	Prices (USD/tonne)				
	S1-S5	S1	S2	S3	S4	S5
I	–12.35	0	0	0	0	0
II (w/tax)	–12.35	–11.52	–11.51	–11.53	–11.54	-12.35

Table A.8.2. Profits for the landfill supply of plastic.

Scenario II(w/tax)	Profits (USD)				
	R1	R2	R3	R4	R5
S1	7755.69	5214.61	304.60	1100.86	2226.55
S2	3625.88	2442.43	142.68	516.18	1040.76
S3	965.27	647.06	37.79	136.61	276.49
S4	1699.68	1146.67	66.93	242.35	489.57

Table A.8.3. Bids and clearing prices for the landfill demand of plastic.

Scenario	Bids (USD/tonne)	Prices (USD/tonne)				
	D1-D5	D1	D2	D3	D4	D5
I	–12.35	–12.29	–12.31	–12.28	–12.28	–12.26
II (w/tax)	–12.35	–23.82	–23.82	–23.82	–23.82	–24.61

Table A.8.4. Profits for the landfill demand of plastic.

Scenario II (w/tax)	Profits (USD)				
	R1	R2	R3	R4	R5
D1	107435.65	72235.29	4219.46	15249.67	30843.32
D2	49532.77	33365.82	1949.21	7051.54	14217.74
D3	13529.78	9069.54	529.73	1914.81	3875.48
D4	23963.77	16166.94	943.65	3416.84	6902.48
D5	138555.99	93065.48	5431.89	19741.16	39727.56

Table A.8.5. Bids and clearing prices for the transportation providers of plastic (enteringlandfill).

Scenario	Bids (USD/tonne)	Prices (USD/tonne)				
	K1–K5	K1	K2	K3	K4	K5
I	0.0165	–12.29	–12.31	–12.28	–12.28	–12.26
II(w/tax)	0.0165	34.64	34.64	34.64	34.64	33.84

Table A.8.6. Profits for the transportation providers of plastic (entering landfill).

Scenario II(w/ tax)	Profits (USD)				
	R1	R2	R3	R4	R5
K1	324382.59	218101.42	12739.89	46043.63	93125.85
K2	149555.26	100742.09	5885.28	21290.85	42927.90
K3	40850.74	27383.85	1599.41	5781.42	11701.31
K4	72354.28	48813.18	2849.17	10316.54	20840.80
K5	382206.68	256721.13	14983.85	54456.00	109588.46

A.3.2 Results of metal waste

Tables A.8.7–A.8.18 show the solutions for the distinct suppliers (*S1–S5*), consumers (*D1–D5*), transportation providers (*K1–K5*), and technology providers (*M1–M5*) of metal. The bids andclearing prices for the landfill supply are shown in **Table A.8.7.** Their profits are presented in **Table A.8.8.** Then, **Table A.8.9** presents the prices for the landfill demand. **Tables A.8.10** and **A.8.11** show the prices and profits for the plant supply, respectively. Similarly, **Tables A.8.12** and **A.8.13** present the prices and profits for the plant demand. The results for the transportation providers of the metal entering the landfill are presented in **Table A.8.14.** The bids and prices for the transportation providers of the metal entering the plant are presented in **Table A.8.15.** The clearing prices for the transportation providers of the metal leaving the plant are equal to the ones obtained for the transportation providers of the metal entering the plant (**Table A.8.15).** Then, the total profits for the transportation providers of metal are presented in **Table A.8.16.** Finally, for the metal waste, the prices and profits for the transformation providers are shown in **Tables A.8.17** and **A.8.18.**

Table A.8.7. Bids and clearing prices for the landfill supply of metal.

Scenario	Bids (USD/tonne)	Prices (USD/tonne)				
	S1–S5	S1	S2	S3	S4	S5
I	–12.35	0	0	0	0	0
II (w/tax)	–12.35	–0.14	–0.11	–0 16	–0.17	–0.21

Table A.8.8. Profits for the landfill supply of met.

Scenario II (w/tax)	Profit (USD)
S1	6460.96
S2	2986.43
S3	806.07
S4	1438.59
S5	7720.98

Table A.8.9. Bids and clearing prices for the landfill demand of metal.

Scenario	Bids (USD/tonne)	Prices (USD/tonne)				
	D1-D5	D1	D2	D3	D4	D5
I	–12.35	–12.21	–12.24	–12.19	–12.18	–12.14
II (w/tax)	–12.35	–12.35	–12.35	–12.35	–12.35	–12.35

Table A.8.10. Bids and clearing prices for the plant supply of metal.

Scenario	Bids (USD/tonne)	Prices (USD/tonne)				
	S1–S5	S1	S2	S3	S4	S5
I	223.5	1204.2	1198.4	1196.6	1198.5	1196
II (w/tax)	223.5	1204.2	1198.4	1196.6	1198.5	1196

Table A.8.11. Profits for the plant supply of metal.

Scenario I	Profits(USD)	Scenario II(w/ tax)	Profits(USD)
S1	289298.65	S1	289298.65
S2	132619.95	S2	132619.95
S3	35880.61	S3	35880.61
S4	64223.88	S4	64223.88
S5	345007.41	S5	345007.41

Table A.8.12. Bids and clearing prices for the plant demand of metal.

Scenario	Bids (USD/tonne)	Prices (USD/tonne)				
	D1–D5	D1	D2	D3	D4	D5
I	1300	1290	1300	1300	1300	1300
II (w/tax)	1300	1290	1300	1300	1300	1300

Table A.8.13. Profits for the plant demand of metal.

Scenario I	Profits(USD)	Scenario II(w/ tax)	Profits(USD)
D1	4713.28	D1	4713.28

Table A.8.14. Bids and clearing prices for the transportation providers of metal (entering landfill).

Scenario	Bids (USD/tonne)	Prices (USD/tonne)				
	K1–K5	K1	K2	K3	K4	K5
I	0.04	–12.21	–12.24	–12.19	–12.18	–12.14
II (w/tax)	0.04	–11.56	–11.56	–11.56	–11.56	–11.56

Table A.8.15. Bids and clearing prices for the transportation providers of metal (enteringplant).

Scenario	Bids (USD/tonne)	Prices (USD/tonne)				
	K1–K5	K1	K2	K3	K4	K5
I	0.04	90.1	101.64	103.44	101.52	104
II (w/tax)	0.04	90.1	101.64	103.44	101.52	104

Table A.8.16. Total profits for the transportation providers of metal.

Scenario I	Profits(USD)	ScenarioII (w/ tax)	Profits(USD)
K1	100754.04	K1	94618.25
K2	13821.66	K2	10992.05
K3	3812.77	K3	3045.80
K4	13235.83	K4	11865.71
K5	36896.91	K5	29517.59

Table A.8.17. Bids and clearing prices for the technology providers of metal.

Scenario	Bids (USD/ton ne)	Prices (USD/tonne)	
	M1–M5	M1	M2–M5
I	90	1294.2	1300
II (w/tax)	90	1294.2	1300

Table A.8.18. Profits for the technology providers of metal.

Scenario I	Profits (USD)	Scenario II (w/tax)	Profits (USD)
M1	1070034.68	M1	1070034.68

A.3.3 Results of glass waste

The information for the distinct types of glass (*G1*, *G2*, and *G3*) are shown in **Tables A.8.19–A.8.24.** The results for the landfill supply (**Tables A.8.19** and **A.8.20**) are like the ones for plastic. Only in scenario II for suppliers *S1*-*S4*, the prices are greater than the bids and positiveprofits are attained. Also, *S5* participates in the market with a profit equal to zero. Then, for the landfill demand, the results are shown in **Tables A.8.21** and **A.8.22.** Again, the prices are like the case of plastic. The bids and clearing prices of the transportation providers arepresented in **Table A.8.23,** while their profits are shown in **Table A.8.24.**

Table A.8.19. Bids and clearing prices for the landfill supply of glass.

Scenario	Bids (USD/tonne)	Prices (USD/tonne)				
	S1–S5	S1	S2	S3	S4	S5
I	–12.35	0	0	0	0	0
II (w/tax)	–12.35	–10.34	–10.32	–10.37	–10.38	–12.35

Table A.8.20. Profits for the landfill supply of glass.

Scenario II (w/tax)	Profits (USD)		
	G1	G2	G3
S1	10454.30	12240.15	34914.54
S2	4883.33	5717.57	16298.13
S3	1296.94	1519.05	4323.14
S4	2306.66	2681.24	7669.14

Table A.8.21. Bids and clearing prices for the landfill demand of glass.

Scenario	Bids (USD/tonne)	Prices (USD/tonne)				
	D1–D5	D1	D2	D3	D4	D5
I	–12.35	–12.21	–12.24	–12.19	–12.18	–12.14
II (w/tax)	–12.35	–22.56	–22.56	–22.56	–22.56	–24.49

Table A.8.22. Profits for the landfill demand of glass.

Scenario II(w/ tax)	Profits (USD)		
	G1	G2	G3
D1	53177.28	62261.31	177597.83
D2	24496.25	28681.03	81756.24
D3	6675.23	7818.39	22250.76
D4	11941.92	13881.21	39704.34
D5	76090.70	89075.87	253635.65

Table A.8.23. Bids and clearing prices for the transportation providers of glass (enteringlandfill).

Scenario	Bids (USD/tonne)	Prices (USD/tonne)				
	K1–K5	K1	K2	K3	K4	K5
I	0.04	–12.21	–12.24	–12.19	–12.18	–12.14
II (w/tax)	0.04	31.20	31.20	31.20	31.20	29.27

Table A.8.24. Profits for the transportation providers of glass (entering landfill).

Scenario II(w/ tax)	Profits (USD)		
	G1	G2	G3
K1	162329.69	190059.71	542137.54
K2	74777.59	87552.10	249570.21
K3	20376.89	23866.51	67922.98
K4	36454.08	42373.97	121202.01
K5	183262.22	214536.64	610874.07

A.3.4 Results of organic waste and non-recyclables

The solutions for organic waste and non-recyclables are presented in **Tables A.8.25–A.8.30**. For this waste, their prices for the landfill supply are equal to the ones for the landfill supply of plastic (**Table A.8.1**). The same occurs with their prices for the landfill demand (**Table A.8.3**).Nevertheless, their profits change because of the flow rates involved. For organic waste, **Tables A.8.25** and **A.8.26** present these profits. **Tables A.8.28** and **A.8.29** show the profits for the non-recyclables. The prices of the transportation providers of the organic and non-recyclable waste entering the landfill are also equal to the ones for plastic (**Table A.8.5**). Their corresponding profits are presented in **Table A.8.27** and **A.8.30.**

Table A.8.25. Profits for the landfill supply of organic waste.

Scenario II(w/ tax)	Profits(USD)
S1	21355.06
S2	9987.94
S3	2642.22
S4	4684.29

Table A.8.26. Profits for the landfill demand of organic waste.

Scenario II(w/ tax)	Profits(USD)
D1	295820.69
D2	136444.43
D3	37034.92
D4	66043.69
D5	380109.35

Table A.8.27. Profits for the transportation providers of organic waste (entering landfill).

Scenario II (w/tax)	Profits(USD)
K1	893177.26
K2	411969.36
K3	111820.25
K4	199407.02
K5	1048531.60

Table A.8.28. Profits for the landfill supply of non-recyclables.

Scenario II(w/ tax)	Profits(USD)
S1	253280.90
S2	118344.55
S3	31412.18
S4	55544.58

Table A.8.29. Profits for the landfill demand of non-recyclables.

Scenario II (w/tax)	Profits(USD)
D1	3508571
D2	1616694.48
D3	440291.26
D4	783122.22
D5	4512265.79

Table A.8.30. Profits for the transportation providers of non-recyclables (entering landfill).

Scenario II(w/ tax)	Profits(USD)
K1	10593497.7
K2	4881317.59
K3	1329380.11
K4	2364496.39
K5	12447084.9

A.4 Chapter 8–GAMS Code

The implementation of this problem in GAMS can be found in Appendix A.4 of this book.

```
$TITLE REFINERY BASED ON MUNICIPAL WASTES

    $ONEMPTY

    ********************************************
    **************SETS*************************
    ********************************************

    SETS

    C      INDEX FOR CITIES AND MARKETS                 /1*5/

    *1.- Morelia
    *2.- Celaya
    *3.- Apatzingan
    *4.- Lazaro Cardenas
    *5.- Leon

    L      INDEX FOR Landfills at each city             /1*5/

    *1.- Morelia
    *2.- Celaya
    *3.- Apatzingan
    *4.- Lazaro Cardenas
    *5.- Leon

    *PROCESSING PLANTS
    PP     INDEX FOR PROCESSING PLANTS FOR PLASTIC      /1*10/
    *1 and 2 Morelia
    *3 and 4 Celaya
    *5 and 6 Apatzingan
    *7 and 8 Lazaro Cardenas
    *9 and 10 Leon
    PM     INDEX FOR PROCESSING PLANTS FOR METAL        /1*10/
    *1 and 2 Morelia
    *3 and 4 Celaya
    *5 and 6 Apatzingan
    *7 and 8 Lazaro Cardenas
    *9 and 10 Leon
    PG     INDEX FOR PROCESSING PLANTS FOR GLASS        /1*10/
    *1 and 2 Morelia
    *3 and 4 Celaya
    *5 and 6 Apatzingan
    *7 and 8 Lazaro Cardenas
    *9 and 10 Leon
    PO     INDEX FOR PROCESSING PLANTS FOR ORGANIC      /1*10/
```

**1 and 2 Morelia*
**3 and 4 Celaya*
**5 and 6 Apatzingan*
**7 and 8 Lazaro Cardenas*
**9 and 10 Leon*
PW INDEX FOR PROCESSING PLANTS FOR NO RECYCLABLES /1*10/
**1 and 2 Morelia*
**3 and 4 Celaya*
**5 and 6 Apatzingan*
**7 and 8 Lazaro Cardenas*
**9 and 10 Leon*

**PROCESSING TECHNOLOGIES*
TP INDEX FOR PROCESSING TECHNOLOGIES FOR PLASTIC /1*6/
**1.- 600 a flujo Normal*
**2.- 600 a flujo doble*
**3.- Material Recycling 1*
**4.- Thermal Recycling*
**5.- Chemical Recycling*
**6.- Pyrolysis*
TM INDEX FOR PROCESSING TECHNOLOGIES FOR METAL /1/
**Material Recycling 2*
TG INDEX FOR PROCESSING TECHNOLOGIES FOR GLASS /1/
**Material Recycling 3*
TO INDEX FOR PROCESSING TECHNOLOGIES FOR ORGANIC /1/
**Material Recycling 4*
TW INDEX FOR PROCESSING TECHNOLOGIES FOR NO RECYCLABLES /1*5/
**1.- INCINERATION*
**2.- PYROLYSIS*
**3.- PYROLYSIS/GASIFICATION*
**4.- PLASMA ARC GASIFICATION*
**5.- CONVENTIONAL GASIFICATION*

**TYPES OF WASTES*
R INDEX FOR TYPES OF PLASTICS /1*5/
**PP*
**PE*
**PET*
**PS*
**HDPE*
S INDEX FOR TYPES OF METAL /1/
**Aluminum*
T INDEX FOR TYPES OF GLASS /1*3/
**Clear Glass*
**Green Glass*
**Brown Glass*
U INDEX FOR TYPES OF ORGANIC /1/
**Paper*
W INDEX FOR TYPES OF NO RECYCLABLES /1/

```
*ECONOMIES OF SCALE
Q      INDEX FOR INTERVALS IN ECONOMIES OF SCALE      /1*2/;

PARAMETERS



*T1P(PP,TP,R,C)
*T1M(PM,TM,S,C)

T1P(PP,TP,R)
T1M(PM,TM,S)
T1G(PG,TG,T)
T1O(PO,TO,U)
T1W(PW,TW,W)
PT(L,R)
PT1(L,S)
PT2(L,T)
PT3(L,U)
PT4(L,W)
PTC(TP,C,R)
TR(TP,C,R)
TR1(TM,C,S)
TR2(TG,C,T)
TR3(TO,C,U)
TR4(TW,C,W)

*PT(L,C,R)

*/PEND
*/PTL(C,L,R)
*PTL1
*/PTL2(L,T)
*/PTL3(L,U)
*/PTL4(L,W)
*/PTL1(TM,C,S)
*/PTL1B(L,S)

PTL(TP,C,R)
PTL1(TM,C,S)
PTL2(TG,C,T)
PTL3(TO,C,U)
PTL4(TW,C,W)


PRP(TP,C,R)
PRM(TM,C,S)
PRG(TG,C,T)
PRO(TO,C,U)
PRW(TW,C,W)
```

```
PRSP(C,R)
PRSM(C,S)
PRSG(C,T)
PRSO(C,U)
PRSW(C,W)


PLP(C,R)
PLM(C,S)
PLG(C,T)
PLO(C,U)
PLW(C,W)

PLDP(L,R)
PLDM(L,S)
PLDG(L,T)
PLDO(L,U)
PLDW(L,W)


A(C,R)  PROFIT FOR PLASTIC SUPPLIER
B(C,S)  PROFIT FOR METAL SUPPLIER
CC(C,T)  PROFIT FOR GLASS SUPPLIER
DD(C,U)  PROFIT FOR ORGANIC SUPPLIER
E(C,W)  PROFIT FOR NR SUPPLIER

FTP(C,R)       TOTAL PLASTIC TYPE R IN THE CITY C
/
1.1    9.37E+06
2.1    4.32E+06
3.1    1.18E+06
4.1    2.09E+06
5.1    1.13E+07

1.2    6.30E+06
2.2    2.91E+06
3.2    7.91E+05
4.2    1.41E+06
5.2    7.59E+06

1.3    3.68E+05
2.3    1.70E+05
3.3    4.62E+04
4.3    8.23E+04
5.3    4.43E+05

1.4    1.33E+06
2.4    6.15E+05
3.4    1.67E+05
4.4    2.98E+05
```

```
5.4     1.61E+06

1.5     2.69E+06
2.5     1.24E+06
3.5     3.38E+05
4.5     6.02E+05
5.5     3.24E+06

/,

LDEMSP(L,R)        DEMAND OF WASTE TYPE R IN LANDFILL L
/
1.1     9.37E+06
2.1     4.32E+06
3.1     1.18E+06
4.1     2.09E+06
5.1     1.13E+07

1.2     6.30E+06
2.2     2.91E+06
3.2     7.91E+05
4.2     1.41E+06
5.2     7.59E+06

1.3     3.68E+05
2.3     1.70E+05
3.3     4.62E+04
4.3     8.23E+04
5.3     4.43E+05

1.4     1.33E+06
2.4     6.15E+05
3.4     1.67E+05
4.4     2.98E+05
5.4     1.61E+06

1.5     2.69E+06
2.5     1.24E+06
3.5     3.38E+05
4.5     6.02E+05
5.5     3.24E+06

/,

ALPHAPSEP(C,R)  SEPARATION FACTOR FOR THE PLASTIC TYPE R IN THE CITY C
/
1*5.1*5  0.064
/,
```

```
ALPHAP(TP,R)   CONVERSION FACTOR FOR THE TECHNOLOGY TP AND THE
TYPE OF PLASTIC R
/
1.1    0.11
1.2    0.16
1.3    0.19
2.1    0.51
2.2    0.32
3.1    0.672
3.2    0.672
3.4    0.672
4.2    0.571
5.2    19.44
*THIS FACTOR HAS UNITS OF MJ/KG WASTE
6.5    0.432
6.1    0.333
/,
PDEMANDP(TP,C,R) DEMAND OF PRODUCT IN CITY C FROM THE WASTE TYPE
R AND TECH TP
/
1*2.1.1   0.5
1*2.2.1   0.5
1*2.4.1   0.5

3*6.1.1    1.27E+09
3*6.2.1    4.72E+08
3*6.3.1    1.28E+08
3*6.4.1    2.29E+08
3*6.5.1    1.23E+09

3*6.1.2    6.90E+08
3*6.2.2    3.01E+08
3*6.3.2    8.61E+07
3*6.4.2    1.51E+08
3*6.5.2    8.29E+08

3*6.1.3    3.98E+07
3*6.2.3    1.90E+07
3*6.3.3    5.02E+06
3*6.4.3    9.03E+06
3*6.5.3    4.83E+07

3*6.1.4    1.43E+08
3*6.2.4    6.75E+07
3*6.3.4    1.77E+07
3*6.4.4    3.18E+07
3*6.5.4    1.71E+08

3*6.1.5    2.89E+08
3*6.2.5    1.24E+08
```

```
    3*6.3.5    3.68E+07
    3*6.4.5    6.62E+07
    3*6.5.5    3.54E+08

    /,
    CSEPP(C,R)    UNITARY SEPARATION COST FOR PLASTIC TYPE R IN THE CITY C
    /
    1*5.1*5  0.2235
    /,
    CNSEP(C,R)       UNITARY NO SEPARATION COST FOR PLASTIC TYPE R IN THE
CITY C
    /
    1*5.1*5  -0.01235
    *-0.01240
    /

    DNSEP(L,R)       UNITARY NO SEPARATION COST FOR PLASTIC TYPE R IN THE
CITY C
    /
    1*5.1*5  -0.01235
    /

    DPP(C,PP)       DISTANCE FROM CITY C TO PROCESSING PLANT PP
    /
    1.1*2   1
    1.3*4   147
    1.5*6   192
    1.7*8   144
    1.9*10  207

    2.1*2   147
    2.3*4   1
    2.5*6   326
    2.7*8   447
    2.9*10  135

    3.1*2   192
    3.3*4   326
    3.5*6   1
    3.7*8   186
    3.9*10  393

    4.1*2   144
    4.3*4   447
    4.5*6   186
    4.7*8   1
    4.9*10  514

    5.1*2   207
    5.3*4   135
```

```
5.5*6   393
5.7*8   514
5.9*10  1
/,

*1.- Morelia
*2.- Celaya
*3.- Apatzingan
*4.- Lazaro Cardenas
*5.- Leon

DCL(C,L)      DISTANCE FROM CITY C TO LANDFILL AT EACH CITY
/
1.1   3.4161417982054
1.2*5  34.161417982054
2.1   27.1264635411009
2.2   2.71264635411009
2.3*5  27.1264635411009
3.1*2  40.0340840483019
3.3   4.00340840483019
3.4*5  40.0340840483019
4.1*3  42.9318223127522
4.4   4.29318223127522
4.5   42.9318223127522
5.1*4  53.5807087856782
5.5   5.35807087856782
/,


CTRANSP(R)     UNITARY TRANSPORTATION COST FOR THE PLASTIC TYPE R
/
1*5    1.65E-5
/,
CTRANSPL(R)    UNITARY TRANSPORTATION COST FOR THE PLASTIC TYPE R
/
1*5    1.65E-5
*1*5    1.65E-5
/,
CTRANSPP(TP,R)   UNITARY TRANSPORTATION COST FOR PRODUCT FROM
TYPE R
/
1*6.1*5 1.65E-5
/,
CPROCP(TP,R) UNITARY COST FOR OPERATIONAL COST OF PLASTIC PLANTS
/
*
1.1      2.6038
1.2      2.6038
1.3      2.6038
```

```
    2.1     2.6038
    2.2     2.6038

    3.2     0.3880
    3.4     0.3880
    3.1     0.3880
    4.2     0.2057
    5.2     0.6705
    *
    6.5     0.2057
    6.1     0.2057

    /,
    ACAPP(PP,TP,R,Q) FIXED CAPITAL COST FOR PLASTIC WHICH DEPENDS OF
THE ECONOMIES OF SCALE
    /
    *
    1*10.1.1.1     2.6038e7
    1*10.1.2.1     2.6038e7
    1*10.1.3.1     2.6038e7
    1*10.2.1.1     2.6038e7
    1*10.2.2.1     2.6038e7

    1*10.3.2.1     0.3880e7
    1*10.3.4.1     0.3880e7
    1*10.3.1.1     0.3880e7
    1*10.4.2.1     0.2057e7
    1*10.5.2.1     0.6705e7
    *
    1*10.6.5.1     0.2057e7
    1*10.6.1.1     0.2057e7
    /,
    BCAPP(PP,TP,R,Q) UNITARY VARIABLE CAPITAL COST FOR PLASTIC WHICH
DEPENDS OF THE ECONOMIES OF SCALE
    /
    *
    1*10.1.1.1     0.52076e-1
    1*10.1.2.1     0.52076e-1
    1*10.1.3.1     0.52076e-1
    1*10.2.1.1     0.52076e-1
    1*10.2.2.1     0.52076e-1

    1*10.3.2.1     0.0051165
    1*10.3.4.1     0.0051165
    1*10.3.1.1     0.0051165
    1*10.4.2.1     0.0059503
    1*10.5.2.1     0.0026819
    *
    1*10.6.5.1     0.0026819
    1*10.6.1.1     0.0026819
```

```
/,
UPFTECHP(PP,TP,R,Q)  UPPER DISAGGREGATED DISTRIBUTED WASTE TO THE
PROCESSING TECHNOLOGIES FOR PLASTIC
/
1*10.1*6.1*5.1        1e8
/,
LOFTECHP(PP,TP,R,Q)  LOWER DISAGGREGATED DISTRIBUTED WASTE TO THE
PROCESSING TECHNOLOGIES FOR PLASTIC
/
1*10.1*6.1*5.1        0
* 1e5
/,
CPP(TP,C)       UNITARY COST OF PRODUCT TP IN CITY C
/
*nanotubes
1.1    100
1.2    100
1.5    100
2.1    70
2.2    70
2.5    70

3.1*5     0.3645

4.1*5     0.3320
5.1*5     0.052632

*
6.1*5     0.18

/,


FTM(C,S)        TOTAL METAL TYPE S IN THE CITY C
/
*data are kg
1.1    8.24E+05
2.1    3.80E+05
3.1    1.03E+05
4.1    1.84E+05
5.1    9.91E+05

/,

LDEMSM(L,S)         DEMAND OF WASTE TYPE S IN LANDFILL L
/
*data are kg
1.1    8.24E+05
2.1    3.80E+05
3.1    1.03E+05
```

```
4.1    1.84E+05
5.1    9.91E+05

/,
ALPHAMSEP(C,S)  SEPARATION FACTOR FOR THE METAL TYPE S IN THE CITY C
/
1*5.1   0.358
/,
ALPHAM(TM,S)   CONVERSION FACTOR FOR THE TECHNOLOGY TM AND THE TYPE OF METAL S
/
1.1    1
/,
PDEMANDM(TM,C,S) DEMAND OF PRODUCT IN CITY C FROM THE WASTE TYPE S AND TECH TM
/
*DATA ARE GIVEN IN KG
1.1.1    8.24E+05
1.2.1    3.80E+05
1.3.1    1.03E+05
1.4.1    1.84E+05
1.5.1    9.91E+05
/,
CSEPM(C,S)     UNITARY SEPARATION COST FOR METAL TYPE S IN THE CITY C
/
*DATA ARE GIVEN IN $US/KG
1*5.1   0.2235
/,

CNSEM(C,S)     UNITARY COST FOR NOT SEPARATION OF METAL
/
1*5.1   -0.01235
*-0.01240
/

DNSEM(L,S)     UNITARY COST FOR NOT SEPARATION OF METAL
/
1*5.1   -0.01235
/


DPM(C,PM)      DISTANCE FROM CITY C TO PROCESSING PLANT PM
/
*DATA ARE GIVEN IN KM
1.1*2   1
1.3*4   147
1.5*6   192
1.7*8   144
1.9*10  207
```

```
    2.1*2  147
    2.3*4  1
    2.5*6  326
    2.7*8  447
    2.9*10  135

    3.1*2  192
    3.3*4  326
    3.5*6  1
    3.7*8  186
    3.9*10  393

    4.1*2  144
    4.3*4  447
    4.5*6  186
    4.7*8  1
    4.9*10  514

    5.1*2  207
    5.3*4  135
    5.5*6  393
    5.7*8  514
    5.9*10  1
    /,
    CTRANSM(S)     UNITARY TRANSPORTATION COST FOR THE METAL TYPE S
    /
    1      4E-5
    /,
    CTRANSML(S)     UNITARY TRANSPORTATION COST FOR THE METAL TYPE S
    /
    1      4E-5
    *1      4E-5
    /,
    CTRANSPM(TM,S)  UNITARY TRANSPORTATION COST FOR PRODUCT FROM
TYPE S
    /
    1.1    4E-5
    /,
    CPROCM(TM,S) UNITARY COST FOR OPERATIONAL COST OF METAL PLANTS
    /
    *
    1.1      0.090
    /,
    ACAPM(PM,TM,S,Q) FIXED CAPITAL COST FOR METAL WHICH DEPENDS OF
THE ECONOMIES OF SCALE
    /
    1*10.1.1.1      5E6
    /,
    BCAPM(PM,TM,S,Q) UNITARY VARIABLE CAPITAL COST FOR METAL WHICH
DEPENDS OF THE ECONOMIES OF SCALE
```

```
/
1*10.1.1.1    0.00090
/,
UPFTECHM(PM,TM,S,Q) UPPER DISAGGREGATED DISTRIBUTED WASTE TO THE
PROCESSING TECHNOLOGIES FOR METAL
/

1*10.1.1.1    1.91E+08

/,
LOFTECHM(PM,TM,S,Q)  LOWER DISAGGREGATED DISTRIBUTED WASTE TO
THE PROCESSING TECHNOLOGIES FOR METAL
/

1*10.1.1.1    0
*1E5
/,
CPM(TM,C)      UNITARY COST OF PRODUCT TM IN CITY C
/
1.1*5   1.3
/,


FTG(C,T)       TOTAL GLASS TYPE T IN THE CITY C
/
1.1    5.21E+06
2.1    2.40E+06
3.1    6.54E+05
4.1    1.17E+06
5.1    6.27E+06

1.2    6.10E+06
2.2    2.81E+06
3.2    7.66E+05
4.2    1.36E+06
5.2    7.34E+06

1.3    1.74E+07
2.3    8.01E+06
3.3    2.18E+06
4.3    3.89E+06
5.3    2.09E+07

/,
LDEMSG(L,T)      DEMAND OF WASTE TYPE T IN LANDFILL L
/
1.1    5.21E+06
2.1    2.40E+06
3.1    6.54E+05
```

```
4.1    1.17E+06
5.1    6.27E+06

1.2    6.10E+06
2.2    2.81E+06
3.2    7.66E+05
4.2    1.36E+06
5.2    7.34E+06

1.3    1.74E+07
2.3    8.01E+06
3.3    2.18E+06
4.3    3.89E+06
5.3    2.09E+07

/,

ALPHAGSEP(C,T)  SEPARATION FACTOR FOR THE GLASS TYPE T IN THE CITY C
/
1*5.1   0.247
1*5.2   0.414
1*5.3   0.181

/,
ALPHAG(TG,T)    CONVERSION FACTOR FOR THE TECHNOLOGY TG AND THE
TYPE OF GLASS T
/
1.1*3       1
/,
PDEMANDG(TG,C,T) DEMAND OF PRODUCT IN CITY C FROM THE WASTE TYPE
T AND TECH TG
/
1.1.1    5.21E+09
1.2.1    2.40E+09
1.3.1    6.54E+08
1.4.1    1.17E+09
1.5.1    6.27E+09

1.1.2    6.10E+09
1.2.2    2.81E+09
1.3.2    7.66E+08
1.4.2    1.36E+09
1.5.2    7.34E+09

1.1.3    1.74E+10
1.2.3    8.01E+09
1.3.3    2.18E+09
1.4.3    3.89E+09
1.5.3    2.09E+10
/,
```

```
CSEPG(C,T)    UNITARY SEPARATION COST FOR GLASS TYPE T IN THE CITY C
/
1*5.1*3  0.2235
/,

CNSEG(C,T)    UNITARY NO SEPARATION COST FOR METAL TYPE T IN THE
CITY C
/
1*5.1*3  -0.01235
*-0.01240
/

DNSEG(L,T)    UNITARY NO SEPARATION COST FOR METAL TYPE T IN THE
CITY C
/
1*5.1*3  -0.01235
/

DPG(C,PG)    DISTANCE FROM CITY C TO PROCESSING PLANT PG
/
1.1*2   1
1.3*4   147
1.5*6   192
1.7*8   144
1.9*10  207

2.1*2   147
2.3*4   1
2.5*6   326
2.7*8   447
2.9*10  135

3.1*2   192
3.3*4   326
3.5*6   1
3.7*8   186
3.9*10  393

4.1*2   144
4.3*4   447
4.5*6   186
4.7*8   1
4.9*10  514

5.1*2   207
5.3*4   135
5.5*6   393
5.7*8   514
5.9*10  1
/,
```

```
    CTRANSG(T)      UNITARY TRANSPORTATION COST FOR THE GLASS TYPE T
    /
    1*3     4e-5
    /,
    CTRANSGL(T)     UNITARY TRANSPORTATION COST FOR THE GLASS TYPE T
    /
    *1*3    4e-5
    1*3     4e-5
    /,
    CTRANSPG(TG,T)  UNITARY TRANSPORTATION COST FOR PRODUCT FROM
TYPE T
    /
    1.1*3   4e-5
    /,
    CPROCG(TG,T) UNITARY COST FOR OPERATIONAL COST OF GLASS PLANTS
    /
    1.1*3      0.010
    /,
    ACAPG(PG,TG,T,Q) FIXED CAPITAL COST FOR GLASS WHICH DEPENDS OF THE
ECONOMIES OF SCALE
    /
    1*10.1.1*3.1      4e6
    /,
    BCAPG(PG,TG,T,Q) UNITARY VARIABLE CAPITAL COST FOR GLASS WHICH
DEPENDS OF THE ECONOMIES OF SCALE
    /
    1*10.1.1*3.1      0.00025
    /,
    UPFTECHG(PG,TG,T,Q) UPPER DISAGGREGATED DISTRIBUTED WASTE TO THE
PROCESSING TECHNOLOGIES FOR GLASS
    /
    1*10.1.1*3.1    40000e4
    /,
    LOFTECHG(PG,TG,T,Q) LOWER DISAGGREGATED DISTRIBUTED WASTE TO THE
PROCESSING TECHNOLOGIES FOR GLASS
    /
    1*10.1.1*3.1    0
    *200e3
    /,
    CPG(TG,C)       UNITARY COST OF PRODUCT TG IN CITY C
    /
    1.1*5   0.039057
    *2.1*5  0.01171
    *3.1*5  0.02994

    /,


    FTO(C,U)        TOTAL ORGANIC TYPE U IN THE CITY C
    /
```

```
    1.1    2.58E+07
    2.1    1.19E+07
    3.1    3.23E+06
    4.1    5.76E+06
    5.1    3.10E+07
    /,
    LDEMSO(L,U)        DEMAND OF WASTE TYPE U IN LANDFILL L
    /
    1.1    2.58E+07
    2.1    1.19E+07
    3.1    3.23E+06
    4.1    5.76E+06
    5.1    3.10E+07
    /,

    ALPHAOSEP(C,U)  SEPARATION FACTOR FOR THE ORGANIC TYPE U IN THE
CITY C
    /
    1*5.1   0.716
    /,
    ALPHAO(TO,U)   CONVERSION FACTOR FOR THE TECHNOLOGY TO AND THE
TYPE OF ORGANIC U
    /
    1.1    0.84
    /,
    PDEMANDO(TO,C,U) DEMAND OF PRODUCT IN CITY C FROM THE WASTE TYPE
U AND TECH TO
    /
    1.1.1    2.58E+07
    1.2.1    1.19E+07
    1.3.1    3.23E+06
    1.4.1    5.76E+06
    1.5.1    3.10E+07
    /,
    CSEPO(C,U)   UNITARY SEPARATION COST FOR ORGANIC TYPE U IN THE CITY C
    /
    1*5.1   0.2235
    /,

    CNSEO(C,U)       UNITARY NO SEPARATION COST FOR METAL TYPE T IN THE
CITY C
    /
    1*5.1   -0.01235
    *-0.01240
    /

    DNSEO(L,U)       UNITARY NO SEPARATION COST FOR METAL TYPE T IN THE
CITY C
    /
    1*5.1   -0.01235
```

```
/

DPO(C,PO)      DISTANCE FROM CITY C TO PROCESSING PLANT PO
/
1.1*2   1
1.3*4   147
1.5*6   192
1.7*8   144
1.9*10  207

2.1*2   147
2.3*4   1
2.5*6   326
2.7*8   447
2.9*10  135

3.1*2   192
3.3*4   326
3.5*6   1
3.7*8   186
3.9*10  393

4.1*2   144
4.3*4   447
4.5*6   186
4.7*8   1
4.9*10  514

5.1*2   207
5.3*4   135
5.5*6   393
5.7*8   514
5.9*10  1
/,
CTRANSO(U)     UNITARY TRANSPORTATION COST FOR THE ORGANIC TYPE U
/
1      1.65E-5
/,
CTRANSOL(U)    UNITARY TRANSPORTATION COST FOR THE ORGANIC TYPE U
/
1      1.65E-5
*1      1.65E-5
/,
CTRANSPO(TO,U)   UNITARY TRANSPORTATION COST FOR PRODUCT FROM
TYPE U
/
1.1     1.65E-5
/,
CPROCO(TO,U) UNITARY COST FOR OPERATIONAL COST OF PLASTIC PLANTS
/
```

```
    1.1      0.00444
    /,
    ACAPO(PO,TO,U,Q) FIXED CAPITAL COST FOR ORGANIC WHICH DEPENDS OF
THE ECONOMIES OF SCALE
    /
    1*10.1.1.1    5E6
    /,
    BCAPO(PO,TO,U,Q) UNITARY VARIABLE CAPITAL COST FOR ORGANIC WHICH
DEPENDS OF THE ECONOMIES OF SCALE
    /
    1*10.1.1.1    0.000444
    /,
    UPFTECHO(PO,TO,U,Q) UPPER DISAGGREGATED DISTRIBUTED WASTE TO THE
PROCESSING TECHNOLOGIES FOR ORGANIC
    /
    1*10.1.1.1   1.095E8
    /,
    LOFTECHO(PO,TO,U,Q) LOWER DISAGGREGATED DISTRIBUTED WASTE TO THE
PROCESSING TECHNOLOGIES FOR ORGANIC
    /
    1*10.1.1.1   0
    *1.095E5
    /,
    CPO(TO,C)      UNITARY COST OF PRODUCT TO IN CITY C
    /
    1.1*5   0.2109
    /

    ****
    ****
    ****
    ****

    FTW(C,W)       TOTAL NO RECYCLABLE TYPE W IN THE CITY C
    /
    1.1     3.06E8
    2.1     1.41E8
    3.1     3.84E7
    4.1     6.83E7
    5.1     3.68E8
    /,


    LDEMSW(L,W)      DEMAND OF WASTE TYPE W IN LANDFILL L
    /
    1.1     3.06E8
    2.1     1.41E8
    3.1     3.84E7
    4.1     6.83E7
    5.1     3.68E8
```

```
/,
ALPHAWSEP(C,W)  SEPARATION FACTOR FOR THE ORGANIC TYPE U IN THE
CITY C
/
1*5.1   0.1
/,
ALPHAW(TW,W)    CONVERSION FACTOR FOR THE TECHNOLOGY TO AND
THE TYPE OF ORGANIC U
/
1.1     1.9584
2.1     2.0556
3.1     2.466
4.1     2.9376
5.1     2.466
/,
PDEMANDW(TW,C,W) DEMAND OF PRODUCT IN CITY C FROM THE WASTE
TYPE U AND TECH WASTE
/
1*5.1.1     2.58E+10
1*5.2.1     1.19E+10
1*5.3.1     3.23E+09
1*5.4.1     5.76E+09
1*5.5.1     3.10E+10
/,
CSEPW(C,W)      UNITARY SEPARATION COST FOR ORGANIC TYPE W IN THE
CITY C
/
1*5.1   0.2235
/,

CNSEW(C,W)      UNITARY NO SEPARATION COST FOR METAL TYPE W IN THE
CITY C
/
1*5.1   -0.01235
*-0.01240
/
DNSEW(L,W)      UNITARY NO SEPARATION COST FOR METAL TYPE W IN THE
CITY C
/
1*5.1   -0.01235
/
DPW(C,PW)       DISTANCE FROM CITY C TO PROCESSING PLANT PW
/
1.1*2   1
1.3*4   147
1.5*6   192
1.7*8   144
1.9*10  207

2.1*2   147
```

```
2.3*4  1
2.5*6  326
2.7*8  447
2.9*10  135

3.1*2  192
3.3*4  326
3.5*6  1
3.7*8  186
3.9*10  393

4.1*2  144
4.3*4  447
4.5*6  186
4.7*8  1
4.9*10  514

5.1*2  207
5.3*4  135
5.5*6  393
5.7*8  514
5.9*10  1
/,
CTRANSW(W)    UNITARY TRANSPORTATION COST FOR THE ORGANIC TYPE W
/
1      1.65E-5
/,
CTRANSWL(W)    UNITARY TRANSPORTATION COST FOR THE ORGANIC TYPE W
/
1      1.65E-5
*1      1.65E-5
/,
CTRANSPW(TW,W)  UNITARY TRANSPORTATION COST FOR PRODUCT FROM
TYPE U
/
1*5.1     1.65E-5
/,
CPROCW(TW,W) UNITARY COST FOR OPERATIONAL COST OF PLASTIC PLANTS
/
1.1       0.04
2.1       0.04
3.1       0.04
4.1       0.04
5.1       0.04

/,
ACAPW(PW,TW,W,Q) FIXED CAPITAL COST FOR ORGANIC WHICH DEPENDS
OF THE ECONOMIES OF SCALE
/
1*10.1.1.1     115997700
```

```
    1*10.2.1.1    86936900
    1*10.3.1.1    102593400
    1*10.4.1.1    101583800
    1*10.5.1.1    80337800
    /,
    BCAPW(PW,TW,W,Q) UNITARY VARIABLE CAPITAL COST FOR ORGANIC WHICH
DEPENDS OF THE ECONOMIES OF SCALE
    /
    1*10.1*5.1.1    0.004
    /,
    UPFTECHW(PW,TW,W,Q)  UPPER DISAGGREGATED DISTRIBUTED WASTE TO
THE PROCESSING TECHNOLOGIES FOR ORGANIC
    /
    1*10.1*5.1.1   1.095E8
    /,
    LOFTECHW(PW,TW,W,Q)  LOWER DISAGGREGATED DISTRIBUTED WASTE TO
THE PROCESSING TECHNOLOGIES FOR ORGANIC
    /
    1*10.1*5.1.1   0
    *1.095E5
    /,
    CPW(TW,C)      UNITARY COST OF PRODUCT TO IN CITY C
    /
    1*5.1*5   0.052632
    /
    ;


    POSITIVE VARIABLES
    *CONSUMETOTAL    TOTAL AMOUNT OF CONSUMED WASTE
    MCT
    FWV(C,W)    WASTE LEFT AT THE CITY
    FOV(C,U)    WASTE LEFT AT THE CITY
    FGV(C,T)    WASTE LEFT AT THE CITY
    FMV(C,S)    WASTE LEFT AT THE CITY
    FPV(C,R)    WASTE LEFT AT THE CITY

    CM(C,R)       MANAGEMENT COST OF WASTE LEFT AT THE CITY
    CM1(C,S)      MANAGEMENT COST OF WASTE LEFT AT THE CITY
    CM2(C,T)      MANAGEMENT COST OF WASTE LEFT AT THE CITY
    CM3(C,U)      MANAGEMENT COST OF WASTE LEFT AT THE CITY
    CM4(C,W)      MANAGEMENT COST OF WASTE LEFT AT THE CITY

    SUP(C,L,R)       DEMAND NEW VARIABLE FOR DUAL VALUES,
    SUP1(C,L,S)      DEMAND NEW VARIABLE FOR DUAL VALUES,
    SUP2(C,L,T)      DEMAND NEW VARIABLE FOR DUAL VALUES,
    SUP3(C,L,U)      DEMAND NEW VARIABLE FOR DUAL VALUES,
    SUP4(C,L,W)      DEMAND NEW VARIABLE FOR DUAL VALUES,

    DEMSP(L,R)
```

DEMSM(L,S)
DEMSG(L,T)
DEMSO(L,U)
DEMSW(L,W)

FTPV(C,R)	variable amount of waste sent to the landfill
FTMV(C,S)	variable amount of waste sent to the landfill
FTGV(C,T)	variable amount of waste sent to the landfill
FTOV(C,U)	variable amount of waste sent to the landfill
FTWV(C,W)	variable amount of waste sent to the landfill

FP(C,R)	USED PLASTIC TYPE R IN THE CITY C,
FPR(C,R)	PLASTIC DELIVERED TO THE LANDFILL,
FPPLANT(C,PP,R)	DELIVERED WASTE FROM THE CITY C TO PLASTIC PLANT PP,
FTECHP(PP,TP,R)	DISTRIBUTED WASTE TO THE PROCESSING TECHNOLOGIES FOR PLASTIC,
PRODP(PP,TP,R)	PRODUCED PRODUCT IN PROCESSING PLANT PP THROUGH TECH TP AND TYPE R,
PRODCITYP(PP,TP,C,R)	DELIVERED PRODUCT TO ALL CITIES FROM PROCESSING PLANTS PP TYPE R AND TECH TP,
PTCITYP(TP,C,R)	TOTAL PRODUCT IN CITY C FROM THE TYPE R AND TECH TP,

FM(C,S)	USED METAL TYPE S IN THE CITY C,
FMPLANT(C,PM,S)	DELIVERED WASTE FROM THE CITY C TO METAL PLANT PM,
FTECHM(PM,TM,S)	DISTRIBUTED WASTE TO THE PROCESSING TECHNOLOGIES FOR METAL,
PRODM(PM,TM,S)	PRODUCED PRODUCT IN PROCESSING PLANT PM THROUGH TECH TM AND TYPE S,
PRODCITYM(PM,TM,C,S)	DELIVERED PRODUCT TO ALL CITIES FROM PROCESSING PLANTS PM TYPE S AND TECH TM,
PTCITYM(TM,C,S)	TOTAL PRODUCT IN CITY C FROM THE TYPE S AND TECH TM,

FG(C,T)	USED GLASS TYPE T IN THE CITY C,
FGPLANT(C,PG,T)	DELIVERED WASTE FROM THE CITY C TO GLASS PLANT PG,
FTECHG(PG,TG,T)	DISTRIBUTED WASTE TO THE PROCESSING TECHNOLOGIES FOR GLASS,
PRODG(PG,TG,T)	PRODUCED PRODUCT IN PROCESSING PLANT PP THROUGH TECH TG AND TYPE T,
PRODCITYG(PG,TG,C,T)	DELIVERED PRODUCT TO ALL CITIES FROM PROCESSING PLANTS PG TYPE T AND TECH TG,
PTCITYG(TG,C,T)	TOTAL PRODUCT IN CITY C FROM THE TYPE T AND TECH TG,

FO(C,U)	USED ORGANIC TYPE U IN THE CITY C,

FOPLANT(C,PO,U)	DELIVERED WASTE FROM THE CITY C TO ORGANIC PLANT PO,
FTECHO(PO,TO,U)	DISTRIBUTED WASTE TO THE PROCESSING TECHNOLOGIES FOR ORGANIC,
PRODO(PO,TO,U)	PRODUCED PRODUCT IN PROCESSING PLANT PO THROUGH TECH TO AND TYPE U,
PRODCITYO(PO,TO,C,U)	DELIVERED PRODUCT TO ALL CITIES FROM PROCESSING PLANTS PO TYPE U AND TECH TO,
PTCITYO(TO,C,U)	TOTAL PRODUCT IN CITY C FROM THE TYPE U AND TECH TO
FW(C,W)	USED ORGANIC TYPE U IN THE CITY C,
FWPLANT(C,PW,W)	DELIVERED WASTE FROM THE CITY C TO ORGANIC PLANT PO,
FTECHW(PW,TW,W)	DISTRIBUTED WASTE TO THE PROCESSING TECHNOLOGIES FOR ORGANIC,
PRODW(PW,TW,W)	PRODUCED PRODUCT IN PROCESSING PLANT PO THROUGH TECH TO AND TYPE U,
PRODCITYW(PW,TW,C,W)	DELIVERED PRODUCT TO ALL CITIES FROM PROCESSING PLANTS PO TYPE U AND TECH TO,
PTCITYW(TW,C,W)	TOTAL PRODUCT IN CITY C FROM THE TYPE U AND TECH TO

CONSUMEP PERCENTAGE OF CONSUMED PLASTIC,
CONSUMEM PERCENTAGE OF CONSUMED METAL,
CONSUMEG PERCENTAGE OF CONSUMED GLASS,
CONSUMEO PERCENTAGE OF CONSUMED ORGANIC
CONSUMEW PERCENTAGE OF CONSUMED ORGANIC;

VARIABLES

TRANSCOSTCL transport from cities to landfills at each city

**DE(TP,C) D,*
**DE1(TM,C) D1,*
**DE2(TG,C) D2,*
**DE3(TO,C) D3,*
**DE4(TW,C) D4,*

**NEW*
D(C,R) DEMAND NEW VARIABLE FOR DUAL VALUES,
D1(C,S) DEMAND NEW VARIABLE FOR DUAL VALUES,
D2(C,T) DEMAND NEW VARIABLE FOR DUAL VALUES,
D3(C,U) DEMAND NEW VARIABLE FOR DUAL VALUES,
D4(C,W) DEMAND NEW VARIABLE FOR DUAL VALUES,

***TR(C,R) TRANSPORT NEW VARIABLE FOR DUAL VALUES,*
**TR1(C,S) TRANSPORT NEW VARIABLE FOR DUAL VALUES,*
**TR2(C,T) TRANSPORT NEW VARIABLE FOR DUAL VALUES,*
**TR3(C,U) TRANSPORT NEW VARIABLE FOR DUAL VALUES,*

```
    *TR4(C,W)        TRANSPORT NEW VARIABLE FOR DUAL VALUES,

    COSTSEPP        TOTAL SEPARATION COST FOR PLASTIC,
    COSTDISP        TOTAL DISPOSITION COST FOR PLASTIC,
    COSTDISPSUP


CAPCOSTP(PP,TP,R)   CAPITAL COST FOR THE PLANT PP TECH TP AND TYPE R,
APROCPP(PP,TP,R)    FIXED CAPITAL COST FOR THE PLANT PP TECH TP AND
                    TYPE R,
BPROCPP(PP,TP,R)    VARIABLE CAPITAL COST FOR THE PLANT PP TECH TP AND
                    TYPE R,
DFTECHP(PP,TP,R,Q)  DISAGGREGATED DISTRIBUTED WASTE TO THE PROCESSING
                    TECHNOLOGIES FOR PLASTIC,

COSTSEPM            TOTAL SEPARATION COST FOR METAL,
CAPCOSTM(PM,TM,S)   CAPITAL COST FOR THE PLANT PM TECH TM AND TYPE S,
APROCPM(PM,TM,S)    FIXED CAPITAL COST FOR THE PLANT PM TECH TM AND
                    TYPE S,
BPROCPM(PM,TM,S)    VARIABLE CAPITAL COST FOR THE PLANT PM TECH TM
                    AND TYPE S,
DFTECHM(PM,TM,S,Q)  DISAGGREGATED DISTRIBUTED WASTE TO THE PROCESSING
                    TECHNOLOGIES FOR METAL,

COSTSEPG            TOTAL SEPARATION COST FOR GLASS,
CAPCOSTG(PG,TG,T)   CAPITAL COST FOR THE PLANT PG TECH TG AND TYPE T,
APROCPG(PG,TG,T)    FIXED CAPITAL COST FOR THE PLANT PG TECH TG AND
                    TYPE T,
BPROCPG(PG,TG,T)    VARIABLE CAPITAL COST FOR THE PLANT PG TECH TG AND
                    TYPE T,
DFTECHG(PG,TG,T,Q)  DISAGGREGATED DISTRIBUTED WASTE TO THE PROCESSING
                    TECHNOLOGIES FOR GLASS,

COSTSEPO            TOTAL SEPARATION COST FOR ORGANIC,
CAPCOSTO(PO,TO,U)   CAPITAL COST FOR THE PLANT PO TECH TO AND TYPE U,
APROCPO(PO,TO,U)    FIXED CAPITAL COST FOR THE PLANT PO TECH TO AND
                    TYPE U,
BPROCPO(PO,TO,U)    VARIABLE CAPITAL COST FOR THE PLANT PO TECH TO AND
                    TYPE U,
DFTECHO(PO,TO,U,Q)  DISAGGREGATED DISTRIBUTED WASTE TO THE PROCESSING
                    TECHNOLOGIES FOR ORGANIC,

COSTSEPW            TOTAL SEPARATION COST FOR ORGANIC,
CAPCOSTW(PW,TW,W)   CAPITAL COST FOR THE PLANT PO TECH TO AND TYPE U,
APROCPW(PW,TW,W)    FIXED CAPITAL COST FOR THE PLANT PO TECH TO AND
                    TYPE U,
BPROCPW(PW,TW,W)    VARIABLE CAPITAL COST FOR THE PLANT PO TECH TO AND
                    TYPE U,
DFTECHW(PW,TW,W,Q)  DISAGGREGATED DISTRIBUTED WASTE TO THE
                    PROCESSING TECHNOLOGIES FOR ORGANIC,
```

```
TOTALCOSTSEP   TOTAL SEPARATION COST
TOTALCOSTDIS        TOTAL DISPOSITION COST
TRANSCOSTCP         TRANSPORTATION COST FROM CITIES TO PROCESSING
                    PLANTS,
TRANSCOSTPC         TRANSPORTATION COST FROM PROCESSING PLANTS TO
                    CITIES,
OPERACOST           TOTAL OPERATIONAL PROCESSING COST,
CAPITALCOST         TOTAL CAPITAL PROCESSING COST,
CSALES              PROFIT FROM THE SOLD PRODUCTS,
NETPROFIT           TOTAL NET PROFIT
TAX                 NET VALUE OF THE TAX CREDIT;

    *%%%%%%/
    PARAMETERS
    *BINARY VARIABLES
    YPROCP(PP,TP,R,Q)      BINARY VARIABLE FOR THE PROCESSING OF PLASTIC
AND ES
    /
    1.1.1.1    0
    1.1.1.2    1
    1.1.2.1    0
    1.1.2.2    1
    1.1.3.1    0
    1.1.3.2    1
    1.1.4.1    0
    1.1.4.2    1
    1.1.5.1    0
    1.1.5.2    1
    1.2.1.1    0
    1.2.1.2    1
    1.2.2.1    0
    1.2.2.2    1
    1.2.3.1    0
    1.2.3.2    1
    1.2.4.1    0
    1.2.4.2    1
    1.2.5.1    0
    1.2.5.2    1
    1.3.1.1    0
    1.3.1.2    1
    1.3.2.1    0
    1.3.2.2    1
    1.3.3.1    0
    1.3.3.2    1
    1.3.4.1    0
    1.3.4.2    1
    1.3.5.1    0
    1.3.5.2    1
    1.4.1.1    0
```

1.4.1.2	1
1.4.2.1	0
1.4.2.2	1
1.4.3.1	0
1.4.3.2	1
1.4.4.1	0
1.4.4.2	1
1.4.5.1	0
1.4.5.2	1
1.5.1.1	0
1.5.1.2	1
1.5.2.1	0
1.5.2.2	1
1.5.3.1	0
1.5.3.2	1
1.5.4.1	0
1.5.4.2	1
1.5.5.1	0
1.5.5.2	1
1.6.1.1	0
1.6.1.2	1
1.6.2.1	0
1.6.2.2	1
1.6.3.1	0
1.6.3.2	1
1.6.4.1	0
1.6.4.2	1
1.6.5.1	0
1.6.5.2	1
2.1.1.1	0
2.1.1.2	1
2.1.2.1	0
2.1.2.2	1
2.1.3.1	0
2.1.3.2	1
2.1.4.1	0
2.1.4.2	1
2.1.5.1	0
2.1.5.2	1
2.2.1.1	0
2.2.1.2	1
2.2.2.1	0
2.2.2.2	1
2.2.3.1	0
2.2.3.2	1
2.2.4.1	0
2.2.4.2	1
2.2.5.1	0
2.2.5.2	1
2.3.1.1	0

2.3.1.2	1
2.3.2.1	0
2.3.2.2	1
2.3.3.1	0
2.3.3.2	1
2.3.4.1	0
2.3.4.2	1
2.3.5.1	0
2.3.5.2	1
2.4.1.1	0
2.4.1.2	1
2.4.2.1	1
2.4.2.2	0
2.4.3.1	0
2.4.3.2	1
2.4.4.1	0
2.4.4.2	1
2.4.5.1	0
2.4.5.2	1
2.5.1.1	0
2.5.1.2	1
2.5.2.1	0
2.5.2.2	1
2.5.3.1	0
2.5.3.2	1
2.5.4.1	0
2.5.4.2	1
2.5.5.1	0
2.5.5.2	1
2.6.1.1	0
2.6.1.2	1
2.6.2.1	0
2.6.2.2	1
2.6.3.1	0
2.6.3.2	1
2.6.4.1	0
2.6.4.2	1
2.6.5.1	0
2.6.5.2	1
3.1.1.1	0
3.1.1.2	1
3.1.2.1	0
3.1.2.2	1
3.1.3.1	0
3.1.3.2	1
3.1.4.1	0
3.1.4.2	1
3.1.5.1	0
3.1.5.2	1
3.2.1.1	0

3.2.1.2	1
3.2.2.1	0
3.2.2.2	1
3.2.3.1	0
3.2.3.2	1
3.2.4.1	0
3.2.4.2	1
3.2.5.1	0
3.2.5.2	1
3.3.1.1	0
3.3.1.2	1
3.3.2.1	0
3.3.2.2	1
3.3.3.1	0
3.3.3.2	1
3.3.4.1	0
3.3.4.2	1
3.3.5.1	0
3.3.5.2	1
3.4.1.1	0
3.4.1.2	1
3.4.2.1	0
3.4.2.2	1
3.4.3.1	0
3.4.3.2	1
3.4.4.1	0
3.4.4.2	1
3.4.5.1	0
3.4.5.2	1
3.5.1.1	0
3.5.1.2	1
3.5.2.1	0
3.5.2.2	1
3.5.3.1	0
3.5.3.2	1
3.5.4.1	0
3.5.4.2	1
3.5.5.1	0
3.5.5.2	1
3.6.1.1	0
3.6.1.2	1
3.6.2.1	0
3.6.2.2	1
3.6.3.1	0
3.6.3.2	1
3.6.4.1	0
3.6.4.2	1
3.6.5.1	0
3.6.5.2	1
4.1.1.1	0

4.1.1.2	1
4.1.2.1	0
4.1.2.2	1
4.1.3.1	0
4.1.3.2	1
4.1.4.1	0
4.1.4.2	1
4.1.5.1	0
4.1.5.2	1
4.2.1.1	0
4.2.1.2	1
4.2.2.1	0
4.2.2.2	1
4.2.3.1	0
4.2.3.2	1
4.2.4.1	0
4.2.4.2	1
4.2.5.1	0
4.2.5.2	1
4.3.1.1	0
4.3.1.2	1
4.3.2.1	0
4.3.2.2	1
4.3.3.1	0
4.3.3.2	1
4.3.4.1	0
4.3.4.2	1
4.3.5.1	0
4.3.5.2	1
4.4.1.1	0
4.4.1.2	1
4.4.2.1	0
4.4.2.2	1
4.4.3.1	0
4.4.3.2	1
4.4.4.1	0
4.4.4.2	1
4.4.5.1	0
4.4.5.2	1
4.5.1.1	0
4.5.1.2	1
4.5.2.1	0
4.5.2.2	1
4.5.3.1	0
4.5.3.2	1
4.5.4.1	0
4.5.4.2	1
4.5.5.1	0
4.5.5.2	1
4.6.1.1	0

4.6.1.2	1
4.6.2.1	0
4.6.2.2	1
4.6.3.1	0
4.6.3.2	1
4.6.4.1	0
4.6.4.2	1
4.6.5.1	0
4.6.5.2	1
5.1.1.1	0
5.1.1.2	1
5.1.2.1	0
5.1.2.2	1
5.1.3.1	0
5.1.3.2	1
5.1.4.1	0
5.1.4.2	1
5.1.5.1	0
5.1.5.2	1
5.2.1.1	0
5.2.1.2	1
5.2.2.1	0
5.2.2.2	1
5.2.3.1	0
5.2.3.2	1
5.2.4.1	0
5.2.4.2	1
5.2.5.1	0
5.2.5.2	1
5.3.1.1	0
5.3.1.2	1
5.3.2.1	0
5.3.2.2	1
5.3.3.1	0
5.3.3.2	1
5.3.4.1	0
5.3.4.2	1
5.3.5.1	0
5.3.5.2	1
5.4.1.1	0
5.4.1.2	1
5.4.2.1	0
5.4.2.2	1
5.4.3.1	0
5.4.3.2	1
5.4.4.1	0
5.4.4.2	1
5.4.5.1	0
5.4.5.2	1
5.5.1.1	0

5.5.1.2	1
5.5.2.1	0
5.5.2.2	1
5.5.3.1	0
5.5.3.2	1
5.5.4.1	0
5.5.4.2	1
5.5.5.1	0
5.5.5.2	1
5.6.1.1	0
5.6.1.2	1
5.6.2.1	0
5.6.2.2	1
5.6.3.1	0
5.6.3.2	1
5.6.4.1	0
5.6.4.2	1
5.6.5.1	0
5.6.5.2	1
6.1.1.1	0
6.1.1.2	1
6.1.2.1	0
6.1.2.2	1
6.1.3.1	0
6.1.3.2	1
6.1.4.1	0
6.1.4.2	1
6.1.5.1	0
6.1.5.2	1
6.2.1.1	0
6.2.1.2	1
6.2.2.1	0
6.2.2.2	1
6.2.3.1	0
6.2.3.2	1
6.2.4.1	0
6.2.4.2	1
6.2.5.1	0
6.2.5.2	1
6.3.1.1	0
6.3.1.2	1
6.3.2.1	0
6.3.2.2	1
6.3.3.1	0
6.3.3.2	1
6.3.4.1	0
6.3.4.2	1
6.3.5.1	0
6.3.5.2	1
6.4.1.1	0

6.4.1.2	1
6.4.2.1	0
6.4.2.2	1
6.4.3.1	0
6.4.3.2	1
6.4.4.1	0
6.4.4.2	1
6.4.5.1	0
6.4.5.2	1
6.5.1.1	0
6.5.1.2	1
6.5.2.1	0
6.5.2.2	1
6.5.3.1	0
6.5.3.2	1
6.5.4.1	0
6.5.4.2	1
6.5.5.1	0
6.5.5.2	1
6.6.1.1	0
6.6.1.2	1
6.6.2.1	0
6.6.2.2	1
6.6.3.1	0
6.6.3.2	1
6.6.4.1	0
6.6.4.2	1
6.6.5.1	0
6.6.5.2	1
7.1.1.1	0
7.1.1.2	1
7.1.2.1	0
7.1.2.2	1
7.1.3.1	0
7.1.3.2	1
7.1.4.1	0
7.1.4.2	1
7.1.5.1	0
7.1.5.2	1
7.2.1.1	0
7.2.1.2	1
7.2.2.1	0
7.2.2.2	1
7.2.3.1	0
7.2.3.2	1
7.2.4.1	0
7.2.4.2	1
7.2.5.1	0
7.2.5.2	1
7.3.1.1	0

7.3.1.2	1
7.3.2.1	0
7.3.2.2	1
7.3.3.1	0
7.3.3.2	1
7.3.4.1	0
7.3.4.2	1
7.3.5.1	0
7.3.5.2	1
7.4.1.1	0
7.4.1.2	1
7.4.2.1	0
7.4.2.2	1
7.4.3.1	0
7.4.3.2	1
7.4.4.1	0
7.4.4.2	1
7.4.5.1	0
7.4.5.2	1
7.5.1.1	0
7.5.1.2	1
7.5.2.1	0
7.5.2.2	1
7.5.3.1	0
7.5.3.2	1
7.5.4.1	0
7.5.4.2	1
7.5.5.1	0
7.5.5.2	1
7.6.1.1	0
7.6.1.2	1
7.6.2.1	0
7.6.2.2	1
7.6.3.1	0
7.6.3.2	1
7.6.4.1	0
7.6.4.2	1
7.6.5.1	0
7.6.5.2	1
8.1.1.1	0
8.1.1.2	1
8.1.2.1	0
8.1.2.2	1
8.1.3.1	0
8.1.3.2	1
8.1.4.1	0
8.1.4.2	1
8.1.5.1	0
8.1.5.2	1
8.2.1.1	0

8.2.1.2	1
8.2.2.1	0
8.2.2.2	1
8.2.3.1	0
8.2.3.2	1
8.2.4.1	0
8.2.4.2	1
8.2.5.1	0
8.2.5.2	1
8.3.1.1	0
8.3.1.2	1
8.3.2.1	0
8.3.2.2	1
8.3.3.1	0
8.3.3.2	1
8.3.4.1	0
8.3.4.2	1
8.3.5.1	0
8.3.5.2	1
8.4.1.1	0
8.4.1.2	1
8.4.2.1	0
8.4.2.2	1
8.4.3.1	0
8.4.3.2	1
8.4.4.1	0
8.4.4.2	1
8.4.5.1	0
8.4.5.2	1
8.5.1.1	0
8.5.1.2	1
8.5.2.1	0
8.5.2.2	1
8.5.3.1	0
8.5.3.2	1
8.5.4.1	0
8.5.4.2	1
8.5.5.1	0
8.5.5.2	1
8.6.1.1	0
8.6.1.2	1
8.6.2.1	0
8.6.2.2	1
8.6.3.1	0
8.6.3.2	1
8.6.4.1	0
8.6.4.2	1
8.6.5.1	0
8.6.5.2	1
9.1.1.1	0

9.1.1.2	1
9.1.2.1	0
9.1.2.2	1
9.1.3.1	0
9.1.3.2	1
9.1.4.1	0
9.1.4.2	1
9.1.5.1	0
9.1.5.2	1
9.2.1.1	0
9.2.1.2	1
9.2.2.1	0
9.2.2.2	1
9.2.3.1	0
9.2.3.2	1
9.2.4.1	0
9.2.4.2	1
9.2.5.1	0
9.2.5.2	1
9.3.1.1	0
9.3.1.2	1
9.3.2.1	0
9.3.2.2	1
9.3.3.1	0
9.3.3.2	1
9.3.4.1	0
9.3.4.2	1
9.3.5.1	0
9.3.5.2	1
9.4.1.1	0
9.4.1.2	1
9.4.2.1	0
9.4.2.2	1
9.4.3.1	0
9.4.3.2	1
9.4.4.1	0
9.4.4.2	1
9.4.5.1	0
9.4.5.2	1
9.5.1.1	0
9.5.1.2	1
9.5.2.1	0
9.5.2.2	1
9.5.3.1	0
9.5.3.2	1
9.5.4.1	0
9.5.4.2	1
9.5.5.1	0
9.5.5.2	1
9.6.1.1	0

9.6.1.2	1
9.6.2.1	0
9.6.2.2	1
9.6.3.1	0
9.6.3.2	1
9.6.4.1	0
9.6.4.2	1
9.6.5.1	0
9.6.5.2	1
10.1.1.1	0
10.1.1.2	1
10.1.2.1	0
10.1.2.2	1
10.1.3.1	0
10.1.3.2	1
10.1.4.1	0
10.1.4.2	1
10.1.5.1	0
10.1.5.2	1
10.2.1.1	0
10.2.1.2	1
10.2.2.1	0
10.2.2.2	1
10.2.3.1	0
10.2.3.2	1
10.2.4.1	0
10.2.4.2	1
10.2.5.1	0
10.2.5.2	1
10.3.1.1	0
10.3.1.2	1
10.3.2.1	0
10.3.2.2	1
10.3.3.1	0
10.3.3.2	1
10.3.4.1	0
10.3.4.2	1
10.3.5.1	0
10.3.5.2	1
10.4.1.1	0
10.4.1.2	1
10.4.2.1	0
10.4.2.2	1
10.4.3.1	0
10.4.3.2	1
10.4.4.1	0
10.4.4.2	1
10.4.5.1	0
10.4.5.2	1
10.5.1.1	0

```
    10.5.1.2    1
    10.5.2.1    0
    10.5.2.2    1
    10.5.3.1    0
    10.5.3.2    1
    10.5.4.1    0
    10.5.4.2    1
    10.5.5.1    0
    10.5.5.2    1
    10.6.1.1    0
    10.6.1.2    1
    10.6.2.1    0
    10.6.2.2    1
    10.6.3.1    0
    10.6.3.2    1
    10.6.4.1    0
    10.6.4.2    1
    10.6.5.1    0
    10.6.5.2    1
    /,

    YPROCM(PM,TM,S,Q)        BINARY VARIABLE FOR THE PROCESSING OF METAL
AND ES
    /
    1.1.1.1     1
    1.1.1.2     0
    2.1.1.1     0
    2.1.1.2     1
    3.1.1.1     0
    3.1.1.2     1
    4.1.1.1     0
    4.1.1.2     1
    5.1.1.1     0
    5.1.1.2     1
    6.1.1.1     0
    6.1.1.2     1
    7.1.1.1     0
    7.1.1.2     1
    8.1.1.1     0
    8.1.1.2     1
    9.1.1.1     0
    9.1.1.2     1
    10.1.1.1     0
    10.1.1.2     1
    /,

    YPROCG(PG,TG,T,Q)        BINARY VARIABLE FOR THE PROCESSING OF GLASS
AND ES
    /
    1.1.1.1     0
```

1.1.1.2	1
1.1.2.1	0
1.1.2.2	1
1.1.3.1	1
1.1.3.2	0
2.1.1.1	1
2.1.1.2	0
2.1.2.1	1
2.1.2.2	0
2.1.3.1	0
2.1.3.2	1
3.1.1.1	0
3.1.1.2	1
3.1.2.1	0
3.1.2.2	1
3.1.3.1	0
3.1.3.2	1
4.1.1.1	0
4.1.1.2	1
4.1.2.1	0
4.1.2.2	1
4.1.3.1	0
4.1.3.2	1
5.1.1.1	0
5.1.1.2	1
5.1.2.1	0
5.1.2.2	1
5.1.3.1	0
5.1.3.2	1
6.1.1.1	0
6.1.1.2	1
6.1.2.1	0
6.1.2.2	1
6.1.3.1	0
6.1.3.2	1
7.1.1.1	0
7.1.1.2	1
7.1.2.1	0
7.1.2.2	1
7.1.3.1	0
7.1.3.2	1
8.1.1.1	0
8.1.1.2	1
8.1.2.1	0
8.1.2.2	1
8.1.3.1	0
8.1.3.2	1
9.1.1.1	0
9.1.1.2	1
9.1.2.1	0

```
    9.1.2.2     1
    9.1.3.1     0
    9.1.3.2     1
    10.1.1.1    0
    10.1.1.2    1
    10.1.2.1    0
    10.1.2.2    1
    10.1.3.1    0
    10.1.3.2    1
    /,

    YPROCO(PO,TO,U,Q)    BINARY VARIABLE FOR THE PROCESSING OF ORGANIC
AND ES
    /
    1.1.1.1     1
    1.1.1.2     0
    2.1.1.1     0
    2.1.1.2     1
    3.1.1.1     0
    3.1.1.2     1
    4.1.1.1     0
    4.1.1.2     1
    5.1.1.1     0
    5.1.1.2     1
    6.1.1.1     0
    6.1.1.2     1
    7.1.1.1     0
    7.1.1.2     1
    8.1.1.1     0
    8.1.1.2     1
    9.1.1.1     0
    9.1.1.2     1
    10.1.1.1    0
    10.1.1.2    1
    /,

    YPROCW(PW,TW,W,Q)    BINARY VARIABLE FOR THE PROCESSING OF ORGANIC
AND ES
    /
    1.1.1.1     0
    1.1.1.2     1
    1.2.1.1     0
    1.2.1.2     1
    1.3.1.1     0
    1.3.1.2     1
    1.4.1.1     0
    1.4.1.2     1
    1.5.1.1     0
    1.5.1.2     1
    2.1.1.1     0
```

2.1.1.2	1
2.2.1.1	0
2.2.1.2	1
2.3.1.1	0
2.3.1.2	1
2.4.1.1	1
2.4.1.2	0
2.5.1.1	0
2.5.1.2	1
3.1.1.1	0
3.1.1.2	1
3.2.1.1	0
3.2.1.2	1
3.3.1.1	0
3.3.1.2	1
3.4.1.1	0
3.4.1.2	1
3.5.1.1	0
3.5.1.2	1
4.1.1.1	0
4.1.1.2	1
4.2.1.1	0
4.2.1.2	1
4.3.1.1	0
4.3.1.2	1
4.4.1.1	0
4.4.1.2	1
4.5.1.1	0
4.5.1.2	1
5.1.1.1	0
5.1.1.2	1
5.2.1.1	0
5.2.1.2	1
5.3.1.1	0
5.3.1.2	1
5.4.1.1	0
5.4.1.2	1
5.5.1.1	0
5.5.1.2	1
6.1.1.1	0
6.1.1.2	1
6.2.1.1	0
6.2.1.2	1
6.3.1.1	0
6.3.1.2	1
6.4.1.1	0
6.4.1.2	1
6.5.1.1	0
6.5.1.2	1
7.1.1.1	0

```
7.1.1.2    1
7.2.1.1    0
7.2.1.2    1
7.3.1.1    0
7.3.1.2    1
7.4.1.1    0
7.4.1.2    1
7.5.1.1    0
7.5.1.2    1
8.1.1.1    0
8.1.1.2    1
8.2.1.1    0
8.2.1.2    1
8.3.1.1    0
8.3.1.2    1
8.4.1.1    0
8.4.1.2    1
8.5.1.1    0
8.5.1.2    1
9.1.1.1    0
9.1.1.2    1
9.2.1.1    0
9.2.1.2    1
9.3.1.1    0
9.3.1.2    1
9.4.1.1    0
9.4.1.2    1
9.5.1.1    0
9.5.1.2    1
10.1.1.1   0
10.1.1.2   1
10.2.1.1   0
10.2.1.2   1
10.3.1.1   0
10.3.1.2   1
10.4.1.1   0
10.4.1.2   1
10.5.1.1   0
10.5.1.2   1
/;
*%%%%%/
```

PARAMETERS

```
*F1    TO DECIDE IF PLASTIC IS CONSIDERED,
*F2    TO DECIDE IF METAL IS CONSIDERED,
*F3    TO DECIDE IF GLASS IS CONSIDERED,
*F4    TO DECIDE IF ORGANIC IS CONSIDERED
*F5    TO DECIDE IF NO RECYCLABLE ARE CONSIDERED
LIMIT  THIS IS THE LIMIT FOR THE PERCENTAGE OF THE WASTE CONSUME
```

X VALUE OF THE TAX CREDIT GIVEN FROM THE GOVERNMENT;

EQUATIONS

**CONSTRAINT OF THE SOURCES IN THE CITIES*
CSOURCEP(C,R) CONSTRAINT OF THE SOURCES IN THE CITIES FOR PLASTIC
CSOURCEM(C,S) CONSTRAINT OF THE SOURCES IN THE CITIES FOR METAL
CSOURCEG(C,T) CONSTRAINT OF THE SOURCES IN THE CITIES FOR GLASS
CSOURCEO(C,U) CONSTRAINT OF THE SOURCES IN THE CITIES FOR ORGANIC
CSOURCEW(C,W) CONSTRAINT OF THE SOURCES IN THE CITIES FOR ORGANIC

**DISTRIBUTION FROM CITIES TO PROCESSING PLANTS*
DCPP(C,R) DISTRIBUTION FROM CITIES TO PROCESSING PLANTS FOR PLASTIC
DCPM(C,S) DISTRIBUTION FROM CITIES TO PROCESSING PLANTS FOR METAL
DCPG(C,T) DISTRIBUTION FROM CITIES TO PROCESSING PLANTS FOR GLASS
DCPO(C,U) DISTRIBUTION FROM CITIES TO PROCESSING PLANTS FOR ORGANIC
DCPW(C,W) DISTRIBUTION FROM CITIES TO PROCESSING PLANTS FOR ORGANIC

**DISTRIBUTION OF WASTE INTO THE PLANT TO DIFFERENT TECHNOLOGIES*
DPTCHP(PP,R) DISTRIBUTION OF WASTE INT THE PLANT TO DIFFERENT TECHNOLOGIES FOR PLASTIC
DPTCHM(PM,S) DISTRIBUTION OF WASTE INT THE PLANT TO DIFFERENT TECHNOLOGIES FOR METAL
DPTCHG(PG,T) DISTRIBUTION OF WASTE INT THE PLANT TO DIFFERENT TECHNOLOGIES FOR GLASS
DPTCHO(PO,U) DISTRIBUTION OF WASTE INT THE PLANT TO DIFFERENT TECHNOLOGIES FOR ORGANIC
DPTCHW(PW,W) DISTRIBUTION OF WASTE INT THE PLANT TO DIFFERENT TECHNOLOGIES FOR ORGANIC

**TO ENSURE THAT FTECH IS GREATER THAN ZERO ONLY IF ALPHA IS GREATER THAN ZERO*
BMP1(PP,TP,R)
BMM1(PM,TM,S)
BMG1(PG,TG,T)
BMO1(PO,TO,U)
BMW1(PW,TW,W)

**BALANCES MATERIAL IN PLANTS*
BMP(PP,TP,R) BALANCES MATERIAL IN PLANTS FOR PLASTIC
BMM(PM,TM,S) BALANCES MATERIAL IN PLANTS FOR METAL
BMG(PG,TG,T) BALANCES MATERIAL IN PLANTS FOR GLASS
BMO(PO,TO,U) BALANCES MATERIAL IN PLANTS FOR ORGANIC
BMW(PW,TW,W) BALANCES MATERIAL IN PLANTS FOR ORGANIC

**BALANCES IN SPLITTERS AFTER THE PROCESSING PLANTS*

BSPP(PP,TP,R) BALANCES MATERIAL IN SPLITTERS AFTER PROCESSING PLANTS FOR PLASTIC
BSPM(PM,TM,S) BALANCES MATERIAL IN SPLITTERS AFTER PROCESSING PLANTS FOR METAL
BSPG(PG,TG,T) BALANCES MATERIAL IN SPLITTERS AFTER PROCESSING PLANTS FOR GLASS
BSPO(PO,TO,U) BALANCES MATERIAL IN SPLITTERS AFTER PROCESSING PLANTS FOR ORGANIC
BSPW(PW,TW,W) BALANCES MATERIAL IN SPLITTERS AFTER PROCESSING PLANTS FOR ORGANIC

**BALANCES IN MIXERS BEFORE CITIES*
BMBCP(TP,C,R) BALANCES IN MIXERS BEFORE CITIES FOR PLASTIC
BMBCM(TM,C,S) BALANCES IN MIXERS BEFORE CITIES FOR METAL
BMBCG(TG,C,T) BALANCES IN MIXERS BEFORE CITIES FOR GLASS
BMBCO(TO,C,U) BALANCES IN MIXERS BEFORE CITIES FOR ORGANIC
BMBCW(TW,C,W) BALANCES IN MIXERS BEFORE CITIES FOR ORGANIC

**CONSTRAINTS FOR THE DEMAND OF PRODUCTS IN CITIES*
DEMP(TP,C,R) CONSTRAINTS FOR THE DEMAND OF PRODUCTS IN CITIES FOR PLASTIC
DEMM(TM,C,S) CONSTRAINTS FOR THE DEMAND OF PRODUCTS IN CITIES FOR METAL
DEMG(TG,C,T) CONSTRAINTS FOR THE DEMAND OF PRODUCTS IN CITIES FOR GLASS
DEMO(TO,C,U) CONSTRAINTS FOR THE DEMAND OF PRODUCTS IN CITIES FOR ORGANIC
DEMW(TW,C,W) CONSTRAINTS FOR THE DEMAND OF PRODUCTS IN CITIES FOR ORGANIC

**SEPARATION COST OF WASTES*
CSP SEPARATION COST OF PLASTIC
CSM SEPARATION COST OF METAL
CSG SEPARATION COST OF GLASS
CSO SEPARATION COST OF ORGANIC
CSW SEPARATION COST OF ORGANIC

**DISPOSITION COST OF WASTES*
CDP DEMAND WITH BIDDING NEGATIVE (LANDFILL)

CDPsup SUPPLY WITH BIDDING POSITIVE (PAGAN PARA QUE SE LLEVEN ESE RESIDUO)

**TOTAL SEPARATION COST*
TSC TOTAL SEPARATION COST

**TOTAL DISPOSITION COST*
TDS TOTAL DISPOSITION COST

TOTAL TRANSPORTATION COST FROM CITIES TO PROCESSING PLANTS
TTCCP TOTAL TRANSPORTATION COST FROM CITIES TO PROCESSING PLANTS

TOTAL TRANSPORTATION COST FROM PROCESSING PLANTS TO CITIES
TTCPC TOTAL TRANSPORTATION COST FROM PROCESSING PLANTS TO CITIES

TCL transport from cities to landfills ar each city

PROCESSING COST INTO PROCESSING FACILITIES
PCOP PROCESSING COST INTO PROCESSING FACILITIES (OPERATIONAL COST)

CAPITAL COST FOR EACH WASTE
CAPP(PP,TP,R) CAPITAL COST FOR PLASTIC
CAPM(PM,TM,S) CAPITAL COST FOR METAL
CAPG(PG,TG,T) CAPITAL COST FOR GLASS
CAPO(PO,TO,U) CAPITAL COST FOR ORGANIC
CAPW(PW,TW,W) CAPITAL COST FOR ORGANIC

TOTAL CAPITAL COST FOR PROCESSING FACILITIES
TCAP TOTAL CAPITAL COST FOR PROCESSING FACILITIES

ECONOMIC OBJECTIVE FUNCTION
ECOOBJ NET PROFIT IS THE ECONOMIC OBJECTIVE FUNCTION

ASIGNATION FO VALUES OF THE VARIABLES TO CAPITAL COST PLASTIC
A1P(PP,TP,R) ASSIGNATION OF THE FIXED VALUE IN CAPITAL COST PLASTIC
A2P(PP,TP,R) ASSIGNATION OF THE VARIABLE COST IN CAPITAL COST PLASTIC
A3P(PP,TP,R) ASSIGNATION OF PROCESSED PLASTIC FOR ECONOMIES OF SCALE
A4P(PP,TP,R,Q) LOWER LIMIT FOR PROCESSED PLASTIC AT THE INTERVAL Q FOR PLASTIC
A5P(PP,TP,R,Q) UPPER LIMIT FOR PROCESSED PLASTIC AT THE INTERVAL Q FOR PLASTIC
* %%%%%%%%%%%
A6P(PP,TP,R) CONSTRAINT FOR ONLY ONE SELECTION OF INTERVAL Q FOR PLASTIC
* %%%%%%%%%%%

ASSIGNATION FOR VALUES OF THE VARIABLES TO CAPITAL COST METAL
A1M(PM,TM,S) ASSIGNATION OF THE FIXED VALUE IN CAPITAL COST METAL
A2M(PM,TM,S) ASSIGNATION OF THE VARIABLE COST IN CAPITAL COST METAL
A3M(PM,TM,S) ASSIGNATION OF PROCESSED METAL FOR ECONOMIES OF SCALE
A4M(PM,TM,S,Q) LOWER LIMIT FOR PROCESSED METAL AT THE INTERVAL Q FOR METAL
A5M(PM,TM,S,Q) UPPER LIMIT FOR PROCESSED METAL AT THE INTERVAL Q FOR METAL
* %%%%%%%%%%%

A6M(PM,TM,S) CONSTRAINT FOR ONLY ONE SELECTION OF INTERVAL Q FOR METAL
* %%%%%%%%%%%%

ASSIGNATION FOR VALUES OF THE VARIABLES TO CAPITAL COST GLASS
A1G(PG,TG,T) ASSIGNATION OF THE FIXED VALUE IN CAPITAL COST GLASS
A2G(PG,TG,T) DESIGNATION OF THE VARIABLE COST IN CAPITAL COST GLASS
A3G(PG,TG,T) ASSIGNATION OF PROCESSED GLASS FOR ECONOMIES OF SCALE
A4G(PG,TG,T,Q) LOWER LIMIT FOR PROCESSED GLASS AT THE INTERVAL Q FOR GLASS
A5G(PG,TG,T,Q) UPPER LIMIT FOR PROCESSED GLASS AT THE INTERVAL Q FOR GLASS
* %%%%%%%%%%%%
A6G(PG,TG,T) CONSTRAINT FOR ONLY ONE SELECTION OF INTERVAL Q FOR GLASS
* %%%%%%%%%%%%

ASSIGNATION FOR VALUES OF THE VARIABLES TO CAPITAL COST ORGANIC
A1O(PO,TO,U) ASSIGNATION OF THE FIXED VALUE IN CAPITAL COST ORGANIC
A2O(PO,TO,U) ASSIGNATION OF THE VARIABLE COST IN CAPITAL COST ORGANIC
A3O(PO,TO,U) ASSIGNATION OF PROCESSED ORGANIC FOR ECONOMIES OF SCALE
A4O(PO,TO,U,Q) LOWER LIMIT FOR PROCESSED ORGANIC AT THE INTERVAL Q FOR ORGANIC
A5O(PO,TO,U,Q) UPPER LIMIT FOR PROCESSED ORGANIC AT THE INTERVAL Q FOR ORGANIC
* %%%%%%%%%%%%
A6O(PO,TO,U) CONSTRAINT FOR ONLY ONE SELECTION OF INTERVAL Q FOR ORGANIC
* %%%%%%%%%%%%

ASSIGNATION FOR VALUES OF THE VARIABLES TO CAPITAL COST NO RECYCLABLE
A1W(PW,TW,W) ASSIGNATION OF THE FIXED VALUE IN CAPITAL COST ORGANIC
A2W(PW,TW,W) ASSIGNATION OF THE VARIABLE COST IN CAPITAL COST ORGANIC
A3W(PW,TW,W) ASSIGNATION OF PROCESSED ORGANIC FOR ECONOMIES OF SCALE
A4W(PW,TW,W,Q) LOWER LIMIT FOR PROCESSED ORGANIC AT THE INTERVAL Q FOR ORGANIC
A5W(PW,TW,W,Q) UPPER LIMIT FOR PROCESSED ORGANIC AT THE INTERVAL Q FOR ORGANIC
* %%%%%%%%%%%%

```
A6W(PW,TW,W)      CONSTRAINT FOR ONLY ONE SELECTION OF INTERVAL Q FOR
                  ORGANIC
   * %%%%%%%%%%%%

   *PROFIT
CSALESP           PROFIT FOR SELLING OF PRODUCTS

*TAXCREDIT
TAXCREDIT         TAX CREDIT TO CONSUME WASTE

*ECONOMIC OBJECTIVE FUNCTION
ECOOBJ            ECONOMIC OBJECTIVE FUNCTION

SUPP(C,R)         TO DEFINE NEW VARIABLE FOR DUAL VALUES
   SUPP1(C,S)
   SUPP2(C,T)
   SUPP3(C,U)
   SUPP4(C,W)

   DEMSa(L,R)
   DEMSb(L,R)
   DEMS1a(L,S)
   DEMS1b(L,S)
   DEMS2a(L,T)
   DEMS2b(L,T)
   DEMS3a(L,U)
   DEMS3b(L,U)
   DEMS4a(L,W)
   DEMS4b(L,W)

   VARS(C,R)
   VARS1(C,S)
   VARS2(C,T)
   VARS3(C,U)
   VARS4(C,W)

   FV(C,R)
   FV1(C,S)
   FV2(C,T)
   FV3(C,U)
   FV4(C,W)

   MC(C,R)
   MC1(C,S)
   MC2(C,T)
   MC3(C,U)
   MC4(C,W)
   SUMC

   *WCONSP
```

```
*WCONSM
*WCONSG
*WCONSO
*WCONSW

*PARETO
*CONS

*CONST

;

*CONSTRAINT OF THE SOURCES IN THE CITIES
CSOURCEP(C,R)..  FP(C,R)=L=FTP(C,R)*ALPHAPSEP(C,R);
CSOURCEM(C,S)..  FM(C,S)=L=FTM(C,S)*ALPHAMSEP(C,S);
CSOURCEG(C,T)..  FG(C,T)=L=FTG(C,T)*ALPHAGSEP(C,T);
CSOURCEO(C,U)..  FO(C,U)=L=FTO(C,U)*ALPHAOSEP(C,U);
CSOURCEW(C,W)..FW(C,W)=L=FTW(C,W)*ALPHAWSEP(C,W);

*DISTRIBUTION FROM CITIES TO PROCESSING PLANTS
DCPP(C,R)..    FP(C,R)=E=SUM((PP),FPPLANT(C,PP,R));
DCPM(C,S)..    FM(C,S)=E=SUM((PM),FMPLANT(C,PM,S));
DCPG(C,T)..    FG(C,T)=E=SUM((PG),FGPLANT(C,PG,T));
DCPO(C,U)..    FO(C,U)=E=SUM((PO),FOPLANT(C,PO,U));
DCPW(C,W)..    FW(C,W)=E=SUM((PW),FWPLANT(C,PW,W));

*DISTRIBUTION OF WASTE INTO THE PLANT TO DIFFERENT TECHNOLOGIES
DPTCHP(PP,R)..    SUM((C),FPPLANT(C,PP,R))=E=SUM((TP),FTECHP(PP,TP,R));
DPTCHM(PM,S)..    SUM((C),FMPLANT(C,PM,S))=E=SUM((TM),FTECHM(PM,TM
,S));
DPTCHG(PG,T)..    SUM((C),FGPLANT(C,PG,T))=E=SUM((TG),FTECHG(PG,TG,T));
DPTCHO(PO,U)..    SUM((C),FOPLANT(C,PO,U))=E=SUM((TO),FTECHO(PO,TO,U));
DPTCHW(PW,W)..  SUM((C),FWPLANT(C,PW,W))=E=SUM((TW),FTECHW(PW,TW
,W));

*TO ENSURE THAT FTECH IS GREATER THAN ZERO ONLY IF ALPHA IS GREATER
THAN ZERO
BMP1(PP,TP,R)..     ALPHAP(TP,R)*5E20=G=FTECHP(PP,TP,R);
BMM1(PM,TM,S)..  ALPHAM(TM,S)*5E20=G=FTECHM(PM,TM,S);
BMG1(PG,TG,T)..    ALPHAG(TG,T)*5E20=G=FTECHG(PG,TG,T);
BMO1(PO,TO,U)..   ALPHAO(TO,U)*5E20=G=FTECHO(PO,TO,U);
BMW1(PW,TW,W)..ALPHAW(TW,W)*5E20=G=FTECHW(PW,TW,W);


*BALANCES MATERIAL IN PLANTS
BMP(PP,TP,R)..      PRODP(PP,TP,R)=E=ALPHAP(TP,R)*FTECHP(PP,TP,R);
BMM(PM,TM,S)..    PRODM(PM,TM,S)=E=ALPHAM(TM,S)*FTECHM(PM,TM,S);
BMG(PG,TG,T)..      PRODG(PG,TG,T)=E=ALPHAG(TG,T)*FTECHG(PG,TG,T);
BMO(PO,TO,U)..     PRODO(PO,TO,U)=E=ALPHAO(TO,U)*FTECHO(PO,TO,U);
```

```
    BMW(PW,TW,W)..          PRODW(PW,TW,W)=E=ALPHAW(TW,W)*FTECHW(PW
,TW,W);

    *BALANCES IN SPLITTERS AFTER THE PROCESSING PLANTS
    BSPP(PP,TP,R).. PRODP(PP,TP,R)=E=SUM((C),PRODCITYP(PP,TP,C,R));
    BSPM(PM,TM,S).. PRODM(PM,TM,S)=E=SUM((C),PRODCITYM(PM,TM,C,S));
BSPG(PG,TG,T)..     PRODG(PG,TG,T)=E=SUM((C),PRODCITYG(PG,TG,C,T));
BSPO(PO,TO,U)..     PRODO(PO,TO,U)=E=SUM((C),PRODCITYO(PO,TO,C,U));
BSPW(PW,TW,W)..     PRODW(PW,TW,W)=E=SUM((C),PRODCITYW(PW,TW,C,W));

*BALANCES IN MIXERS BEFORE CITIES
BMBCP(TP,C,R)..     SUM((PP),PRODCITYP(PP,TP,C,R))=E=PTCITYP(TP,C,R);
BMBCM(TM,C,S)..     SUM((PM),PRODCITYM(PM,TM,C,S))=E=PTCITYM(TM,C,S);
BMBCG(TG,C,T)..     SUM((PG),PRODCITYG(PG,TG,C,T))=E=PTCITYG(TG,C,T);
BMBCO(TO,C,U)..     SUM((PO),PRODCITYO(PO,TO,C,U))=E=PTCITYO(TO,C,U);
BMBCW(TW,C,W)..     SUM((PW),PRODCITYW(PW,TW,C,W))=E=PTCITYW(TW,C,W);

*CONSTRAINTS FOR THE DEMAND OF PRODUCTS IN CITIES
DEMP(TP,C,R)..      PTCITYP(TP,C,R)=L=PDEMANDP(TP,C,R);
DEMM(TM,C,S)..      PTCITYM(TM,C,S)=L=PDEMANDM(TM,C,S);
DEMG(TG,C,T)..      PTCITYG(TG,C,T)=L=PDEMANDG(TG,C,T);
DEMO(TO,C,U)..      PTCITYO(TO,C,U)=L=PDEMANDO(TO,C,U);
DEMW(TW,C,W)..      PTCITYW(TW,C,W)=L=PDEMANDW(TW,C,W);

*SEPARATION COST OF WASTES
CSP..       COSTSEPP=E=SUM((C,R),CSEPP(C,R)*FP(C,R));
CSM..       COSTSEPM=E=SUM((C,S),CSEPM(C,S)*FM(C,S));
CSG..       COSTSEPG=E=SUM((C,T),CSEPG(C,T)*FG(C,T));
CSO..       COSTSEPO=E=SUM((C,U),CSEPO(C,U)*FO(C,U));
CSW..       COSTSEPW=E=SUM((C,W),CSEPW(C,W)*FW(C,W));

*DISPOSITION COST OF WASTES
*CDP..      COSTDISP=E=SUM((C,R),CNSEP(C,R)*(FTP(C,R)-FP(C,R)))+SUM((C,S),CNS
                      EM(C,S)*(FTM(C,S)-FM(C,S)))+
*           SUM((C,T),CNSEG(C,T)*(FTG(C,T)-FG(C,T)))+SUM((C,U),CNSEO(C,U)*(FTO(
                      C,U)-FO(C,U)))+SUM((C,W),CNSEW(C,W)*(FTW(C,W)-FW(C,W)));


*DEMAND WITH NEGATIVE BIDDING (LANDFILL)
CDP..       COSTDISP=E=SUM((L,R),DNSEP(L,R)*(DEMSP(L,R)))+SUM((L,S),DNSEM(
                      L,S)*(DEMSM(L,S)))+
          SUM((L,T),DNSEG(L,T)*(DEMSG(L,T)))+SUM((L,U),DNSEO(L,U)*(DEMSO(L,
                      U)))+SUM((L,W),DNSEW(L,W)*(DEMSW(L,W)));

*SUPPLY WITH NEGATIVE BIDDING (PAY TO GET WASTE COLLECTED)
*CDPsup..       COSTDISPSUP=E=SUM((C,R),CNSEP(C,R)*(FTP(C,R)-FTPV(C,R)-FP(C,R
                      )))+SUM((C,S),CNSEM(C,S)*(DEMSM(C,S)))+
*         SUM((C,T),CNSEG(C,T)*(FTG(C,T)-FTGV(C,T)-FG(C,T)))+SUM((C,U),CNSEO(C
                      ,U)*(FTO(C,U)-FTOV(C,U)-FO(C,U)))+SUM((C,W),CNSEW(C,W)*(F
                      TW(C,W)-FTWV(C,W)-FW(C,W)));
```

```
*SUPP(C,R)..      SUM(L,SUP(C,L,R))=E=FTPV(C,R);
*SUPP1(C,S)..     SUM(L,SUP1(C,L,S))=E=FTMV(C,S);
*SUPP2(C,T)..     SUM(L,SUP2(C,L,T))=E=FTGV(C,T);
*SUPP3(C,U)..     SUM(L,SUP3(C,L,U))=E=FTOV(C,U);
*SUPP4(C,W)..     SUM(L,SUP4(C,L,W))=E=FTWV(C,W);
    *SUPPLY WITH NEGATIVE BIDDING (PAY TO GET WASTE COLLECTED)
    CDPsup..      COSTDISPSUP=E=SUM((C,R),CNSEP(C,R)*(SUM(L,SUP(C,L,R))))+S
UM((C,S),CNSEM(C,S)*(SUM(L,SUP1(C,L,S))))+
              SUM((C,T),CNSEG(C,T)*(SUM(L,SUP2(C,L,T))))+SUM((C,U),CNSEO(C,U)*
(SUM(L,SUP3(C,L,U))))+SUM((C,W),CNSEW(C,W)*(SUM(L,SUP4(C,L,W))));

    *-(COSTDISPSUP)  +(COSTDISP)

    *ECOOBJ..     NETPROFIT=E=CSALES-COSTDISPSUP+COSTDISP-TOTALCOSTSEP-
OPERACOST-TRANSCOSTCP-TRANSCOSTPC-TRANSCOSTCL;

    *CDPsup..     COSTDISPSUP=E=2*(SUM((C,R),CNSEP(C,R)*(FTPV(C,R)))+SUM((C
,S),CNSEM(C,S)*(DEMSM(C,S)))+
    *             SUM((C,T),CNSEG(C,T)*(FTGV(C,T)))+SUM((C,U),CNSEO(C,U)*(FTOV(C,U
)))+SUM((C,W),CNSEW(C,W)*(FTWV(C,W))));

    *TOTAL DISPOSITION COST
    TDS..         TOTALCOSTDIS=E=COSTDISP;

    *TOTAL SEPARATION COST
    TSC..         TOTALCOSTSEP=E=COSTSEPP+COSTSEPM+COSTSEPG+COSTSEPO+
COSTSEPW;

    *TOTAL TRANSPORTATION COST FROM CITIES TO PROCESSING PLANTS
    TTCCP..       TRANSCOSTCP=E=SUM((C,PP,R),CTRANSP(R)*DPP(C,PP)*FPPLAN
T(C,PP,R))
              +SUM((C,PM,S),CTRANSM(S)*DPM(C,PM)*FMPLANT(C,PM,S))+SUM((C
,PG,T),CTRANSG(T)*DPG(C,PG)*FGPLANT(C,PG,T))
              + SUM((C,PO,U),CTRANSO(U)*DPO(C,PO)*FOPLANT(C,PO,U))+ SUM((C
,PW,W),CTRANSW(W)*DPW(C,PW)*FWPLANT(C,PW,W));

    *added for including transport to landfills
    *TTCCP..      TRANSCOSTCP=E=SUM((C,PP,R),CTRANSP(R)*DPP(C,PP)*FPPLAN
T(C,PP,R))
    *             +SUM((C,PM,S),CTRANSM(S)*DPM(C,PM)*FMPLANT(C,PM,S))+SUM((C,
PG,T),CTRANSG(T)*DPG(C,PG)*FGPLANT(C,PG,T))
    *             + SUM((C,PO,U),CTRANSO(U)*DPO(C,PO)*FOPLANT(C,PO,U))+ SUM((C
,PW,W),CTRANSW(W)*DPW(C,PW)*FWPLANT(C,PW,W))+SUM((C,PP,R),CTRANSP(R)*
DPP(C,PP)*(SUP(C,R)));

    *TOTAL TRANSPORTATION COST FROM PROCESSING PLANTS TO CITIES
    TTCPC..       TRANSCOSTPC=E=SUM((PP,TP,C,R),CTRANSPP(TP,R)*DPP(C,PP)*P
RODCITYP(PP,TP,C,R))
```

```
            +SUM((PM,TM,C,S),CTRANSPM(TM,S)*DPM(C,PM)*PRODCITYM(PM,
TM,C,S))+SUM((PG,TG,C,T),CTRANSPG(TG,T)*DPG(C,PG)*PRODCITYG(PG,TG,C,T))
            + SUM((PO,TO,C,U),CTRANSPO(TO,U)*DPO(C,PO)*PRODCITYO(PO,TO,C
,U))+ SUM((PW,TW,C,W),CTRANSPW(TW,W)*DPW(C,PW)*PRODCITYW(PW,TW,C,W));

    *TOTAL TRANSPORTATION COST FROM CITIES TO LANDFILLS AT EACH CITY
    TCL..      TRANSCOSTCL=E=SUM((C,L,R),CTRANSPL(R)*DCL(C,L)*SUP(C,L,R))
            +SUM((C,L,S),CTRANSML(S)*DCL(C,L)*SUP1(C,L,S))+SUM((C,L,T),CTRA
NSGL(T)*DCL(C,L)*SUP2(C,L,T))
            + SUM((C,L,U),CTRANSOL(U)*DCL(C,L)*SUP3(C,L,U))+ SUM((C,L,W),CT
RANSWL(W)*DCL(C,L)*SUP4(C,L,W));

    *TCL..     TRANSCOSTCL=E=SUM((PP,TP,C,R),CTRANSP(R)*DCL(C)*FTPV(C,R))
    *          +SUM((PM,TM,C,S),CTRANSM(S)*DCL(C)*DEMSM(C,S))+SUM((PG,TG,C,
T),CTRANSG(T)*DCL(C)*FTGV(C,T))
    *          + SUM((PO,TO,C,U),CTRANSO(U)*DCL(C)*FTOV(C,U))+ SUM((PW,TW,C,
W),CTRANSW(W)*DCL(C)*FTWV(C,W));


    *PROCESSING COST INTO PROCESSING FACILITIES
    PCOP..      OPERACOST=E=SUM((PP,TP,R),CPROCP(TP,R)*FTECHP(PP,TP,R))+
            SUM((PM,TM,S),CPROCM(TM,S)*FTECHM(PM,TM,S))+
            SUM((PG,TG,T),CPROCG(TG,T)*FTECHG(PG,TG,T))+
            SUM((PO,TO,U),CPROCO(TO,U)*FTECHO(PO,TO,U))+
            SUM((PW,TW,W),CPROCW(TW,W)*FTECHW(PW,TW,W));

    *CAPITAL COST FOR EACH WASTE
    CAPP(PP,TP,R)..     CAPCOSTP(PP,TP,R)=E=APROCPP(PP,TP,R)+BPROCPP(PP,
TP,R);
    CAPM(PM,TM,S)..     CAPCOSTM(PM,TM,S)=E=APROCPM(PM,TM,S)+BPROCP
M(PM,TM,S);
    CAPG(PG,TG,T)..     CAPCOSTG(PG,TG,T)=E=APROCPG(PG,TG,T)+BPROCPG(
PG,TG,T);
    CAPO(PO,TO,U)..     CAPCOSTO(PO,TO,U)=E=APROCPO(PO,TO,U)+BPROCPO
(PO,TO,U);
    CAPW(PW,TW,W)..     CAPCOSTW(PW,TW,W)=E=APROCPW(PW,TW,W)+BPR
OCPW(PW,TW,W);

    *TOTAL CAPITAL COST FOR PROCESSING FACILITIES
    TCAP..         CAPITALCOST=E=0.1*(SUM((PP,TP,R),CAPCOSTP(PP,TP,R))
              +SUM((PM,TM,S),CAPCOSTM(PM,TM,S))
              +SUM((PG,TG,T),CAPCOSTG(PG,TG,T))
              +SUM((PO,TO,U),CAPCOSTO(PO,TO,U))
              +SUM((PW,TW,W),CAPCOSTW(PW,TW,W)));

    *ASIGNATION FO VALUES OF THE VARIABLES TO CAPITAL COST PLASTIC
    A1P(PP,TP,R)..    APROCPP(PP,TP,R)=E=SUM((Q),ACAPP(PP,TP,R,Q)*YPROCP(P
P,TP,R,Q));
    A2P(PP,TP,R)..    BPROCPP(PP,TP,R)=E=SUM((Q),BCAPP(PP,TP,R,Q)*DFTECHP(P
P,TP,R,Q));
```

A3P(PP,TP,R).. FTECHP(PP,TP,R)=E=**SUM**((Q),DFTECHP(PP,TP,R,Q));
A4P(PP,TP,R,Q).. LOFTECHP(PP,TP,R,Q)*YPROCP(PP,TP,R,Q)=L=DFTECHP(PP,TP,R,Q);
A5P(PP,TP,R,Q).. UPFTECHP(PP,TP,R,Q)*YPROCP(PP,TP,R,Q)=G=DFTECHP(PP,TP,R,Q);
* %%%%%%%%%%%%
A6P(PP,TP,R).. **SUM**((Q),YPROCP(PP,TP,R,Q))=E=1;
* %%%%%%%%%%%%

**ASIGNATION FO VALUES OF THE VARIABLES TO CAPITAL COST METAL*
A1M(PM,TM,S).. APROCPM(PM,TM,S)=E=**SUM**((Q),ACAPM(PM,TM,S,Q)*YPROCM(PM,TM,S,Q));
A2M(PM,TM,S).. BPROCPM(PM,TM,S)=E=**SUM**((Q),BCAPM(PM,TM,S,Q)*DFTECHM(PM,TM,S,Q));
A3M(PM,TM,S).. FTECHM(PM,TM,S)=E=**SUM**((Q),DFTECHM(PM,TM,S,Q));
A4M(PM,TM,S,Q).. LOFTECHM(PM,TM,S,Q)*YPROCM(PM,TM,S,Q)=L=DFTECHM(PM,TM,S,Q);
A5M(PM,TM,S,Q).. UPFTECHM(PM,TM,S,Q)*YPROCM(PM,TM,S,Q)=G=DFTECHM(PM,TM,S,Q);
* %%%%%%%%%%%%
A6M(PM,TM,S).. **SUM**((Q),YPROCM(PM,TM,S,Q))=E=1;
* %%%%%%%%%%%%

**ASIGNATION FO VALUES OF THE VARIABLES TO CAPITAL COST GLASS*
A1G(PG,TG,T).. APROCPG(PG,TG,T)=E=**SUM**((Q),ACAPG(PG,TG,T,Q)*YPROCG(PG,TG,T,Q));
A2G(PG,TG,T).. BPROCPG(PG,TG,T)=E=**SUM**((Q),BCAPG(PG,TG,T,Q)*DFTECHG(PG,TG,T,Q));
A3G(PG,TG,T).. FTECHG(PG,TG,T)=E=**SUM**((Q),DFTECHG(PG,TG,T,Q));
A4G(PG,TG,T,Q).. LOFTECHG(PG,TG,T,Q)*YPROCG(PG,TG,T,Q)=L=DFTECHG(PG,TG,T,Q);
A5G(PG,TG,T,Q).. UPFTECHG(PG,TG,T,Q)*YPROCG(PG,TG,T,Q)=G=DFTECHG(PG,TG,T,Q);
* %%%%%%%%%%%%
A6G(PG,TG,T).. **SUM**((Q),YPROCG(PG,TG,T,Q))=E=1;
* %%%%%%%%%%%%

**ASIGNATION FO VALUES OF THE VARIABLES TO CAPITAL COST ORGANIC*
A1O(PO,TO,U).. APROCPO(PO,TO,U)=E=**SUM**((Q),ACAPO(PO,TO,U,Q)*YPROCO(PO,TO,U,Q));
A2O(PO,TO,U).. BPROCPO(PO,TO,U)=E=**SUM**((Q),BCAPO(PO,TO,U,Q)*DFTECHO(PO,TO,U,Q));
A3O(PO,TO,U).. FTECHO(PO,TO,U)=E=**SUM**((Q),DFTECHO(PO,TO,U,Q));
A4O(PO,TO,U,Q).. LOFTECHO(PO,TO,U,Q)*YPROCO(PO,TO,U,Q)=L=DFTECHO(PO,TO,U,Q);
A5O(PO,TO,U,Q).. UPFTECHO(PO,TO,U,Q)*YPROCO(PO,TO,U,Q)=G=DFTECHO(PO,TO,U,Q);
* %%%%%%%%%%%%
A6O(PO,TO,U).. **SUM**((Q),YPROCO(PO,TO,U,Q))=E=1;
* %%%%%%%%%%%%

```
    *ASIGNATION FO VALUES OF THE VARIABLES TO CAPITAL COST ORGANIC
A1W(PW,TW,W)..     APROCPW(PW,TW,W)=E=SUM((Q),ACAPW(PW,TW,W,Q)*YPRO
                   CW(PW,TW,W,Q));
A2W(PW,TW,W)..     BPROCPW(PW,TW,W)=E=SUM((Q),BCAPW(PW,TW,W,Q)*DFTE
                   CHW(PW,TW,W,Q));
A3W(PW,TW,W)..     FTECHW(PW,TW,W)=E=SUM((Q),DFTECHW(PW,TW,W,Q));
A4W(PW,TW,W,Q)..   LOFTECHW(PW,TW,W,Q)*YPROCW(PW,TW,W,Q)=L=DFTECHW
                   (PW,TW,W,Q);
A5W(PW,TW,W,Q)..   UPFTECHW(PW,TW,W,Q)*YPROCW(PW,TW,W,Q)=G=DFTECHW
                   (PW,TW,W,Q);
* %%%%%%%%%%%%
A6W(PW,TW,W)..     SUM((Q),YPROCW(PW,TW,W,Q))=E=1;
* %%%%%%%%%%%%
*PTCITYP(TP,C,R)
*PROFIT
*CSALESP..         CSALES=E=SUM((PP,TP,C,R),CPP(TP,C)*PRODCITYP(PP,TP,C,R))+
*CSALESP..         CSALES=E=SUM((TP,C,R),CPP(TP,C)*PTCITYP(TP,C,R))+
*                  SUM((PM,TM,C,S),CPM(TM,C)*PRODCITYM(PM,TM,C,S))+
*                  SUM((PG,TG,C,T),CPG(TG,C)*PRODCITYG(PG,TG,C,T))+
*                  SUM((PO,TO,C,U),CPO(TO,C)*PRODCITYO(PO,TO,C,U))+
*                  SUM((PW,TW,C,W),CPW(TW,C)*PRODCITYW(PW,TW,C,W));


CSALESP..          CSALES=E=SUM((TP,C,R),CPP(TP,C)*PTCITYP(TP,C,R))+
                   SUM((TM,C,S),CPM(TM,C)*PTCITYM(TM,C,S))+
                   SUM((TG,C,T),CPG(TG,C)*PTCITYG(TG,C,T))+
                   SUM((TO,C,U),CPO(TO,C)*PTCITYO(TO,C,U))+
                   SUM((TW,C,W),CPW(TW,C)*PTCITYW(TW,C,W));

TAXCREDIT..        TAX=E=X*(SUM((C,R),FP(C,R))+SUM((C,S),FM(C,S))+SUM((C,T),F
                   G(C,T))+SUM((C,U),FO(C,U))+SUM((C,W),FW(C,W)));

    ****************************************************ECONOMIC OBJECTIVE FUNCTION
*%ECOOBJ..         NETPROFIT=E=CSALES-TOTALCOSTSEP-TOTALCOSTDIS-
                   CAPITALCOST-OPERACOST-TRANSCOSTCP-
                   TRANSCOSTPC+TAX;
ECOOBJ..           NETPROFIT=E=CSALES+(COSTDISP)-(COSTDISPSUP)-
                   (TOTALCOSTSEP)-(OPERACOST)-(TRANSCOSTCP)-
                   (TRANSCOSTPC)-(TRANSCOSTCL)-MCT;
*
*-(COSTDISPSUP) +(COSTDISP)
*ECOOBJ..          NETPROFIT=E=CSALES-TOTALCOSTSEP-OPERACOST-
                   TRANSCOSTCP-TRANSCOSTPC;
```

```
*DIS..                SDIS(C,R)=E=FTP(C,R)-FP(C,R);

*SUPPLY LANDFILLS
SUPP(C,R)..           FTPV(C,R)=E=SUM(L,SUP(C,L,R));
SUPP1(C,S)..      FTMV(C,S)=E=SUM(L,SUP1(C,L,S));
SUPP2(C,T)..      FTGV(C,T)=E=SUM(L,SUP2(C,L,T));
SUPP3(C,U)..      FTOV(C,U)=E=SUM(L,SUP3(C,L,U));
SUPP4(C,W)..      FTWV(C,W)=E=SUM(L,SUP4(C,L,W));

    *DEMAND LANDFILLS
DEMSa(L,R)..      DEMSP(L,R)=L=LDEMSP(L,R);
DEMSb(L,R)..      SUM(C,SUP(C,L,R))=E=DEMSP(L,R);
DEMS1a(L,S)..     DEMSM(L,S)=L=LDEMSM(L,S);
DEMS1b(L,S)..     SUM(C,SUP1(C,L,S))=E=DEMSM(L,S);
DEMS2a(L,T)..     DEMSG(L,T)=L=LDEMSG(L,T);
DEMS2b(L,T)..     SUM(C,SUP2(C,L,T))=E=DEMSG(L,T);
DEMS3a(L,U)..     DEMSO(L,U)=L=LDEMSO(L,U);
DEMS3b(L,U)..     SUM(C,SUP3(C,L,U))=E=DEMSO(L,U);
DEMS4a(L,W)..     DEMSW(L,W)=L=LDEMSW(L,W);
DEMS4b(L,W)..     SUM(C,SUP4(C,L,W))=E=DEMSW(L,W);

VARS(C,R)..      FTPV(C,R)=L=FTP(C,R)-FP(C,R);
VARS1(C,S)..     FTMV(C,S)=L=FTM(C,S)-FM(C,S);
VARS2(C,T)..     FTGV(C,T)=L=FTG(C,T)-FG(C,T);
VARS3(C,U)..     FTOV(C,U)=L=FTO(C,U)-FO(C,U);
VARS4(C,W)..     FTWV(C,W)=L=FTW(C,W)-FW(C,W);

*VARS(C,R)..     FTPV(C,R)=E=FTP(C,R)-FP(C,R);
*VARS1(C,S)..    FTMV(C,S)=E=FTM(C,S)-FM(C,S);
*VARS2(C,T)..    FTGV(C,T)=E=FTG(C,T)-FG(C,T);
*VARS3(C,U)..    FTOV(C,U)=E=FTO(C,U)-FO(C,U);
*VARS4(C,W)..    FTWV(C,W)=E=FTW(C,W)-FW(C,W);

*CM(C,R)
FV(C,R)..      FPV(C,R)=E=FTP(C,R)-FP(C,R)-FTPV(C,R);
FV1(C,S)..     FMV(C,S)=E=FTM(C,S)-FM(C,S)-FTMV(C,S);
FV2(C,T)..     FGV(C,T)=E=FTG(C,T)-FG(C,T)-FTGV(C,T);
FV3(C,U)..     FOV(C,U)=E=FTO(C,U)-FO(C,U)-FTOV(C,U);
FV4(C,W)..     FWV(C,W)=E=FTW(C,W)-FW(C,W)-FTWV(C,W);


*MANAGEMENT COST OF THE WASTE LEFT AT THE CITY
MC(C,R)..              CM(C,R)=E=FPV(C,R)*0.01235;
MC1(C,S)..             CM1(C,S)=E=FMV(C,S)*0.01235;
*VERIFICAR PT1 NEGATIVO {
MC2(C,T)..             CM2(C,T)=E=FGV(C,T)*0.01235;
MC3(C,U)..             CM3(C,U)=E=FOV(C,U)*0.01235;
```

```
MC4(C,W)..      CM4(C,W)=E=FWV(C,W)*0.01235;
SUMC..          MCT=E=SUM((C,R),CM(C,R))+SUM((C,S),CM1(C,S))+SUM((C,T),
                CM2(C,T))+SUM((C,U),CM3(C,U))+SUM((C,W),CM4(C,W));

   *WASTE CONSUMING
*WCONSP..       CONSUMEP=E=100*SUM((C,R),FP(C,R))/SUM((C,R),FTP(C,R)*AL
                PHAPSEP(C,R));
*WCONSM..       CONSUMEM=E=100*SUM((C,S),FM(C,S))/SUM((C,S),FTM(C,S)*AL
                PHAMSEP(C,S));
*WCONSG..       CONSUMEG=E=100*SUM((C,T),FG(C,T))/SUM((C,T),FTG(C,T)*AL
                PHAGSEP(C,T));
*WCONSO..       CONSUMEO=E=100*SUM((C,U),FO(C,U))/SUM((C,U),FTO(C,U)*A
                LPHAOSEP(C,U));
*WCONSW..       CONSUMEW=E=100*SUM((C,W),FW(C,W))/SUM((C,W),FTW(C,W
                )*ALPHAWSEP(C,W));

   *PARETO
*CONS..         CONSUMETOTAL=E=100*(SUM((C,R),FP(C,R))+
*               SUM((C,S),FM(C,S))+
*               SUM((C,T),FG(C,T))+
*               SUM((C,U),FO(C,U))+
*               SUM((C,W),FW(C,W)))/(
*               SUM((C,R),FTP(C,R)*ALPHAPSEP(C,R))+
*               SUM((C,S),FTM(C,S)*ALPHAMSEP(C,S))+
*               SUM((C,T),FTG(C,T)*ALPHAGSEP(C,T))+
*               SUM((C,U),FTO(C,U)*ALPHAOSEP(C,U))+
*               SUM((C,W),FTW(C,W)*ALPHAWSEP(C,W)));

   *CONST..     CONSUMETOTAL=E=LIMIT;

   *LIMIT=98;

   X=0;
   LIMIT=0;

   OPTION OPTCR=1E-10;

   MODEL WASTEREFINERY /ALL/;
   SOLVE WASTEREFINERY USING LP MAXIMIZED NETPROFIT;


   *PROFIT FOR SUPPLY
   PRSP(C,R)=(-DCPP.M(C,R)-(CSEPP(C,R)))*FP.L(C,R);
   PRSM(C,S)=(-DCPM.M(C,S)-(CSEPM(C,S)))*FM.L(C,S);
   PRSG(C,T)=(-DCPG.M(C,T)-(CSEPG(C,T)))*FG.L(C,T);
   PRSO(C,U)=(-DCPO.M(C,U)-(CSEPO(C,U)))*FO.L(C,U);
   PRSW(C,W)=(-DCPW.M(C,W)-(CSEPW(C,W)))*FW.L(C,W);

   *PROFIT FOR DEMAND
   PRP(TP,C,R)=(CPP(TP,C)-(-BMBCP.M(TP,C,R)))*PTCITYP.L(TP,C,R);
```

PRM(TM,C,S)=(CPM(TM,C)-(-BMBCM.M(TM,C,S)))*PTCITYM.L(TM,C,S);
PRG(TG,C,T)=(CPG(TG,C)-(-BMBCG.M(TG,C,T)))*PTCITYG.L(TG,C,T);
PRO(TO,C,U)=(CPO(TO,C)-(-BMBCO.M(TO,C,U)))*PTCITYO.L(TO,C,U);
PRW(TW,C,W)=(CPW(TW,C)-(-BMBCW.M(TW,C,W)))*PTCITYW.L(TW,C,W);

**PROFIT FOR SUPPLY: landfill*
PLP(C,R)=(-SUPP.M(C,R)-(CNSEP(C,R)))***SUM**(L,SUP.L(C,L,R));
PLM(C,S)=(-SUPP1.M(C,S)-(CNSEM(C,S)))***SUM**(L,SUP1.L(C,L,S));
PLG(C,T)=(-SUPP2.M(C,T)-(CNSEG(C,T)))***SUM**(L,SUP2.L(C,L,T));
PLO(C,U)=(-SUPP3.M(C,U)-(CNSEO(C,U)))***SUM**(L,SUP3.L(C,L,U));
PLW(C,W)=(-SUPP4.M(C,W)-(CNSEW(C,W)))***SUM**(L,SUP4.L(C,L,W));

**PROFIT FOR DEMAND: landfill*
PLDP(L,R)=(DNSEP(L,R)-(-DEMSb.M(L,R)))*DEMSP.L(L,R);
PLDM(L,S)=(DNSEM(L,S)-(-DEMS1b.M(L,S)))*DEMSM.L(L,S);
PLDG(L,T)=(DNSEG(L,T)-(-DEMS2b.M(L,T)))*DEMSG.L(L,T);
PLDO(L,U)=(DNSEO(L,U)-(-DEMS3b.M(L,U)))*DEMSO.L(L,U);
PLDW(L,W)=(DNSEW(L,W)-(-DEMS4b.M(L,W)))*DEMSW.L(L,W);

** WASTE NOT SENT TO PLANT OR LANDFILL*
**CM(C,R)*
**FPV(C,R)=FTP(C,R)-FP.L(C,R)-FTPV.L(C,R);*
**CM1(C,S)..*
**FMV(C,S)=FTM(C,S)-FM.L(C,S)-FTMV.L(C,S);*
**CM2(C,T)..*
**FGV(C,T)=FTG(C,T)-FG.L(C,T)-FTGV.L(C,T);*
**CM3(C,U)..*
**FOV(C,U)=FTO(C,U)-FO.L(C,U)-FTOV.L(C,U);*
**CM4(C,W)..*
**FWV(C,W)=FTW(C,W)-FW.L(C,W)-FTWV.L(C,W);*

************************PROFIT TECH PROVIDER*
**T1P(PP,TP,R,C)=(((-BMP.M(PP,TP,R))*ALPHAP(TP,R))-CPP(TP,C))*FTECHP.L(PP,TP,R);*
**T1M(PM,TM,S,C)=(((-BMM.M(PM,TM,S))*ALPHAM(TM,S))-CPM(TM,C))*FTECHM.L(PM,TM,S);*
**T1M(PM,TM,S)=(((-BSPM.M(PM,TM,S))*ALPHAM(TM,S))-CPROCM(TM,S))*FTECHM.L(PM,TM,S);*
**BMG(PG,TG,T).. PRODG(PG,TG,T)=E=ALPHAG(TG,T)*FTECHG(PG,TG,T);*
**BMO(PO,TO,U).. PRODO(PO,TO,U)=E=ALPHAO(TO,U)*FTECHO(PO,TO,U);*
**BMW(PW,TW,W).. PRODW(PW,TW,W)=E=ALPHAW(TW,W)*FTECHW(PW,TW,W);*

*********************** PROFIT TECH PROVIDER*
T1P(PP,TP,R)=(((-BSPP.M(PP,TP,R))*ALPHAP(TP,R))-CPROCP(TP,R))*FTECHP.L(PP,TP,R);
T1M(PM,TM,S)=(((-BSPM.M(PM,TM,S))*ALPHAM(TM,S))-CPROCM(TM,S))*FTECHM.L(PM,TM,S);
T1G(PG,TG,T)=(((-BSPG.M(PG,TG,T))*ALPHAG(TG,T))-CPROCG(TG,T))*FTECHG.L(PG,TG,T);

```
    T1O(PO,TO,U)=(((-BSPO.M(PO,TO,U))*ALPHAO(TO,U))-CPROCO(TO,U))*FTECH
O.L(PO,TO,U);
    T1W(PW,TW,W)=(((-BSPW.M(PW,TW,W))*ALPHAW(TW,W))-CPROCW(TW,W))*FT
ECHW.L(PW,TW,W);


    ***1 TRANSPORT ENTERING LANDFILL  (FROM CITIES TO LANDFILLS AT EACH
CITY )
    PT(L,R)=(-DEMSb.M(L,R)+SUM(C,SUPP.M(C,R))-CTRANSPL(R))*DEMSP.L(L,R);
    PT1(L,S)=(-DEMS1b.M(L,S)+SUM(C,SUPP1.M(C,S))-CTRANSML(S))*DEMSM.L(L,S);
    PT2(L,T)=(-DEMS2b.M(L,T)+SUM(C,SUPP2.M(C,T))-CTRANSGL(T))*DEMSG.L(L,T);
    PT3(L,U)=(-DEMS3b.M(L,U)+SUM(C,SUPP3.M(C,U))-CTRANSOL(U))*DEMSO.L(L,U);
    PT4(L,W)=(-DEMS4b.M(L,W)+SUM(C,SUPP4.M(C,W))-
CTRANSWL(W))*DEMSW.L(L,W);


    *2 TRANSPORT ENTERING PLANT  (FROM CITIES TO PROCESSING PLANTS )
    *DEMAND PLANTA
    TR(TP,C,R)=(-BMBCP.M(TP,C,R)+(DCPP.M(C,R))-CTRANSP(R))*FP.L(C,R);
    TR1(TM,C,S)=(-BMBCM.M(TM,C,S)+(DCPM.M(C,S))-CTRANSM(S))*FM.L(C,S);
    TR2(TG,C,T)=(-BMBCG.M(TG,C,T)+(DCPG.M(C,T))-CTRANSG(T))*FG.L(C,T);
    TR3(TO,C,U)=(-BMBCO.M(TO,C,U)+(DCPO.M(C,U))-CTRANSO(U))*FO.L(C,U);
    TR4(TW,C,W)=(-BMBCW.M(TW,C,W)+(DCPW.M(C,W))-CTRANSW(W))*FW.L(C,W);


    ***3 TRANSPORT LEAVING PLANT  (FROM PROCESSING PLANTS TO CITIES)
    *DEMAND CITY
PTL(TP,C,R)=(-BMBCP.M(TP,C,R)+(DCPP.M(C,R))-CTRANSPP(TP,R))*PTCITYP.L(TP,C,R);
PTL1(TM,C,S)            =(-BMBCM.M(TM,C,S)+(DCPM.M(C,S))-CTRANSPM(TM,S))*P
                        TCITYM.L(TM,C,S);
PTL2(TG,C,T)            =(-BMBCG.M(TG,C,T)+(DCPG.M(C,T))-CTRANSPG(TG,T))*PT
                        CITYG.L(TG,C,T);
PTL3(TO,C,U)            =(-BMBCO.M(TO,C,U)+(DCPO.M(C,U))-CTRANSPO(TO,U))*PT
                        CITYO.L(TO,C,U);
PTL4(TW,C,W)            =(-BMBCW.M(TW,C,W)+(DCPW.M(C,W))-CTRANSPW(TW,W)
                        )*PTCITYW.L(TW,C,W);

    *CONSUMETOTAL=100*(SUM((C,R),FP.L(C,R))+
    *          SUM((C,S),FM.L(C,S))+
    *          SUM((C,T),FG.L(C,T))+
    *          SUM((C,U),FO.L(C,U))+
    *          SUM((C,W),FW.L(C,W)))/(
    *          SUM((C,R),FTP(C,R)*ALPHAPSEP(C,R))+
    *          SUM((C,S),FTM(C,S)*ALPHAMSEP(C,S))+
    *          SUM((C,T),FTG(C,T)*ALPHAGSEP(C,T))+
    *          SUM((C,U),FTO(C,U)*ALPHAOSEP(C,U))+
    *          SUM((C,W),FTW(C,W)*ALPHAWSEP(C,W)));


    *PTL1(C,L,S)=(-DEMS1b.M(L,S)-(-SUPP1.M(C,S))-(CTRANSM(S)))*SUP1.L(C,L,S);
```

```
*PTL2(C,L,T)=(-DEMS2b.M(L,T)-(-SUPP2.M(C,T))-(CTRANSG(T)))*SUP2.L(C,L,T);
*PTL3(C,L,U)=(-DEMS3b.M(L,U)-(-SUPP3.M(C,U))-(CTRANSO(U)))*SUP3.L(C,L,U);
*PTL4(C,L,W)=(-DEMS4b.M(L,W)-(-SUPP4.M(C,W))-(CTRANSW(W)))*SUP4.L(C,L,W);

*DEM1-S*F_DEM+DEM2-S*F_DEM

$ONTEXT
PTL1              =SUM((TM,C,S),((-BMBCM.M(TM,C,S)-(-DCPM.M(C,S))-CTRANSP
                  M(TM,S))*PTCITYM.L(TM,C,S)))+SUM((L,C,S),((-DEMS1b.M(L,S)-(-
                  SUPP1.M(C,S))-CTRANSM(S))*DEMSM.L(L,S)));
PTL2              =SUM((TG,C,T),((-BMBCG.M(TG,C,T)-(-DCPG.M(C,T))-CTRANSP
                  G(TG,T))*PTCITYG.L(TG,C,T)))+SUM((L,C,T),((-DEMS2b.M(L,T)-
                  (-SUPP2.M(C,T))-CTRANSG(T))*DEMSG.L(L,T)));
PTL3              =SUM((TO,C,U),((-BMBCO.M(TO,C,U)-(-DCPO.M(C,U))-CTRANSP
                  O(TO,U))*PTCITYO.L(TO,C,U)))+SUM((L,C,U),((-DEMS3b.M(L,U)-
                  (-SUPP3.M(C,U))-CTRANSO(U))*DEMSO.L(L,U)));
PTL4              =SUM((TW,C,W),((-BMBCW.M(TW,C,W)-(-DCPW.M(C,W))-CTR
                  ANSPW(TW,W))*PTCITYW.L(TW,C,W)))+SUM((L,C,W),((-
                  DEMS4b.M(L,W)-(-SUPP4.M(C,W))-CTRANSW(W))*DEMS-
                  W.L(L,W)));
PTL1(TM,C,S,L)    =(-BMBCM.M(TM,C,S)-(-DCPM.M(C,S))-
                  CTRANSPM(TM,S))*PTCITYM.L(TM,C,S)+(-DEMS1b.M(L,S)-(-
                  SUPP1.M(C,S))-CTRANSM(S))*DEMSM.L(L,S);
PTL2(TG,C,T,L)    =(-BMBCG.M(TG,C,T)-(-DCPG.M(C,T))-
                  CTRANSPG(TG,T))*PTCITYG.L(TG,C,T)+(-DEMS2b.M(L,T)-(-
                  SUPP2.M(C,T))-CTRANSG(T))*DEMSG.L(L,T);
PTL3(TO,C,U,L)    =(-BMBCO.M(TO,C,U)-(-DCPO.M(C,U))-
                  CTRANSPO(TO,U))*PTCITYO.L(TO,C,U)+(-DEMS3b.M(L,U)-(-
                  SUPP3.M(C,U))-CTRANSO(U))*DEMSO.L(L,U);
PTL4(TW,C,W,L)    =(-BMBCW.M(TW,C,W)-(-DCPW.M(C,W))-
                  CTRANSPW(TW,W))*PTCITYW.L(TW,C,W)+(-DEMS4b.M(L,W)-(-
                  SUPP4.M(C,W))-CTRANSW(W))*DEMSW.L(L,W);

*TOTAL TRANSPORTATION COST FROM CITIES TO PROCESSING PLANTS
TTCCP..           TRANSCOSTCP=E=SUM((C,PP,R),CTRANSP(R)*DPP(C,PP)*FPPL
                  ANT(C,PP,R))
                  +SUM((C,PM,S),CTRANSM(S)*DPM(C,PM)*FMPLANT(C,PM,S))+
                  SUM((C,PG,T),CTRANSG(T)*DPG(C,PG)*FGPLANT(C,PG,T))
                  + SUM((C,PO,U),CTRANSO(U)*DPO(C,PO)*FOPLANT(C,PO,U))+
                  SUM((C,PW,W),CTRANSW(W)*DPW(C,PW)*FWPLANT(C,PW,W));

*TOTAL TRANSPORTATION COST FROM PROCESSING PLANTS TO CITIES
TTCPC..           TRANSCOSTPC=E=SUM((PP,TP,C,R),CTRANSPP(TP,R)*DPP(C,P
                  P)*PRODCITYP(PP,TP,C,R))
                  +SUM((PM,TM,C,S),CTRANSPM(TM,S)*DPM(C,PM)*PRODCITYM
                  (PM,TM,C,S))+SUM((PG,TG,C,T),CTRANSPG(TG,T)*DPG(C,PG)*P
                  RODCITYG(PG,TG,C,T))
                  +SUM((PO,TO,C,U),CTRANSPO(TO,U)*DPO(C,PO)*PRODCITYO
                  (PO,TO,C,U))+SUM((PW,TW,C,W),CTRANSPW(TW,W)*DPW(C,P
                  W)*PRODCITYW(PW,TW,C,W));
```

```
*TOTAL TRANSPORTATION COST FROM CITIES TO LANDFILLS AT EACH CITY
TCL..           TRANSCOSTCL=E=SUM((C,L,R),CTRANSP(R)*DCL(C,L)*SUP(C,L
                ,R))
                +SUM((C,L,S),CTRANSM(S)*DCL(C,L)*SUP1(C,L,S))+SUM((C,L,T),
                CTRANSG(T)*DCL(C,L)*SUP2(C,L,T))
                + SUM((C,L,U),CTRANSO(U)*DCL(C,L)*SUP3(C,L,U))+ SUM((C,L,W),
                CTRANSW(W)*DCL(C,L)*SUP4(C,L,W));
$OFFTEXT

    EXECUTE_UNLOAD "MSW_10.cl_5_wTAX.gdx"

    **************************************************************************
```

Index

Authors Biographies

José María Ponce-Ortega is a Professor at the Chemical Engineering Department, Universidad Michoacana de San Nicolás de Hidalgo, Mexico. He obtained his PhD (2023) and Master's (2003) degrees in Chemical Engineering from the Institute of Technology of Celaya, Mexico. He was a postdoctoral researcher at Texas A&M University, USA, and as a visiting scholar at Carnegie Mellon University, USA. He is a full professor at the Universidad Michoacana de San Nicolás de Hidalgo, and is a member of the National Research System of Mexico. The research interest of Dr Ponce-Ortega cover optimization of chemical processes, sustainable design, energy, mass, water and property integration and supply chain optimization. He has published more than 300 papers, 3 books and 50 book chapters. He has supervised 25 PhD and 40 Master's students. He also has 15 funded research projects for about $US 1,000,000.00. Dr Ponce-Ortega is a member of the editorial board of the Clean Technologies and Environmental Policy Journal, and of the Process Integration and Optimization for Sustainability Journal. He is also a subject editor in the Journal of Sustainable Production and Consumption.

Rogelio Ochoa-Barragán has a PhD (2023) and Master's Degrees (2020) in Chemical Engineering from the Michoacana University of San Nicolás de Hidalgo, Mexico. His research is focused on process optimization, energy network management, social justice, and machine learning. His research has been presented at 11 international and regional conferences, he currently has 9 published scientific articles, and has participated in the writing of 15 books.

César Ramírez-Márquez is a Postdoctoral Fellow at the Chemical Engineering Department, Universidad Michoacana de San Nicolás de Hidalgo, Mexico. He has a PhD from the University of Guanajuato (Mexico) in 2020. Currently published contributions focus on the production of materials for the solar energy industry and base chemicals in the chemical industry. He has published more than 60 journal papers, 10 book chapters, has presented his work at more than 20 international/regional conferences, and has registered 4 patents.

José Ezequiel Santibañez-Aguilar has a PhD in Chemical Engineering from the Universidad Michoacana de San Nicolás de Hidalgo, México. He is member level 1 of the Mexican Research System. Currently, he is full time Professor at Tecnologico de Monterrey, México. His expertise covers mathematical model developing for the design of supply chains considering various criteria for sustainability such as life cycle analysis, generation of jobs, waste recycling and processing and financial risk. His research focuses on strategies to solve multi-objective formulations and making decision approaches. He has authored and co-authored more than 30 papers in journals such as Applied Energy, Journal of Cleaner Production, ACS Sustainable Chemistry and Engineering, Renewable and Sustainable Energy Reviews, Waste Management, AIChE Journal, etc. Furthermore, he has written several book chapters. His research has got more than 1300 citations with h and i10 indexes equal to 18 and 25 respectively.

Aurora del Carmen Munguía-López Assistant Professor in the Department of Chemical and Biological Engineering at the University at Buffalo, State University of New York. Previously, she served as a postdoctoral researcher in the Department of Chemical and Biological Engineering at the University of Wisconsin-Madison, USA. She holds BSc and MSc degrees from the Technical

Institute of Celaya and a PhD in Chemical Engineering from the University of Michoacan in Mexico. Her research interests include mathematical optimization, sustainability, social justice, and process modeling. Dr Munguía-López has published papers in over 20 journals and conferences and a book chapter. She has presented her work at several recognized conferences. She has supervised 2 graduate and 3 undergraduate students. Aurora is also passionate about fostering diversity, equity, and inclusion in STEM.